Mathematical Modelling of Solids with Nonregular Boundaries

Mathematical Modelling of Solids with Nonregular Boundaries

A.B. Movchan
N.V. Movchan
School of Mathematical Sciences
University of Bath
Bath, U.K.

CRC Press
Boca Raton New York London Tokyo

Library of Congress Cataloging-in-Publication Data

Movchan, A. B. (Alexander B.)
Mathematical modelling of solids with nonregular boundaries / A.B. Movchan and N.V. Movchan
p. cm.
Includes bibliographical references and index.
ISBN 0-8493-8338-2 (alk. paper)
1. Elasticity. 2. Boundary value problems--Asymptotic theory. I. Movchan, N. V. (Nataliya V.). II. Title.
QA931.M685 1995
620.1'126'015118--dc20 95-23090
CIP

Direct all inquiries to CRC Press, Inc., 2000 Corporate Blvd., N.W., Boca Raton, Florida 33431.

International Standard Book Number 0-8493-8338-2
Library of Congress Card Number 95-23090
Printed in the United States of America 1 2 3 4 5 6 7 8 9 0
Printed on acid-free paper

CRC Mathematical Modelling Series

Titles included in the series:

Movchan: ***Mathematical Modelling of Solids with Nonregular Boundaries***

Mittnik: ***System Theoretic Methods in Empirical Economic Modelling***

Villa/Brandimarte: ***Advanced Models for Manufacturing Systems Management***

Kusiak/Bielli: ***Designing Innovations in Industrial Logistics Modelling***

Kliemann/Namachchivaya: ***Nonlinear Dynamics and Stochastic Mechanics***

Bellomo: ***Modelling Mathematical Methods and Scientific Computation***

Contents

Preface

This volume is the advanced textbook which deals with the asymptotic methods of analysis of linear elasticity problems for solids with cracks and inclusions. The problems are natural in fracture mechanics of inhomogeneous media and are important in analysis of strength of elastic structures and in geophysics. The rigorous approach allows one to study the displacement field and the stress–strain state in a continuum with thin inclusions or cavities, in composite materials with small defects and in elastic solids with sharp inclusions. We also describe the asymptotic procedure for the eigenvalue problems in domains with small defects. Problems, posed in domains with singularly perturbed boundary, involve special solutions of the boundary layer type. A boundary layer is specified in an unbounded region, and the analysis requires the information on the asymptotic behaviour of solutions to boundary value problems in noncompact domains with cylindrical or conical extensions at infinity.

The textbook is addressed to applied scientists, engineers and postgraduate students at the MSc and PhD level who wish to use the asymptotic analysis of elasticity problems in domains with nonregular boundary as a powerful tool for the strength evaluation in models of elastic structures and geophysics applications.

The book consists of five chapters.

In the first introductory chapter a short survey, including equations of linear elasticity and formulations of the boundary value problems, is presented.

In Chapter 2 we discuss the asymptotic approach to the mathematical modelling of cracks and inclusions in a linear elastic solid. A crack is represented as a thin cavity. For the case of a singularly perturbed boundary we introduce a boundary layer which describes the displacement field near the ends of the crack. The brittle fracture criteria are described for quasi–static formulations. The displacement field in a neighbourhood of a thin elastic inclusion is studied.

Chapter 3 deals with main definitions and general properties of solutions of linear elasticity problems in domains with conical points on the boundary and in unbounded domains with cylindrical or conical extensions at infinity. We discuss examples of the elasticity formulations in unbounded domains and the asymptotic behaviour of the displacement field at infinity.

Chapter 4 is related to the asymptotic analysis of the displacement and stress fields in the vicinity of sharp inclusions. In a two–dimensional case certain classes of cusp–shaped and thin angular inclusions are analyzed.

For the equations of three-dimensional elasticity, the stress-strain behaviour in the vicinity of cusp-shaped elastic and rigid inclusions have been studied. The dependence of the asymptotics on the geometry of a cusp (in the sense of the integral Dini condition) is emphasized, and a boundary layer effect in a neighbourhood of the irregular point on the boundary is considered.

In Chapter 5 we work with integral characteristics of finite cavities and inclusions in elastic media. The Pólya–Szegö tensor is defined for the elasticity formulations (in the case of the Laplacian similar characteristics were studied by G. Pólya and G. Szegö in [107]). The canonical asymptotic approximation of the energy increment and asymptotic expansions for fundamental frequencies are presented for an elastic solid with a small defect. Simple asymptotic formulae describe the interaction between a small inclusion or cavity and a crack in an elastic plane. The applications to the theory of composite materials with a periodic structure of small defects are presented in Section 5.3.

We express our deep gratitude to Professor J.R. Willis for many valuable discussions, his support and interest to our work.

A.B. Movchan would like to thank Dr. J.R. Ockendon and Dr. A.B. Tayler for their support during his work at Oxford in 1990–1991.

We are grateful to Professors V.G. Maz'ya and V.A. Kozlov and also to Professors S.A. Nazarov and N.F. Morozov, whose critical comments were always helpful and stimulating.

We wish to thank Professor K.Z. Markov for encouraging us to write this book, Professor E. Inan for reading the draft and useful suggestions, and the publishers for their assistance in the preparation of the manuscript.

Finally our thanks are due to Miss S. Love and Miss Nadya Movchan for efficient typing and help with preparation of figures.

ABM
NVM

1

Introduction

In the present introductory chapter we consider the equations of linear elasticity and formulations of the boundary value problems. In particular, the two–dimensional simple formulations are presented.

The model problems are discussed for cracks in an elastic continuum, where the stress components are singular at the crack front. The interface boundary conditions are introduced for problems related to elastic inclusions.

The energy estimates for an elastic region with a small defect (a cavity or an elastic inclusion) are obtained with the use of certain matrix integral characteristics that depend on the shape of the defect, its elastic moduli and elastic coefficients of surrounding medium. Here we discuss the integral characteristics for the case of the Laplacian, which provides the motivation for further analysis related to the case of the Lamé system.

In the final section of this chapter the concept of asymptotic expansions is stated.

1.1 Equations of linear elasticity

1.1.1 The Lamé system. Hooke's law

Within any continuum the following three groups of balance relations hold:

(i) The equations of balance of linear momentum

$$\sum_{j=1}^{3} \frac{\partial \sigma_{ij}}{\partial x_j} + F_i = \rho \ddot{u}_i, \; i = 1, 2, 3, \tag{1.1.1}$$

(ii) the balance of angular momentum

$$\sigma_{ij} = \sigma_{ji}, \; i, j = 1, 2, 3, \tag{1.1.2}$$

(iii) conservation of mass

$$\dot{\rho} + \rho \nabla \cdot \dot{\boldsymbol{u}} = 0. \tag{1.1.3}$$

Here σ_{ij} are components of the stress tensor, $\boldsymbol{u} = (u_1, u_2, u_3)^T$ is the displacement vector, $\mathbf{F} = (F_1, F_2, F_3)^T$ is the body force density, ρ is the density of the material; the dot denotes the derivative with respect to time. For any orthogonal transformation of the Cartesian basis $\{\boldsymbol{e}^{(j)}\}_{j=1}^3$

$$\boldsymbol{e}'^{(i)} = \sum_{j=1}^{3} l_{ij} \boldsymbol{e}^{(j)}, \; j = 1, 2, 3, \tag{1.1.4}$$

the components of the rank–2 tensor σ_{ij} change as follows

$$\sigma'_{km} = \sum_{i,j=1}^{3} l_{ki} l_{mj} \sigma_{ij}. \tag{1.1.5}$$

(Here we present only the basic properties of stresses. For the detailed analysis we refer the reader to the books [117], [6], [120].)

Let us discuss the physical interpretation of components of stress, and the relation to the vector of tractions. It is assumed here and further in the text that an elastic medium is subjected to a small deformation, and, therefore, just a linear analysis will be involved.

Consider an infinitesimal tetrahedron V (see Fig. 1.1) such that the outward unit normal vectors with respect to its faces S_1, S_2, S_3 and S_0 are

$$-\boldsymbol{e}^{(1)}, \; -\boldsymbol{e}^{(2)}, \; -\boldsymbol{e}^{(3)} \text{ and } \boldsymbol{n}.$$

Let $\text{mes}S_1$, $\text{mes}S_2$, $\text{mes}S_3$ and $\text{mes}S_0$ denote the corresponding areas, and let $\text{mes}V$ denote the volume of the tetrahedron. If $\boldsymbol{n}(s)$ is the outward unit normal vector defined on

$$S = \bigcup_{i=0}^{3} S_i,$$

then

$$0 = \int_S \boldsymbol{n}(s)ds = \text{mes}S_0\boldsymbol{n} - \sum_{j=1}^{3} \text{mes}S_j \boldsymbol{e}^{(j)}. \tag{1.1.6}$$

The scalar product of (1.1.6) and a basis vector $\boldsymbol{e}^{(i)}$ gives

$$\text{mes}S_i = \text{mes}S_0\boldsymbol{n} \cdot \boldsymbol{e}^{(i)} = \text{mes}S_0 n_i. \tag{1.1.7}$$

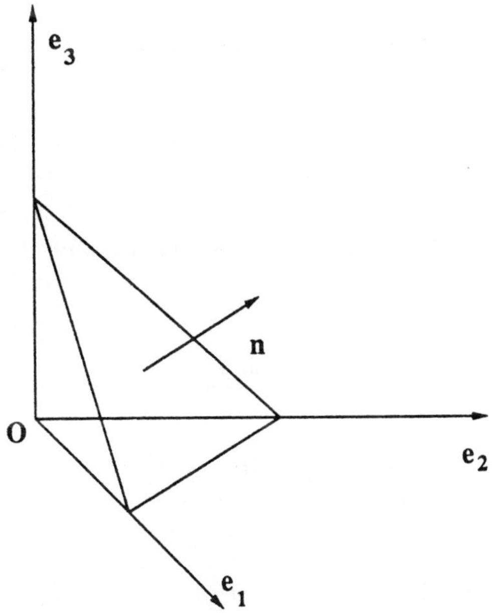

Fig. 1.1: Infinitesimal tetrahedron in an elastic space.

The integral form of the balance of the linear momentum is

$$\int_S \mathbf{t}ds = (\rho\ddot{u} - \mathbf{F})d\boldsymbol{x}, \tag{1.1.8}$$

where $\mathbf{t}$ is the vector of tractions applied on the surface $S = \partial V$. Since the tetrahedron is infinitesimal, the traction vectors on the faces S_i, $i = 0, 1, 2, 3$, may be regarded as constant. To leading order approximation, one can write

$$\sum_{j=1}^{3} \mathbf{t}(-\boldsymbol{e}^{(j)})\text{mes}S_j + \mathbf{t}(\boldsymbol{n})\text{mes}S_0 = (\rho\ddot{\boldsymbol{u}} - \mathbf{F})\text{mes}V,$$

and, therefore,

$$\mathbf{t}(\boldsymbol{n}) = -\sum_{j=1}^{3} \mathbf{t}(-\boldsymbol{e}^{(j)})n_j + (\rho\ddot{\boldsymbol{u}} - \mathbf{F})\frac{\text{mes}V}{\text{mes}S_0}. \tag{1.1.9}$$

As mes$V \to 0$, the last term in (1.1.9) vanishes, and in the limit we have

$$\mathbf{t}(\boldsymbol{n}) = -\sum_{j=1}^{3} \mathbf{t}(-\boldsymbol{e}^{(j)})n_j = \sum_{j=1}^{3} \mathbf{t}(\boldsymbol{e}^{(j)})n_j. \tag{1.1.10}$$

Thus, the traction vector $\mathbf{t}(\boldsymbol{n})$ depends linearly on the components of $\boldsymbol{n}$. In the scalar form the relation (1.1.10) can be written as

$$t_i(\boldsymbol{n}) = \sum_{j=1}^{3} t_i(\boldsymbol{e}^{(j)})n_j. \tag{1.1.11}$$

Then, the Cauchy stress tensor (further it will be simply called the stress tensor) is defined by

$$\sigma_{ij} = t_i(\boldsymbol{e}^{(j)}). \tag{1.1.12}$$

The equality (1.1.11) is equivalent to

$$\sigma_i^{(n)} := t_i(\boldsymbol{n}) = \sum_{j=1}^{3} \sigma_{ij}n_j, \; i = 1, 2, 3. \tag{1.1.13}$$

If the elastic medium is in equilibrium, then

$$0 = \int_V F_i d\boldsymbol{x} + \int_{\partial V} t_i ds = \int_V F_i d\boldsymbol{x} + \int_{\partial V} \sum_{j=1}^{3} \sigma_{ij}n_j ds$$

$$= \int_V (F_i + \sum_{j=1}^{3} \frac{\partial \sigma_{ij}}{\partial x_j})d\boldsymbol{x}. \tag{1.1.14}$$

Assuming that the integrand is continuous and that the relation (1.1.14) holds for any volume V, we deduce

$$\sum_{j=1}^{3} \frac{\partial \sigma_{ij}}{\partial x_j} + F_i = 0, \; i = 1,2,3, \tag{1.1.15}$$

which are the equations of equilibrium. These equations are the same as (1.1.1) when the displacement components are time independent. The balance of the angular momentum provides the symmetry for the stress tensor.

The infinitesimal strain tensor components are defined by

$$\varepsilon_{ij} = \frac{1}{2}(\frac{\partial u_i}{\partial x_j} + \frac{\partial u_j}{\partial x_i}), \; i,j = 1,2,3. \tag{1.1.16}$$

These equalities are called the *Cauchy relations.* The rank–2 strain tensor is symmetric:

$$\varepsilon_{ij} = \varepsilon_{ji} \quad \text{for any} \quad i,j = 1,2,3.$$

For the case of a small deformation a linear relation between stress and strain is postulated

$$\sigma_{ij} = \sum_{k,l=1}^{3} c_{ijkl}\varepsilon_{kl}, \tag{1.1.17}$$

which is called the *Hooke law.* Here (c_{ijkl}) is the rank–4 tensor of linear elastic moduli. The symmetry of the tensors (σ_{ij}), (ε_{kl}) yields the following properties of symmetry for the tensor (c_{ijkl})

$$c_{ijkl} = c_{jikl} = c_{ijlk} = c_{jilk}. \tag{1.1.18}$$

In general, for an inhomogeneous solid, c_{ijkl} may depend on the position vector $\boldsymbol{x}$; for a homogeneous medium they are constant.

In isotropic case c_{ijkl} are specified by

$$c_{ijkl} = \lambda\delta_{ij}\delta_{kl} + \mu(\delta_{ik}\delta_{jl} + \delta_{il}\delta_{jk}), \tag{1.1.19}$$

where δ_{ij} is the Kronecker delta, and the coefficients λ, μ are called the *Lamé elastic moduli.* The Hooke's law (1.1.17) takes then the form

$$\sigma_{ij} = \lambda\delta_{ij} \sum_{k=1}^{3} \varepsilon_{kk} + 2\mu\varepsilon_{ij}. \tag{1.1.20}$$

Relations (1.1.16), (1.1.20) give

$$\sigma_{ij} = \lambda\delta_{ij} \sum_{k=1}^{3} \frac{\partial u_k}{\partial x_k} + \mu(\frac{\partial u_i}{\partial x_j} + \frac{\partial u_j}{\partial x_i}), \tag{1.1.21}$$

and, therefore, the equilibrium equations (1.1.15) can be rewritten in terms of displacements

$$\mu\Delta u_i + (\lambda + \mu)\frac{\partial}{\partial x_i}(\nabla \cdot \boldsymbol{u}) + F_i = 0, \; i = 1, 2, 3, \tag{1.1.22}$$

where Δ denotes the Laplacian operator. The system of equations (1.1.22) is called the *Lamé system*, and the corresponding vector form is

$$\mathbf{L}(\frac{\partial}{\partial \boldsymbol{x}})\boldsymbol{u} := \mu\Delta\boldsymbol{u} + (\lambda + \mu)\nabla\nabla \cdot \boldsymbol{u} + \mathbf{F} = \mathbf{0}. \tag{1.1.23}$$

In addition to the Lamé constants we shall use two more characteristics of an isotropic elastic material

$$E = \mu\frac{2\mu + 3\lambda}{\lambda + \mu} \quad \text{and} \quad \nu = \frac{\lambda}{2(\lambda + \mu)}. \tag{1.1.24}$$

The quantity E is called the *Young modulus*, and ν is the *Poisson ratio*; E represents a uniaxial elastic modulus, and ν is a measure of the lateral contraction compared with the axial extension.

1.1.2 Formulation of static boundary value problems

In the state of equilibrium the displacement vector $\boldsymbol{u}$ should satisfy the Lamé system (1.1.22) at any point within the elastic body. Also one has to specify certain boundary conditions. First, assume that an isotropic elastic material with the Lamé constants λ, μ occupies a bounded region $\Omega \subset \mathbb{R}^m$ where m may take values 2 or 3. Further in the text we shall deal with the following types of boundary conditions:

(i) *displacement boundary conditions* (Dirichlet's type) which require all components of the displacement vector to be specified on $\partial\Omega$

$$\boldsymbol{u}(\boldsymbol{x}) = \boldsymbol{\varphi}(\boldsymbol{x}), \ \boldsymbol{x} \in \partial\Omega, \tag{1.1.25}$$

where the vector–function $\boldsymbol{\varphi}$ is known;
(ii) *traction boundary conditions* (Neumann's type) that give the vector of tractions applied on the surface $\partial\Omega$

$$\sum_{j=1}^{m} \left(\lambda\delta_{ij} \sum_{k=1}^{m} \frac{\partial u_k}{\partial x_k} + \mu\left(\frac{\partial u_i}{\partial x_j} + \frac{\partial u_j}{\partial x_i}\right) \right) n_j = p_i;$$

the corresponding vector form is

$$\boldsymbol{\sigma}^{(n)}(\boldsymbol{u}; \boldsymbol{x}) = \mathbf{p}(\boldsymbol{x}), \ \boldsymbol{x} \in \partial\Omega; \tag{1.1.26}$$

here $\mathbf{p}$ is the given vector function;
(iii) *mixed boundary conditions* (the Dirichlet–Neumann type) that involve displacements prescribed on the subset $\partial\Omega_u$ of the surface, and the tractions given on the remaining part $\partial\Omega_\sigma$ of $\partial\Omega$.

Also, there is a range of boundary conditions that occur in contact problems of solid mechanics. The interface boundary conditions for problems on elastic inclusions will be discussed below.

We introduce the energy functional

$$\epsilon(\boldsymbol{u}; \Omega) = \frac{1}{2} \sum_{i,j=1}^{m} \int_\Omega \varepsilon_{ij}(\boldsymbol{u}; \boldsymbol{x}) \sigma_{ij}(\boldsymbol{u}; \boldsymbol{x}) d\boldsymbol{x}. \tag{1.1.27}$$

The physical displacement fields are characterized by the finite value of elastic energy.

In classical formulations it is assumed that the boundary $\partial\Omega$ and the components of the vector functions $\boldsymbol{\varphi}$ and $\mathbf{F}$, $\mathbf{p}$ are smooth, and, therefore, the functions u_i, $i = 1, 2, 3$, have derivatives of the first and second order in the closure $\overline{\Omega}$. It is known that a boundary value problem with boundary conditions of the type (i) or (iii) is always solvable in the class of functions with the finite elastic energy, and the solution is unique. The solvability conditions for a static problem with prescribed tractions on $\partial\Omega$ (see (ii)) are expressed as the balance relations for the principal force and moment vectors of external load

$$\int_\Omega \mathbf{F} d\boldsymbol{x} + \int_{\partial\Omega} \mathbf{p} ds = \mathbf{0}, \tag{1.1.28}$$

$$\int_{\Omega} \boldsymbol{x} \times \mathbf{F} d\boldsymbol{x} + \int_{\partial\Omega} \boldsymbol{x} \times \mathbf{p} ds = \mathbf{0}. \tag{1.1.29}$$

A solution of the boundary value problem (1.1.23), (1.1.26), subject to the solvability conditions (1.1.28), (1.1.29), is not unique. It is specified up to an arbitrary rigid body displacement which may involve a translation and a rotation

$$\mathbf{a} + \mathbf{b} \times \boldsymbol{x}, \tag{1.1.30}$$

where $\mathbf{a}$ and $\mathbf{b}$ are constant vectors.

Let us introduce the space $E(\Omega)$ of vector functions, defined on Ω, with the norm

$$||\boldsymbol{u}||_{E(\Omega)} = (\epsilon(\boldsymbol{u}; \Omega))^{1/2} < +\infty. \tag{1.1.31}$$

Then, the Neumann problem (1.1.23), (1.1.26) can be reformulated in the variational form: find $\boldsymbol{u} \in E(\Omega)$ such that

$$\sum_{i,j=1}^{m} \int_{\Omega} \sigma_{ij}(\boldsymbol{u}; \boldsymbol{x}) \varepsilon_{ij}(\boldsymbol{v}; \boldsymbol{x}) d\boldsymbol{x} = h(\boldsymbol{v}), \tag{1.1.32}$$

for any $\boldsymbol{v} \in E(\Omega)$, where

$$h(\boldsymbol{v}) = \int_{\Omega} \mathbf{F}(\boldsymbol{x}) \cdot \boldsymbol{v}(\boldsymbol{x}) d\boldsymbol{x} + \int_{\partial\Omega} \mathbf{p}(\boldsymbol{x}) \cdot \boldsymbol{v}(\boldsymbol{x}) ds. \tag{1.1.33}$$

To give the corresponding formulation for the Dirichlet–Neumann problem (and, similarly, for the Dirichlet problem) we define the space

$$E^0(\Omega; S) = \{\boldsymbol{v} \in E(\Omega) : \boldsymbol{v} = \mathbf{0} \text{ on } S\},$$

where $S \subset \partial\Omega$. Then, one has to find a vector function $\boldsymbol{u}$ which satisfies (1.1.32) with

$$h(\boldsymbol{v}) = \int_{\Omega} \mathbf{F}(\boldsymbol{x}) \cdot \boldsymbol{v}(\boldsymbol{x}) d\boldsymbol{x} + \int_{\partial\Omega_\sigma} \mathbf{p}(\boldsymbol{x}) \cdot \boldsymbol{v}(\boldsymbol{x}) ds, \tag{1.1.34}$$

for any $\boldsymbol{v} \in E^0(\Omega, \partial\Omega_u)$, and the boundary condition (1.1.25), with $\partial\Omega$ being replaced by $\partial\Omega_u$. For the Dirichlet boundary value problem one has $\partial\Omega_u = \partial\Omega$, and the last term in (1.1.34) is absent.

Let us note that the variational formulations do not require second-order derivatives of components of the vector function $\boldsymbol{u}$. Further in the text we do not assume the smoothness of the boundary $\partial\Omega$. However, we suppose that the external load is chosen in such a way that it provides a finite elastic energy for the displacement field $\boldsymbol{u}$, and, when necessary, we shall understand the solution of the boundary value problem in the weak sense related to the corresponding variational formulation.

The important place in this book is taken by boundary value problems posed in infinite domains. In these cases we set certain radiation conditions which specify the asymptotic behaviour of components of the displacement vector or their derivatives at infinity.

Special weighted Sobolev spaces are usually introduced for the analysis of boundary value problems posed in domains with conical or cylindrical boundaries. We have no intention to present in details this theory here, and we would like to refer to the classical paper [38] and to the book [59]. When necessary, we shall discuss some elements of this theory in the text.

1.1.3 Compatibility conditions

It is possible to formulate problems of linear elasticity in terms of stress components. One has the equilibrium equations (1.1.15) and the equality (1.1.26) for tractions on the boundary of an elastic body. In order to restore a continuous displacement field, one has to consider additional relations which are called the Saint Venant compatibility equations. In terms of components of the strain tensor these equations can be written as follows:

$$\frac{\partial^2 \varepsilon_{ij}}{\partial x_k \partial x_l} + \frac{\partial^2 \varepsilon_{kl}}{\partial x_i \partial x_j} = \frac{\partial^2 \varepsilon_{ik}}{\partial x_j \partial x_l} + \frac{\partial^2 \varepsilon_{jl}}{\partial x_i \partial x_k}, \quad i, j, k, l = 1, 2, 3. \qquad (1.1.35)$$

With the use of Hooke's law the strain components can be represented in terms of stresses. Then, using the equilibrium equations, one can derive six differential equations of the second order

$$(1+\nu)\Delta\sigma_{ij} + \frac{\partial^2 \Theta}{\partial x_i \partial x_j} = 0, \; i, j = 1, 2, 3; \; i \geq j, \qquad (1.1.36)$$

where $\Theta = \sum_{k=1}^{3} \sigma_{kk}$.

We do not discuss the derivation of equations (1.1.35), (1.1.36); they can be found in any manual on linear elasticity (see, for example [117], [85]).

Thus, we have stated that the components of the stress tensor have to satisfy three first-order differential equations of equilibrium and six differential equations (1.1.36) of the second order. The relations (1.1.36) are called the Beltrami–Mitchell compatibility equations.

1.1.4 Two-dimensional models

Here we describe the cases of plane strain, plane stress and the anti-plane shear where the elasticity problems admit two–dimensional formulations.

Plane strain. A deformation state is called the plane strain if there exists a system of coordinates where the displacement field satisfies the relations

$$u_1 = u_1(x_1, x_2),\ u_2 = u_2(x_1, x_2),\ u_3 = 0. \tag{1.1.37}$$

Thus, the displacement vector is parallel to the plane Ox_1x_2, and the components u_i do not depend on x_3.

For example, one can use the model of plane strain to describe the deformation within an infinite cylindrical bar with tractions $\mathbf{t}$ applied on the lateral surface in such a way that

$$t_i = t_i(x_1, x_2),\ i = 1, 2,$$

$$t_3 = 0, \tag{1.1.38}$$

and with the body force vector

$$\mathbf{F} = (F_1(x_1, x_2), F_2(x_1, x_2), 0)^T. \tag{1.1.39}$$

As an approximation, the plane strain model can be used for a middle region of a long bar of length l and diameter d, $l \gg d$ (see Fig. 1.2). Also, on some part of the boundary one can set the displacement boundary conditions which satisfy (1.1.37).

Basic relations of linear elasticity can be written as follows.

(i) Cauchy's relations:

$$\varepsilon_{11} = \frac{\partial u_1}{\partial x_1},\ \varepsilon_{22} = \frac{\partial u_2}{\partial x_2},\ \varepsilon_{12} = \frac{1}{2}(\frac{\partial u_1}{\partial x_2} + \frac{\partial u_2}{\partial x_1}),$$

$$\varepsilon_{13} = \varepsilon_{23} = \varepsilon_{33} = 0. \tag{1.1.40}$$

(ii) Hooke's law:

$$\sigma_{11} = (2\mu + \lambda)\varepsilon_{11} + \lambda\varepsilon_{22},\ \sigma_{12} = 2\mu\varepsilon_{12},$$

$$\sigma_{22} = \lambda \varepsilon_{11} + (2\mu + \lambda)\varepsilon_{22}, \tag{1.1.41}$$

and

$$\sigma_{33} = \lambda(\varepsilon_{11} + \varepsilon_{22}) = \nu(\sigma_{11} + \sigma_{22}), \ \sigma_{23} = \sigma_{13} = 0. \tag{1.1.42}$$

The linearly independent stress components are σ_{11}, σ_{12} and σ_{22}.

(iii) Equations of equilibrium:

$$\frac{\partial \sigma_{i1}}{\partial x_1} + \frac{\partial \sigma_{i2}}{\partial x_2} + F_i = 0, \ i = 1, 2. \tag{1.1.43}$$

(iv) The components of the traction vector are given by

$$t_i = \sigma_{i1} n_1 + \sigma_{i2} n_2, \ i = 1, 2, \tag{1.1.44}$$

where n_1, n_2 denote the components of the outward unit normal on the boundary of the elastic region.

(v) Compatibility equation (the Saint Venant condition):

$$\frac{\partial^2 \varepsilon_{11}}{\partial x_2^2} + \frac{\partial^2 \varepsilon_{22}}{\partial x_1^2} = 2 \frac{\partial^2 \varepsilon_{12}}{\partial x_1 \partial x_2}. \tag{1.1.45}$$

(vi) Levy's equation (compatibility equation in terms of stress):

$$\Delta(\sigma_{11} + \sigma_{22}) = 0. \tag{1.1.46}$$

In the case of two dimensions the equations are quite simple. In particular, one needs just one compatibility condition instead of six relations that occur in a general three–dimensional situation.

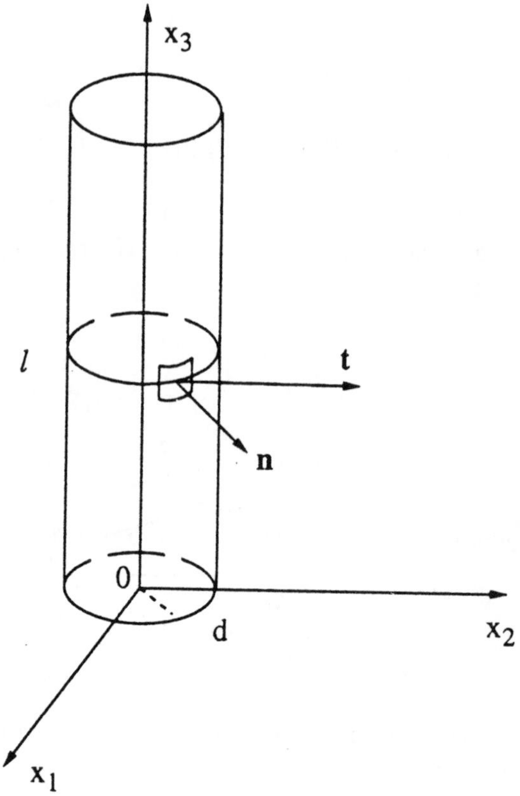

Fig. 1.2: A long cylindrical bar.

Plane stress, generalized plane stress. Here, we consider an example of a formulation which can be used for description of a stress–strain state within a thin elastic layer. A deformation is called the *plane stress,* if the components of the stress tensor satisfy the equalities

$$\sigma_{31} = \sigma_{32} = \sigma_{33} = 0. \tag{1.1.47}$$

Introduce a thin layer $\omega \times (-h, h)$, where ω is a bounded two–dimensional region (see Fig. 1.3). The components of the stress vector

$$\sigma^{(3)} = (\sigma_{13}, \sigma_{23}, \sigma_{33})^T$$

are assumed to be zero on all planes parallel to the flat surfaces $x_3 = \pm h$.

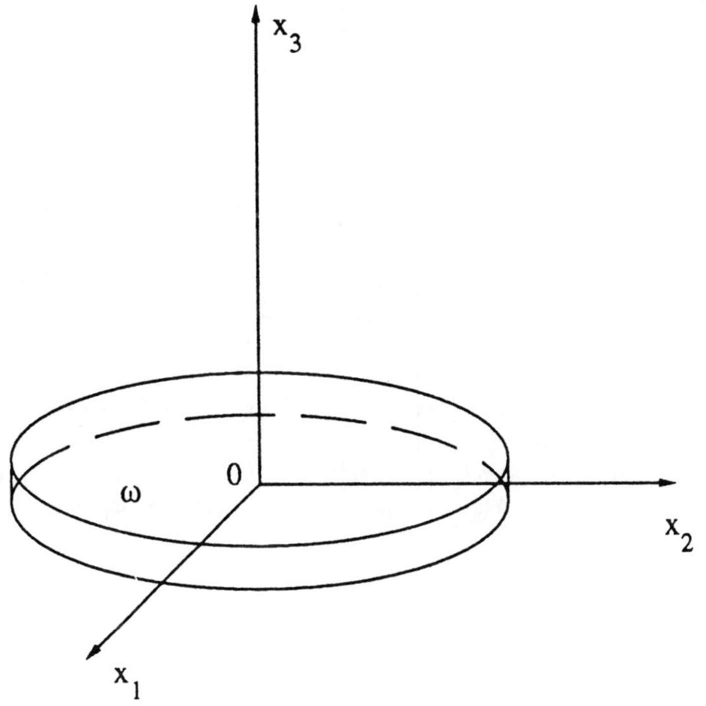

Fig. 1.3: A thin elastic layer.

Consider the stress–strain relations. We have

$$0 = \sigma_{33} = 2\mu\varepsilon_{33} + \lambda(\varepsilon_{11} + \varepsilon_{22} + \varepsilon_{33}).$$

Thus,

$$\varepsilon_{33} = -\frac{\lambda}{2\mu + \lambda}(\varepsilon_{11} + \varepsilon_{22})$$

and

$$\varepsilon_{11} + \varepsilon_{22} + \varepsilon_{33} = \frac{2\mu}{2\mu + \lambda}(\varepsilon_{11} + \varepsilon_{22}).$$

Introducing a new constant

$$\lambda^* = \frac{2\mu\lambda}{2\mu + \lambda},$$

we deduce that

$$\sigma_{11} = 2\mu\varepsilon_{11} + \lambda^*(\varepsilon_{11} + \varepsilon_{22}),$$

$$\sigma_{12} = 2\mu\varepsilon_{12},$$

$$\sigma_{22} = 2\mu\varepsilon_{22} + \lambda^*(\varepsilon_{11} + \varepsilon_{22}),$$

$$\sigma_{23} = \sigma_{13} = \sigma_{33} = 0. \tag{1.1.48}$$

The above equalities look similar to those in the case of plane strain. However, we still have the dependence on the variable x_3.

Assume that tractions are applied on the lateral surface of the thin plane and are symmetric with respect to the plane $x_3 = 0$. Also, suppose that $F_3 = 0$, and F_1, F_2 are even functions of x_3. Then the *averaging* over the thickness of a thin plate removes the dependence on x_3 and provides the *generalized plane stress.* In this case u_3 is an odd function with respect to x_3. The averaged components of the displacement vector are

$$\hat{u}_i = \frac{1}{2h}\int_{-h}^{h} u_i dx_3, \; i = 1, 2, \tag{1.1.49}$$

and

$$\hat{u}_3 = \frac{1}{2h}\int_{-h}^{h} u_3 dx_3 = 0.$$

It follows from the definition of plane stress that the functions u_1, u_2, σ_{11}, σ_{12}, σ_{22} are even with respect to x_3, and u_3, σ_{13}, σ_{23} are odd functions of x_3. Consequently, we have

$$\hat{\sigma}_{i3} = \frac{1}{2h}\int_{-h}^{h} \sigma_{i3} dx_3 = 0, \; i = 1, 2. \tag{1.1.50}$$

The averaged equations of equilibrium take the form

$$\frac{\partial \hat{\sigma}_{i1}}{\partial x_1} + \frac{\partial \hat{\sigma}_{i2}}{\partial x_2} + \hat{F}_i = 0, \; i = 1, 2, \tag{1.1.51}$$

where

$$\hat{F}_i = \frac{1}{2h}\int_{-h}^{h} F_i dx_3, \; i = 1, 2.$$

Hooke's law yields

$$\hat{\sigma}_{ii} = 2\mu\frac{\partial \hat{u}_i}{\partial x_i} + \lambda^*(\frac{\partial \hat{u}_1}{\partial x_1} + \frac{\partial \hat{u}_2}{\partial x_2}),\ i = 1,2,\ \hat{\sigma}_{12} = \mu(\frac{\partial \hat{u}_1}{\partial x_2} + \frac{\partial \hat{u}_2}{\partial x_1}), \quad (1.1.52)$$

with

$$\lambda^* = \frac{2\lambda\mu}{2\mu + \lambda}. \quad (1.1.53)$$

In the sense of formal analysis the generalized plane stress formulation is equivalent to the plane strain, subject to the replacement of the Lamé coefficient λ by λ^*. Further in the text, we shall often refer to the plane strain formulations, and one can see that the results, obtained for the plane strain state, are readily extended to the case of generalized plane stress.

Anti–plane shear. Now, let us consider an infinite elastic cylinder

$$\omega \times \mathbb{R}^1,$$

which axis coincides with the Ox_3 coordinate axis; $\omega \subset \mathbb{R}^2$. Assume that the body force density is equal to zero, and that the deformation of an infinite cylindrical region is determined by tractions directed along the generators of the cylinder in such a way that the displacement vector takes the form

$$\boldsymbol{u} = (0, 0, u_3(x_1, x_2))^T. \quad (1.1.54)$$

The nonzero components of the stress tensor are

$$\sigma_{i3} = \mu\frac{\partial u_3}{\partial x_i},\ i = 1,2, \quad (1.1.55)$$

where μ is the Lamé coefficient, which is usually called the *shear modulus*.

The Lamé system yields that

$$\Delta u_3(x_1, x_2) = 0,\ (x_1, x_2) \in \omega, \quad (1.1.56)$$

with Δ being the Laplacian operator. The traction boundary conditions are represented by

$$\sigma_{13}n_1 + \sigma_{23}n_2 = p(x_1, x_2),\ (x_1, x_2) \in \omega, \quad (1.1.57)$$

or, equivalently,

$$\mu \frac{\partial u_3}{\partial n}(x_1, x_2) = p(x_1, x_2), \ (x_1, x_2) \in \omega. \tag{1.1.58}$$

Here n_i, $i = 1, 2$, are components of the outward unit normal with respect to $\partial\omega$, and p is the given function.

Thus, in the case of an anti-plane shear we have the classical Neumann boundary value problem for the Laplacian equation in the two-dimensional region ω. When ω is bounded, the problem (1.1.56), (1.1.57) is solvable if and only if

$$\int_{\partial\omega} p(x_1, x_2)ds = 0. \tag{1.1.59}$$

Equation (1.1.59) is the balance relation for external load.

If ω is unbounded, then an additional radiation condition should be set in order to specify the behaviour of the displacement field at infinity.

The Airy function. Assume that we deal with the state of plane strain of an elastic region $\omega \in \mathbb{R}^2$, and that the body forces are absent. The equilibrium equations can be written in the form

$$\frac{\partial \sigma_{i1}}{\partial x_1} + \frac{\partial \sigma_{i2}}{\partial x_2} = 0, \ i = 1, 2, \ (x_1, x_2) \in \omega,$$

and it follows that there exist functions $A(x_1, x_2)$ and $B(x_2, x_2)$ which satisfy the equalities

$$\sigma_{11} = \frac{\partial A}{\partial x_2}, \ \sigma_{12} = -\frac{\partial A}{\partial x_1},$$

and

$$\sigma_{22} = \frac{\partial B}{\partial x_1}, \ \sigma_{12} = -\frac{\partial B}{\partial x_2}.$$

The Levy relation (1.1.46) yields

$$\frac{\partial A}{\partial x_1} = \frac{\partial B}{\partial x_2}.$$

Consequently, one can find a function Φ such that

$$\sigma_{11} = \frac{\partial^2 \Phi}{\partial x_2^2}, \ \sigma_{22} = \frac{\partial^2 \Phi}{\partial x_1^2}, \ \sigma_{12} = -\frac{\partial^2 \Phi}{\partial x_1 \partial x_2}. \tag{1.1.60}$$

The function Φ is called the *Airy stress function.* It follows from (1.1.46) that

$$\Delta\Delta\Phi(x_1, x_2) = 0, \ (x_1, x_2) \in \omega, \tag{1.1.61}$$

where

$$\Delta\Delta = \frac{\partial^4}{\partial x_1^4} + 2\frac{\partial^4}{\partial x_1^2 \partial x_2^2} + \frac{\partial^4}{\partial x_2^4}.$$

If tractions are applied on $\partial\omega$, then one can derive the following boundary conditions for Φ

$$\Phi(x_1, x_2) = q(x_1, x_2), \ \frac{\partial\Phi}{\partial n}(x_1, x_2) = g(x_1, x_2), \ (x_1, x_2) \in \omega. \tag{1.1.62}$$

The functions q and g depend on the external load. We do not give the explicit representation of these functions, and for more details we refer to the text [85].

The Airy function can be expressed in terms of two holomorphic functions of complex variable which are called the complex potentials of Kolosov–Muskhelishvili. Further in the text of the book we shall consider some examples that are treated with the use of complex potentials, and we shall show how to apply the conformal mapping theory in this case.

1.2 Cracks and inclusions

In this section we discuss mathematical models that describe stress–strain state of a nonhomogeneous elastic body containing cracks and inclusions.

In the mathematical theory of brittle fracture, cracks are regarded as surfaces of discontinuity of the components of the displacement vector. Near the edge of the crack the stress field has a singularity. Here we describe this singularity and recall the classical concept of the stress–intensity factors. For additional reading we recommend the classical papers [125], [116], [109].

For the case of a rigid inclusion one has to prescribe a displacement, corresponding to a linear combination of the rigid–body translation and rotation, on the boundary of the inclusion. In particular, when the rigid inclusion is fixed one has the homogeneous boundary condition of the Dirichlet type.

The elasticity problems on elastic inclusions involve additional conditions on the interface boundary. The ideal contact of the inclusion and the elastic matrix means continuity of the displacement vector and continuity of tractions across the interface surface.

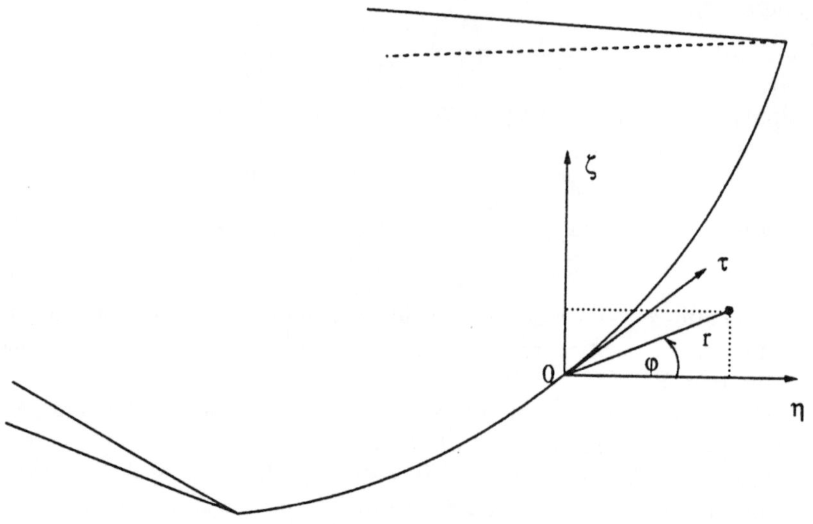

Fig. 1.4: A local system of coordinates on the crack front.

1.2.1 Singularity of stress at the crack front

Consider an infinite elastic medium with a crack occupying the region

$$M = \{\boldsymbol{x} \in \mathbb{R}^3 : (x_1, x_2) \in \omega \subset \mathbb{R}^2,\ x_3 = 0\},$$

where ω is a bounded domain, and $\partial\omega$ is smooth.

Assume that the elastic material is characterized by the Lamé moduli λ, μ, and let the displacement vector satisfy the homogeneous Lamé system

$$\mathbf{L}(\frac{\partial}{\partial \boldsymbol{x}})\boldsymbol{u}(\boldsymbol{x}) = \mathbf{0} \text{ in } \mathbb{R}^3 \setminus M. \tag{1.2.1}$$

The crack faces

$$\Gamma^{\pm} = \{\boldsymbol{x} : (x_1, x_2) \in \omega \subset \mathbb{R}^2,\ x_3 = \pm 0\}$$

are supposed to be free of tractions, which means that

$$\sigma_{i3}(\boldsymbol{u};\boldsymbol{x}) = 0,\ (x_1, x_2) \in \omega,\ x_3 = \pm 0,\ i = 1,2,3. \tag{1.2.2}$$

At infinity the constant load is given:

$$\sigma_{ij} \to \sigma_{ij}^{\infty},\ \text{as } \|\boldsymbol{x}\| \to \infty. \tag{1.2.3}$$

On the crack front we set the local system of coordinates (η, τ, ζ), where η and τ correspond to the outward normal direction and to the tangent direction on $\partial\omega$, respectively (see Fig. 1.4); $\zeta = x_3$. In the plane $O\eta\zeta$ the polar coordinates (r, φ) are introduced. The displacement and stress components near the crack front are given by

$$u_\eta = \frac{1}{\mu}\sqrt{\frac{r}{2\pi}}\left\{ K_I \cos\frac{\varphi}{2}[(1-2\nu) + \sin^2\frac{\varphi}{2}]\right.$$

$$\left. + K_{II} \sin\frac{\varphi}{2}[2(1-\nu) + \cos^2\frac{\varphi}{2}]\right\} + O(r),$$

$$u_\tau = \frac{K_{III}}{\mu}\sqrt{\frac{2r}{\pi}} \sin\frac{\varphi}{2} + O(r),$$

$$u_\zeta = \frac{1}{\mu}\sqrt{\frac{r}{2\pi}}\left\{ K_I \sin\frac{\varphi}{2}[2(1-\nu) - \cos^2\frac{\varphi}{2}]\right.$$

$$\left. - K_{II} \cos\frac{\varphi}{2}[(1-2\nu) - \sin^2\frac{\varphi}{2}]\right\} + O(r), \tag{1.2.4}$$

and

$$\sigma_{\eta\eta} = \frac{1}{4\sqrt{2\pi r}}\left\{ K_I[3\cos\frac{\varphi}{2} + \cos\frac{5\varphi}{2}]\right.$$

$$\left. - K_{II}[7\sin\frac{\varphi}{2} + \sin\frac{5\varphi}{2}]\right\} + O(1),$$

$$
\begin{aligned}
\sigma_{\tau\tau} &= \frac{2\nu}{\sqrt{2\pi r}}\left\{K_I \cos\frac{\varphi}{2} - K_{II}\sin\frac{\varphi}{2}\right\} + O(1),\\
\sigma_{\zeta\zeta} &= \frac{1}{4\sqrt{2\pi r}}\left\{K_I[5\cos\frac{\varphi}{2} - \cos\frac{5\varphi}{2}]\right.\\
&\qquad \left. -K_{II}[\sin\frac{\varphi}{2} - \sin\frac{5\varphi}{2}]\right\} + O(1),\\
\sigma_{\eta\zeta} &= -\frac{1}{4\sqrt{2\pi r}}\left\{K_I[\sin\frac{\varphi}{2} - \sin\frac{5\varphi}{2}]\right.\\
&\qquad \left. -K_{II}[3\cos\frac{\varphi}{2} + \cos\frac{5\varphi}{2}]\right\} + O(1),\\
\sigma_{\tau\zeta} &= \frac{K_{III}}{\sqrt{2\pi r}}\cos\frac{\varphi}{2} + O(1),\\
\sigma_{\tau\eta} &= -\frac{K_{III}}{\sqrt{2\pi r}}\sin\frac{\varphi}{2} + O(1).
\end{aligned}
\tag{1.2.5}
$$

The stress components have a singularity $O(1/\sqrt{r})$, as $r \to 0$. The constant coefficients K_I, K_{II} and K_{III} are called the stress–intensity factors corresponding to the Mode-I, Mode-II and Mode-III loading, respectively.

The derivation of the above formulae is presented in [116], pp. 149–150. Note that in comparison with [116] we have an extra factor $1/\sqrt{\pi}$ in the representations (1.2.4), (1.2.5).

In fact, the solution of the boundary value problem (1.2.1)–(1.2.3) is characterized by the infinite elastic energy. Due to linearity of the formulation, one can subtract a linear polynomial displacement field and set zero stress at infinity and certain tractions on the crack faces. Of course, after these transformations the entire displacement field will change. However, the local representations (1.2.4), (1.2.5) will be the same.

It is important to mention that a variation of the external load contributes to the coefficients K_I, K_{II}, K_{III} only, and the structure of the expansions (1.2.4), (1.2.5) does not change.

In a general three–dimensional situation the Mode–I uncouples from the Modes II and III; however, the stress–intensity factors K_{II} and K_{III} are coupled. For the case of a plane strain state one has just two coefficients K_I and K_{II} which correspond to the normal load and the in–plane shear load. The case of an anti–plane shear of an elastic body with a crack involves the coefficient K_{III} only.

The stress–intensity factors play an important role in fracture mechanics, where they are used in order to set the criteria of fracture. Further in the text we shall consider the examples which involve the fracture criteria. In the present introductory part we would like to cite papers [29], [34], [9], [126] that give the basis for the mathematical concept of the fracture criteria.

1.2.2 Interface conditions for elastic inclusions

We shall describe the formulation of boundary value problems of elasticity posed in an inhomogeneous domain. It is assumed that an elastic body Ω contains an inclusion g with the elastic moduli λ°, μ° which differ from the moduli λ, μ of elastic matrix $\Omega \setminus \bar{g}$. For the sake of simplicity assume that the body force density is equal to zero, so that the displacement vector satisfies the homogeneous Lamé system. The ideal contact is assumed on ∂g.

First, suppose that Ω is bounded, and on the boundary $\partial\Omega$ we prescribe tractions. Then, the displacement vector satisfies the following boundary value problem

$$\mathbf{L}(\frac{\partial}{\partial \boldsymbol{x}})\boldsymbol{u}(\boldsymbol{x}) = \mathbf{0}, \quad \boldsymbol{x} \in \Omega \setminus \bar{g}, \tag{1.2.6}$$

$$\mathbf{L}^\circ(\frac{\partial}{\partial \boldsymbol{x}})\boldsymbol{u}^\circ(\boldsymbol{x}) = \mathbf{0}, \quad \boldsymbol{x} \in g, \tag{1.2.7}$$

where the superscript "o" marks all quantities and differential operators related to the inclusion. The contact conditions on ∂g are

$$\boldsymbol{u}(\boldsymbol{x}) = \boldsymbol{u}^\circ(\boldsymbol{x}), \ \boldsymbol{\sigma}^{(n)}(\boldsymbol{u}; \boldsymbol{x}) = \boldsymbol{\sigma}^{\circ,(n)}(\boldsymbol{u}^\circ; \boldsymbol{x}), \quad \boldsymbol{x} \in \partial g. \tag{1.2.8}$$

On the external boundary we have

$$\boldsymbol{\sigma}^{(n)}(\boldsymbol{u}; \boldsymbol{x}) = \mathbf{p}(\boldsymbol{x}), \ \boldsymbol{x} \in \partial\Omega. \tag{1.2.9}$$

The external load is assumed to be self–balanced, which means that

$$\int_{\partial\Omega} \mathbf{p}(\boldsymbol{x})ds = \mathbf{0}, \int_{\partial\Omega} \boldsymbol{x} \times \mathbf{p}(\boldsymbol{x})ds = \mathbf{0}. \tag{1.2.10}$$

A solution of the problem (1.2.6)–(1.2.10) is not unique. It is defined up to an arbitrary rigid body displacement that may involve translations and rotations of the whole region Ω.

In fact, the relation (1.2.9) can be replaced by the boundary condition of the Dirichlet type or the mixed boundary condition, where the displacement vector is specified on some part of $\partial\Omega$. In this case the boundary value problem will be always solvable, and there will be no need to set the additional solvability conditions (1.2.10).

Also, it should be mentioned that in the case where the region Ω is unbounded one has to set some conditions at infinity. An example of the anti–plane shear problem for an infinite medium with an inclusion is considered in the next section.

1.3 The Pólya–Szegö matrix. Anti-plane shear

We shall talk on the integral characteristics of defects in elastic continuum. By "defects" we mean inclusions or cavities. The idea, established by Pólya and Szegö in [107], is to introduce a matrix which characterizes perturbations of certain polynomial fields at infinity due to the presence of the defect, and allows one to evaluate the energy increment. This matrix will be called the Pólya–Szegö matrix. It turns out that the Pólya–Szegö matrix represents a tensor. In fact, this tensor quantity depends on the shape of the defect and elastic moduli of the media, and it is independent of the external load. The case of the Neumann boundary value problem was considered in [107]. Here, we shall perform similar analysis for the case of an anti–plane shear of a medium with an inclusion. Further in the text of the book we consider the Pólya–Szegö matrix for the Lamé operator. For additional reading we recommend the papers [131], [76], [66], [83] and [75].

1.3.1 Anti-plane shear of a finite elastic region with a small inclusion

Consider a finite region $\Omega \subset \mathbb{R}^2$ which contains a small inclusion $g_\varepsilon = \{\boldsymbol{x} : \varepsilon^{-1}\boldsymbol{x} \in g\}$; g is a finite two–dimensional region with smooth boundary; $O \in g$. A small positive parameter ε is defined by

$$\varepsilon = \text{diam}\, g_\varepsilon / \text{diam}\, \Omega.$$

Also it is assumed that $\text{dist}(g_\varepsilon, \partial\Omega)$ has the same order of magnitude as $\text{diam}\, \Omega$. Tractions are applied on the external boundary, and the body force densities are set to be zero. The ideal contact is assumed on the interface boundary.

The displacements u_3, u_3° in g_ε and $\Omega_\varepsilon := \Omega \setminus \bar{g}_\varepsilon$ satisfy the equations

$$\Delta u_3(\boldsymbol{x}, \varepsilon) = 0, \; \boldsymbol{x} \in \Omega_\varepsilon, \tag{1.3.1}$$

$$\Delta u_3^\circ(\boldsymbol{x}, \varepsilon) = 0, \; \boldsymbol{x} \in g_\varepsilon, \tag{1.3.2}$$

interface boundary conditions

$$u_3 = u_3^\circ, \; \mu \frac{\partial u_3}{\partial n} = \mu^\circ \frac{\partial u_3^\circ}{\partial n} \; \text{ on } \partial g_\varepsilon, \tag{1.3.3}$$

and the Neumann boundary condition

$$\mu \frac{\partial u_3}{\partial n}(\boldsymbol{x}, \varepsilon) = p(\boldsymbol{x}), \; \boldsymbol{x} \in \partial\Omega. \tag{1.3.4}$$

The right–hand side in (1.3.4) is subjected to the orthogonality condition

$$\int_{\partial\Omega} p(\boldsymbol{x}) ds = 0. \tag{1.3.5}$$

The orientation of the normal $\boldsymbol{n}$ has been chosen to be outward with respect to $\Omega \setminus g_\varepsilon$.

A solution of (1.3.1)–(1.3.5) is determined up to an arbitrary additive constant.

Let v denote a solution of the Neumann boundary value problem in the homogeneous domain Ω (without inclusion). Clearly, the field $v(\boldsymbol{x})$ satisfies all the equations (1.3.1)–(1.3.5) except the second interface condition (1.3.3) which provides the continuity of tractions.

In order to compensate the leading part of this error, one can introduce auxiliary fields $w(\boldsymbol{x}/\varepsilon)$ and $w^\circ(\boldsymbol{x}/\varepsilon)$ such that

$$\Delta w^\circ(\boldsymbol{\xi}) = 0, \; \boldsymbol{\xi} \in g, \tag{1.3.6}$$

$$\Delta w(\boldsymbol{\xi}) = 0, \; \boldsymbol{\xi} \in \mathbb{R}^2 \setminus \bar{g}, \tag{1.3.7}$$

$$\mu \frac{\partial w}{\partial n}(\boldsymbol{\xi}) - \mu^{\circ} \frac{\partial w^{\circ}}{\partial n}(\boldsymbol{\xi}) = (\mu^{\circ} - \mu) \frac{\partial}{\partial n} \left(\boldsymbol{\xi} \cdot (\nabla v)(\mathbf{0}) \right),$$

$$w(\boldsymbol{\xi}) = w^{\circ}(\boldsymbol{\xi}), \quad \boldsymbol{\xi} \in \partial g, \tag{1.3.8}$$

and

$$w(\boldsymbol{\xi}) \to 0, \quad \text{as } ||\boldsymbol{\xi}|| \to \infty. \tag{1.3.9}$$

Here $\boldsymbol{n}$ denotes the unit inward normal on ∂g.

If we know the functions w, w°, then the displacements u_3, u_3° can be approximated by

$$u_3(\boldsymbol{x}, \varepsilon) \sim v(\boldsymbol{x}) + \varepsilon w(\frac{\boldsymbol{x}}{\varepsilon}), \quad \boldsymbol{x} \in \Omega_{\varepsilon}, \tag{1.3.10}$$

and

$$u_3^{\circ}(\boldsymbol{x}, \varepsilon) \sim v(\mathbf{0}) + \boldsymbol{x} \cdot (\nabla v)(\mathbf{0}) + \varepsilon w^{\circ}(\frac{\boldsymbol{x}}{\varepsilon}), \quad \boldsymbol{x} \in g_{\varepsilon}. \tag{1.3.11}$$

Next, we shall consider in more details the properties of the fields w, w°, specified in an infinite elastic plane with a finite inclusion.

1.3.2 Auxiliary boundary value problems in an infinite region

Let us represent the field w from (1.3.10) as the linear combination

$$w(\boldsymbol{\xi}) = \sum_{j=1}^{2} \frac{\partial v}{\partial x_j}(\mathbf{0}) w_j(\boldsymbol{\xi}), \tag{1.3.12}$$

where the functions w_j are given as solutions of the following problem

$$-\mu^{\circ} \Delta w_j^{\circ}(\boldsymbol{\xi}) = 0, \ \boldsymbol{\xi} \in g, \tag{1.3.13}$$

$$-\mu \Delta w_j(\boldsymbol{\xi}) = 0, \ \boldsymbol{\xi} \in \mathbb{R}^2 \setminus \overline{g}, \tag{1.3.14}$$

$$w_j(\boldsymbol{\xi}) = w_j^{\circ}(\boldsymbol{\xi}),$$

$$\mu \frac{\partial w_j}{\partial n}(\boldsymbol{\xi}) - \mu^{\circ} \frac{\partial w_j^{\circ}}{\partial n}(\boldsymbol{\xi}) = (\mu^{\circ} - \mu) n_j(\boldsymbol{\xi}), \ \boldsymbol{\xi} \in \partial g, \tag{1.3.15}$$

$$w_j(\boldsymbol{\xi}) \to 0, \quad \text{as} \quad \|\boldsymbol{\xi}\| \to \infty. \tag{1.3.16}$$

Here the index j may take values 1 or 2.

The right–hand side (1.3.15) of the jump in tractions is self–balanced

$$\int_{\partial g} n_j(\boldsymbol{\xi}) ds = 0, \; j = 1, 2,$$

and therefore, a solution of (1.3.13)–(1.3.16) decays at infinity like a linear combination of first-order derivatives of the fundamental solution of the equation (1.3.14) in two dimensions (it means $O(\|\boldsymbol{\xi}\|^{-1})$). Thus, the radiation condition (1.3.16) can be written in a more precise form

$$w_j = -\frac{1}{2\pi\mu} \sum_{i=1}^{2} m_{ji} \frac{\xi_i}{\|\boldsymbol{\xi}\|^2} + O(\|\boldsymbol{\xi}\|^{-2}), \; \|\boldsymbol{\xi}\| \to \infty. \tag{1.3.17}$$

Note that the function W_j, W_j°, defined by

$$W_j(\boldsymbol{\xi}) = \xi_j + w_j(\boldsymbol{\xi}), \; \boldsymbol{\xi} \in \mathbb{R}^2 \setminus \overline{g}, \tag{1.3.18}$$

$$W_j^\circ(\boldsymbol{\xi}) = \xi_j + w_j^\circ(\boldsymbol{\xi}), \; \boldsymbol{\xi} \in g, \tag{1.3.19}$$

satisfy the homogeneous boundary value problem (1.3.13)–(1.3.15) and behave like linear functions at infinity.

1.3.3 Definition of the Pólya–Szegö matrix

The symmetric matrix $\boldsymbol{m} = (m_{ij})_{i,j=1}^2$ of coefficients from the sum (1.3.17) is called *the Pólya–Szegö matrix* of the inclusion g. This matrix characterizes the perturbation of the linear displacement field due to the presence of the inclusion.

It will be shown that the coefficients m_{ij} can be related to the scalar product of the functions w_i and w_j in the energy space. The Green formula, applied to the sets of functions $\{W_j, W_j^\circ\}$ and $\{w_k, w_k^\circ\}$ in a large disk $D_r = \{\boldsymbol{\xi} : \|\boldsymbol{\xi}\| < r\}$ containing g, gives

$$0 = \mu \int_{D_r \setminus \overline{g}} (w_k \Delta W_j - W_j \Delta w_k) d\boldsymbol{\xi} + \mu \int_g (w_k^\circ \Delta W_j^\circ - W_j^\circ \Delta w_k^\circ) d\boldsymbol{\xi}$$

$$= \mu \int_{\partial D_r} (\frac{\partial W_j}{\partial r} w_k - \frac{\partial w_k}{\partial r} W_j) ds + \int_{\partial g} \left\{ \mu (\frac{\partial W_j}{\partial n} w_k - \frac{\partial w_k}{\partial n} W_j) \right.$$

$$- \mu^\circ \Big(\frac{\partial W_j^\circ}{\partial n} w_k^\circ - \frac{\partial w_k^\circ}{\partial n} W_j^\circ\Big) \Big\} ds = I_{jk}^{(1)} + I_{jk}^{(2)}, \tag{1.3.20}$$

where

$$I_{jk}^{(1)} = -\frac{1}{\pi} \int_0^{2\pi} \Big\{ \xi_j \sum_{i=1}^{2} m_{ki} \frac{\xi_i}{||\xi||^3} \Big\} ||\xi|| d\varphi + o(1) \to -m_{jk},$$

as $r \to \infty$. (1.3.21)

$$I_{jk}^{(2)} = (\mu - \mu^\circ) \int_{\partial g} n_k W_j^\circ ds = (\mu - \mu^\circ) \int_{\partial g} \frac{\partial \xi_k}{\partial n} \xi_j ds$$
$$- \int_{\partial g} (\mu \frac{\partial w_k}{\partial n} w_j - \mu^\circ \frac{\partial w_k^\circ}{\partial n} w_j^\circ) ds. \tag{1.3.22}$$

The second integral in (1.3.22) can be interpreted as the scalar product (w_k, w_j) of solutions of the problem (1.3.13)–(1.3.16) in the energy space. Then the quantity (1.3.22) can be represented by

$$I_{jk}^{(2)} = -(w_k, w_j) + (\mu^\circ - \mu) \int_g \nabla \xi_k \cdot \nabla \xi_j d\boldsymbol{\xi}. \tag{1.3.23}$$

Introduce the positive definite matrix $(\mathcal{M}_{jk})_{j,k=1}^2$ with components $\mathcal{M}_{jk} = (w_j, w_k)$. Then, taking the limit $r \to \infty$ in (1.3.20), one has

$$\mathcal{M}_{jk} = -m_{jk} + \delta_{jk}(\mu^\circ - \mu) \text{mes } g, \; j, k = 1, 2, \tag{1.3.24}$$

where δ_{jk} is the Kronecker delta, and mes g denotes the area of the plane region g.

When $\mu^\circ \to 0$ we obtain the case of a cavity g, and the interface conditions (1.3.8) should be replaced by the homogeneous Neumann boundary condition. It follows from (1.3.24) that in the case of a cavity the Pólya–Szegö matrix $\boldsymbol{m}$ is negative definite.

1.3.4 The energy evaluation

It turns out that the matrix $\boldsymbol{m}$ provides the canonical representation for the energy increment when a defect is introduced in the elastic medium.

The potential energy of the inhomogeneous elastic body Ω is defined by

$$\mathcal{E}_\varepsilon(u; \Omega) := - \int_{\partial \Omega} p(\boldsymbol{x}) u_3(\boldsymbol{x}, \varepsilon) ds. \tag{1.3.25}$$

Unfortunately, the first-order approximations (1.3.10), (1.3.11) do not allow one to evaluate the energy change due to the presence of the small inclusion g_ε, and, hence, the more precise relation should be used:

$$u_3(\boldsymbol{x}, \varepsilon) \sim v(\boldsymbol{x}) + \varepsilon w(\boldsymbol{x}/\varepsilon) + \varepsilon^2 \mathcal{V}(\boldsymbol{x}), \ \boldsymbol{x} \in \Omega \setminus \bar{g}_\varepsilon. \tag{1.3.26}$$

The function $\mathcal{V}$ is defined in Ω and compensates the error produced by w in the Neumann boundary condition (1.3.4). It satisfies the boundary value problem

$$\Delta \mathcal{V}(\boldsymbol{x}) = 0, \ \boldsymbol{x} \in \Omega, \tag{1.3.27}$$

$$\mu \frac{\partial \mathcal{V}}{\partial n}(\boldsymbol{x}) = \frac{1}{2\pi} \frac{\partial}{\partial n} \sum_{j,k=1}^{2} m_{jk} \frac{x_k}{\|\boldsymbol{x}\|^2} \frac{\partial v}{\partial x_j}(\mathbf{0}), \ \boldsymbol{x} \in \partial\Omega. \tag{1.3.28}$$

The function $\mathcal{V}$ can be written in the form

$$\mathcal{V}(\boldsymbol{x}) = \sum_{j,k=1}^{2} m_{jk} \left(T^{(k)}(\boldsymbol{x}) + \frac{1}{2\pi\mu} \frac{x_k}{\|\boldsymbol{x}\|^2} \right) \frac{\partial v}{\partial x_j}(\mathbf{0}). \tag{1.3.29}$$

where $T^{(k)}$, $k = 1, 2$, are singular at the origin, and the second term in brackets compensates the singularity. The functions $T^{(k)}$ are defined as solutions of the Neumann boundary value problems

$$\mu \Delta T^{(k)}(\boldsymbol{x}) + \frac{\partial \delta}{\partial x_k}(\boldsymbol{x}) = 0, \ \boldsymbol{x} \in \Omega, \tag{1.3.30}$$

$$\frac{\partial T^{(k)}}{\partial n}(\boldsymbol{x}) = 0, \ \boldsymbol{x} \in \partial\Omega. \tag{1.3.31}$$

Here $\delta(\boldsymbol{x})$ is the Dirac delta function, and solutions of (1.3.30), (1.3.31) should be understood in the sense of distributions.

We recall that the function v satisfies the Neumann boundary value problem in a homogeneous region Ω

$$\Delta v(\boldsymbol{x}) = 0, \ \boldsymbol{x} \in \Omega; \ \ \mu \frac{\partial v}{\partial n}(\boldsymbol{x}) = p(\boldsymbol{x}), \ \boldsymbol{x} \in \partial\Omega. \tag{1.3.32}$$

On the boundary of the region Ω we have

$$u_3(\boldsymbol{x}, \varepsilon) - v(\boldsymbol{x}) \sim \varepsilon^2 \sum_{j,k=1}^{2} m_{jk} T^{(k)}(\boldsymbol{x}) \frac{\partial v}{\partial x_j}(\mathbf{0}), \ \ \boldsymbol{x} \in \partial\Omega. \tag{1.3.33}$$

Consequently, the leading part of the potential energy increment is specified by

$$\mathcal{E}_\varepsilon(u;\Omega) - \mathcal{E}_0(v;\Omega) \sim -\varepsilon^2 \sum_{j,k=1}^{2} m_{jk} \frac{\partial v}{\partial x_j}(0) \int_{\partial\Omega} p(\boldsymbol{x}) T^{(k)}(\boldsymbol{x}) ds. \quad (1.3.34)$$

Using the Green formula, one can simplify the integral from (1.3.34)

$$\int_{\partial g} p(\boldsymbol{x}) T^{(k)}(\boldsymbol{x}) ds = \mu \int_{\partial\Omega} (T^{(k)}(\boldsymbol{x}) \Delta v(\boldsymbol{x}) - v(\boldsymbol{x}) \Delta T^{(k)}(\boldsymbol{x})) d\boldsymbol{x}$$

$$= \int_{\Omega} v(\boldsymbol{x}) \frac{\partial \delta}{\partial x_k}(\boldsymbol{x}) d\boldsymbol{x} = -\frac{\partial v}{\partial x_k}(0). \quad (1.3.35)$$

The formulae (1.3.34), (1.3.35) imply

$$\mathcal{E}_\varepsilon(u;\Omega) - \mathcal{E}_0(v;\Omega) \sim \varepsilon^2 \sum_{j,k=1}^{2} \frac{\partial v}{\partial x_j}(0) m_{jk} \frac{\partial v}{\partial x_k}(0)$$

$$= \varepsilon^2 \nabla v(0) \cdot \boldsymbol{m} \nabla v(\boldsymbol{x})(0), \quad (1.3.36)$$

i.e. the leading part of the potential energy increment is represented as a quadratic form with the matrix $\boldsymbol{m}$, and in order to evaluate the energy change one needs a solution of the boundary value problem (1.3.32), corresponding to the homogeneous elastic region Ω, and the Pólya–Szegö matrix of the inclusion. The quantity (1.3.36) is invariant with respect to a rotation of the coordinate system, and it is readily verified that $(m_{jk})_{j,k=1}^2$ is the rank-2 Cartesian tensor.

In the last chapter of the book we shall consider the definition of the Pólya–Szegö matrix for the case of the Lamé system. In general, it is possible to establish a correspondence between its components and a certain Cartesian tensor of the rank 4.

1.4 The concept of the asymptotic expansions

In the previous section an example of a finite elastic body, containing a small inclusion, was considered. The reader can notice that we did not construct an exact solution. The objective was to evaluate the displacement to certain order in ε, and to approximate the change of the potential energy due to the presence of a small defect. In fact we have used here the technique which is called the method of compound asymptotic expansions. It gives a certain motivation to recall the definition of an asymptotic expansion and to look at the classification of the asymptotic problems presented in this book. Also, we can recommend the classical texts on asymptotic expansions [123], [86] that can be used for additional reading.

Suppose that one has a pair of functions $\alpha(\boldsymbol{x},\varepsilon)$ and $\beta(\boldsymbol{x},\varepsilon)$ defined in a domain $\Omega \subset \mathbb{R}^3$, and ε is a small positive parameter. The domain Ω is assumed to be independent of ε. We write

$$\alpha(\boldsymbol{x},\varepsilon) = O(\beta(\boldsymbol{x},\varepsilon)) \text{ in } \Omega, \tag{1.4.1}$$

as $\varepsilon \to 0$, if for any $\boldsymbol{x} \in \Omega$ there exist positive quantities $A(\boldsymbol{x})$, $C(\boldsymbol{x})$ such that

$$|\alpha(\boldsymbol{x},\varepsilon)| \leq C(\boldsymbol{x})|\beta(\boldsymbol{x},\varepsilon)|, \tag{1.4.2}$$

for any ε from the interval $(0, A)$. We say that the relation (1.4.1) is uniformly valid in Ω if the quantities A and C do not change within Ω.

If for any $\boldsymbol{x} \in \omega$ and any $C > 0$ there exists a quantity $A(\boldsymbol{x}, C)$ such that

$$|\alpha(\boldsymbol{x},\varepsilon)| \leq C|\beta(\boldsymbol{x},\varepsilon)|, \tag{1.4.3}$$

for any ε from the interval $(0, A)$, then we write

$$\alpha(\boldsymbol{x},\varepsilon) = o(\beta(\boldsymbol{x},\varepsilon)) \text{ in } \Omega, \tag{1.4.4}$$

as $\varepsilon \to 0$. If for sufficiently small positive ε the function β does not vanish, then the relation (1.4.4) yields that

$$\lim_{\varepsilon \to 0} \frac{\alpha}{\beta} = 0.$$

In particular, if $\alpha(\boldsymbol{x},\varepsilon) = o(\beta(\boldsymbol{x},\varepsilon))$, then also $\alpha(\boldsymbol{x},\varepsilon) = O(\beta(\boldsymbol{x},\varepsilon))$. We say that the relation (1.4.4) is uniformly valid in Ω, if $A = A(C)$ and it is the same for all points $\boldsymbol{x} \in \Omega$.

The series

$$\sum_{i \geq k} \alpha_i(\varepsilon) U_i(\boldsymbol{x}),$$

with $k \geq 0$ and a sequence of functions α_i chosen in such a way that

$$\alpha_{i+1}(\varepsilon) = o(\alpha_i(\varepsilon)), \text{ as } \varepsilon \to 0,$$

is called the *asymptotic expansion* of the function $u(\boldsymbol{x}, \varepsilon)$ if

$$u(\boldsymbol{x}, \varepsilon) - \sum_{i=k}^{K} \alpha_i(\varepsilon) U_i(\boldsymbol{x}) = O(\alpha_{K+1}), \quad \text{as } \varepsilon \to 0, \tag{1.4.5}$$

or

$$\|u(\boldsymbol{x}, \varepsilon) - \sum_{i=k}^{K} \alpha_i(\varepsilon) U_i(\boldsymbol{x})\|_{\mathcal{H}} = O(\alpha_{K+1}), \quad \text{as } \varepsilon \to 0, \tag{1.4.6}$$

where $\mathcal{H}$ is the given functional space (for example, it can be the energy space).

The so–called *regular perturbation problem* has a solution that can be represented by a single asymptotic expansion which is uniformly valid in the entire domain. In the next chapter we shall consider an example of the regular perturbation problem for a crack with smoothly closed edges.

However, in the previous section on the Pólya–Szegö matrix we were able to see that for a domain with a small defect some auxiliary fields, which solve certain boundary value problems in an infinite region, may be required. These functions, defined in the scaled variables, are called the terms of the *boundary layer* type, and they describe the asymptotics of the displacement in the vicinity of the small inclusion. Such problems, which solution is characterized by different asymptotic expansions in certain distinguished subsets of the original region Ω, are called the *singular perturbation problems.* Most of formulations considered in this book are posed in domains which depend on a small parameter, and the solution is approximated by an asymptotic series where the coefficients are given as solutions of some auxiliary boundary value problems related to model (limit) domains independent of ε. Two asymptotic techniques will be used. The first is the method of *matched asymptotic expansions* established by Van Dyke [123], where the "outer" and the "inner" expansions hold in different subdomains of the entire region Ω, and these expansions are matched in a certain "layer" which surrounds a singular point of the boundary of the domain. An example of this technique is presented in the section on the stress–strain state around a thin rectangular cavity. The second is the method of *compound asymptotic expansions* which was demonstrated already in the previous section; the algorithm of redistribution of discrepancies is used in such a way that each term of the asymptotic series compensates discrepancies left by the previous terms in equations and boundary conditions. In particular, the method of compound asymptotic expansions involves the boundary layer which decays at infinity; the latter is not required by the method of matched asymptotic expansions. The book [59] can be recommended for detailed mathematical analysis based on the method of compound asymptotic expansions in domains with singular perturbations of the boundary.

1.5 Exercises

1. Consider a semi–infinite cut in two dimensions

$$M = \{\boldsymbol{x} \in \mathbb{R}^2 : x_1 < 0,\ x_2 = 0\}.$$

By direct calculation verify that the field (u_r, u_φ) with components

$$u_r = \frac{1}{2(1+\varkappa)\sqrt{2\pi r}} \left\{ \left((2\varkappa + 1)\cos\frac{3\varphi}{2} - 3\cos\frac{\varphi}{2} \right) c_1 \right.$$

$$\left. + \left((2\varkappa + 1)\sin\frac{3\varphi}{2} - \sin\frac{\varphi}{2} \right) c_2 \right\},$$

$$u_\varphi = \frac{1}{2(1+\varkappa)\sqrt{2\pi r}} \left\{ \left(-(2\varkappa - 1)\sin\frac{3\varphi}{2} + 3\sin\frac{\varphi}{2} \right) c_1 \right.$$

$$\left. + \left((2\varkappa - 1)\cos\frac{3\varphi}{2} - \cos\frac{\varphi}{2} \right) c_2 \right\},$$

satisfies the homogeneous Lamé system in $\mathbb{R}^2 \setminus M$ and the homogeneous traction boundary conditions

$$\sigma_{r\varphi} = \sigma_{\varphi\varphi} = 0, \quad \text{as } \varphi = \pm\pi.$$

Here $\varkappa = 3 - 4\nu$ and (r, φ) are the polar coordinates with the centre at the tip of the crack. Note that the above field has an infinite elastic energy. This displacement cannot be related to any physical phenomena. However, its components are used as weight functions, and then the Green formula for the Lamé operator (it is also called the Betti formula) allows one to evaluate the stress intensity factors for a given semi–infinite crack.

2. In Section 1.3 we used the boundary layer w which solves the problem (1.3.6)–(1.3.9) and compensates a discrepancy of v in the traction boundary condition (second relation in (1.3.3)). Now, we change the procedure and extend the field $v(\boldsymbol{x})$ inside the inclusion in such a way that the interface traction condition is satisfied. However, a jump will occur in the first equality (1.3.3). In order to compensate this jump, we introduce the fields z, z° which solve the boundary value problem

$$\Delta z(\boldsymbol{\xi}) = 0,\ \boldsymbol{\xi} \in \mathbb{R}^2 \setminus \overline{g},\quad \Delta z^{\circ}(\boldsymbol{\xi}) = 0,\ \boldsymbol{\xi} \in g,$$

and

$$z - z^{\circ} = (\frac{\mu}{\mu^{\circ}} - 1)\boldsymbol{\xi} \cdot (\nabla v)(\mathbf{0}),\ \ \mu \frac{\partial z}{\partial n} = \mu^{\circ} \frac{\partial z^{\circ}}{\partial n},\ \boldsymbol{\xi} \in \partial g.$$

$$z(\boldsymbol{\xi}) \to 0,\ \text{ as } \boldsymbol{\xi} \to \infty.$$

Then, similar to (1.3.12), we can write

$$z(\boldsymbol{\xi}) = \sum_{j=1}^{2} \frac{\partial v}{\partial x_j}(\mathbf{0}) z_j(\boldsymbol{\xi}),$$

where the functions z_j, $j = 1, 2$, and the corresponding functions z_j° within the inclusion satisfy the boundary value problems

$$\Delta z_j(\boldsymbol{\xi}) = 0,\ \boldsymbol{\xi} \in \mathbb{R}^2 \setminus \overline{g},\quad \Delta z_j^{\circ}(\boldsymbol{\xi}) = 0,\ \boldsymbol{\xi} \in g,$$

$$z_j - z_j^{\circ} = (\frac{\mu}{\mu^{\circ}} - 1)\boldsymbol{\xi}_j,\ \ \mu \frac{\partial z_j}{\partial n} = \mu^{\circ} \frac{\partial z_j^{\circ}}{\partial n},\ \boldsymbol{\xi} \in \partial g.$$

As $\boldsymbol{\xi} \to \infty$, the fields z_j admit the asymptotic representations

$$z_j = -\frac{1}{2\pi\mu} \sum_{k=1}^{2} m_{jk} \frac{\xi_k}{||\boldsymbol{\xi}||^2} + O(||\boldsymbol{\xi}||^{-2}).$$

Show that the symmetric matrix $\boldsymbol{m} = (m_{jk})_{j,k=1}^2$ and the 2×2 matrix $\mathbf{P}$ with components

$$P_{jk} = \mu \int_{g} \frac{\partial z_j}{\partial n}(z_k - z_k^{\circ}) ds,$$

where $\boldsymbol{n}$ is the inward unit normal with respect to g, are related by

$$\mathbf{P} = \boldsymbol{m} + \mu(\frac{\mu}{\mu^{\circ}} - 1)\text{mes } g\ \mathbf{I},$$

with $\mathbf{I}$ being the identity matrix. In particular, for the case of a rigid inclusion ($\mu^\circ \to \infty$) the above relation reduces to

$$m = \mathbf{P} + \mu \operatorname{mes} g \, \mathbf{I},$$

and one can see that for the case of a rigid inclusion the matrix m is positive definite.

2

Modelling of Cracks and Thin Inclusions in Elastic Media

In the present chapter we consider mathematical models of cracks and inclusions in elastic media. Each problem analyzed here involves a small positive parameter characterizing the geometry of the domain. The objective is to construct the asymptotic representation of the stress and displacement fields and, in some cases, to evaluate the critical load which provides the brittle fracture of the inhomogeneous elastic medium. Section 2.1 includes the analysis of the regular perturbation problem on a crack which is modelled as a thin cavity with smoothly closed edges. This model is quite simple to tackle, and it allows one to take into account the effect of the longitudinal compression along the crack; the analysis is based on the results of the papers [67], [68]. An example of a boundary value problem in a domain with a singular perturbation of the boundary is considered in Section 2.2, where we deal with a thin rectangular cavity, and a boundary layer is required to describe the stress–strain state near the ends of the cavity. We refer to the papers [65] and [78]. Also, we should mention the papers [128], [129] that can be used for additional reading. Sections 2.3 and 2.4, where we use the results [72] and [77], deal with the fracture criteria in elastic media. Also, it is highly recommended to read in addition the classical papers [29], [30], [126] and [97], [98]. A singular perturbation problem for a thin elastic inclusion is discussed in Section 2.5 (also see [71]).

2.1 Cracks with smoothly closed edges

In this section we describe a simple model that presents a crack as a thin cavity with smoothly closed edges. This model is treated within the regular perturbations approach and does not require a solution of the boundary layer type. It allows one to take into account the longitudinal (with respect to the crack orientation) external load. For this particular example we show how, with the use of the asymptotic methods, to calculate the stress–intensity factor and obtain the representation for the potential energy.

2.1.1 Two-dimensional cracks

Let $\Omega \in \mathbb{R}^2$ be a domain containing a segment $M = \{\boldsymbol{x} : x_2 = 0, |x_1| \leq a\}$, $h^0_\pm$ be smooth functions on $[-a, a]$ satisfying the equalities

$$h^0_\pm(a) = h^0_\pm(-a) = {h^0_\pm}'(a) = {h^0_\pm}'(-a) = 0, \qquad (2.1.1)$$

and $h_\pm(x_1) = a h^0_\pm(x_1)$.

Introduce a small nondimensional parameter $\varepsilon > 0$ and consider a "thin" domain (see Fig. 2.1)

$$G_\varepsilon = \{\boldsymbol{x} : |x_1| < a, -\varepsilon h_-(x_1) < x_2 < \varepsilon h_+(x_1)\},\ \Omega_\varepsilon = \Omega \setminus \overline{G}_\varepsilon.$$

Suppose that the displacement vector $\boldsymbol{u}(\varepsilon, \boldsymbol{x}) = (u_1(\varepsilon, \boldsymbol{x}), u_2(\varepsilon, \boldsymbol{x}))$ satisfies the problem

$$\mu \Delta \boldsymbol{u}(\varepsilon, \boldsymbol{x}) + (\lambda + \mu) \nabla \nabla \cdot \boldsymbol{u}(\varepsilon, \boldsymbol{x}) = \boldsymbol{0}, \quad \boldsymbol{x} \in \Omega_\varepsilon, \qquad (2.1.2)$$

$$\boldsymbol{\sigma}^{(n)}(\boldsymbol{u}; \varepsilon, \boldsymbol{x}) = \boldsymbol{p}^\pm(\varepsilon, x_1), \quad \boldsymbol{x} \in \gamma^\pm_\varepsilon = \{\boldsymbol{x} \in \partial G_\varepsilon : \pm x_2 > 0\}, \qquad (2.1.3)$$

$$\boldsymbol{\sigma}^{(n)}(\boldsymbol{u}; \varepsilon, \boldsymbol{x}) = \boldsymbol{p}(\boldsymbol{x}), \quad \boldsymbol{x} \in \partial \Omega, \qquad (2.1.4)$$

where $\boldsymbol{p}, \boldsymbol{p}^\pm$ are smooth vector–valued functions corresponding to external surface traction; λ, μ are the Lamé constants; $\boldsymbol{n} = (n_1, n_2)^T$ is the unit outward normal vector; $\sigma^{(n)}_i = n_j \sigma_{ij}$, $i, j = 1, 2, 3$; σ_{ij} are components of the stress tensor $\boldsymbol{\sigma}$.

Assume that the external load is self–balanced (the right–hand sides in the traction boundary conditions are orthogonal to the rigid–body displacements which include translations and a rotation)

$$\int_{\partial\Omega} \boldsymbol{p}(\boldsymbol{x})ds = -\sum_{\pm} \int_{\gamma_\varepsilon^\pm} \boldsymbol{p}^\pm(\varepsilon, x_1)ds,$$

$$\int_{\partial\Omega} (p_1x_2 - p_2x_1)ds = -\sum_{\pm} \int_{\gamma_\varepsilon^\pm} (p_1^\pm x_2 - p_2^\pm x_1)ds.$$

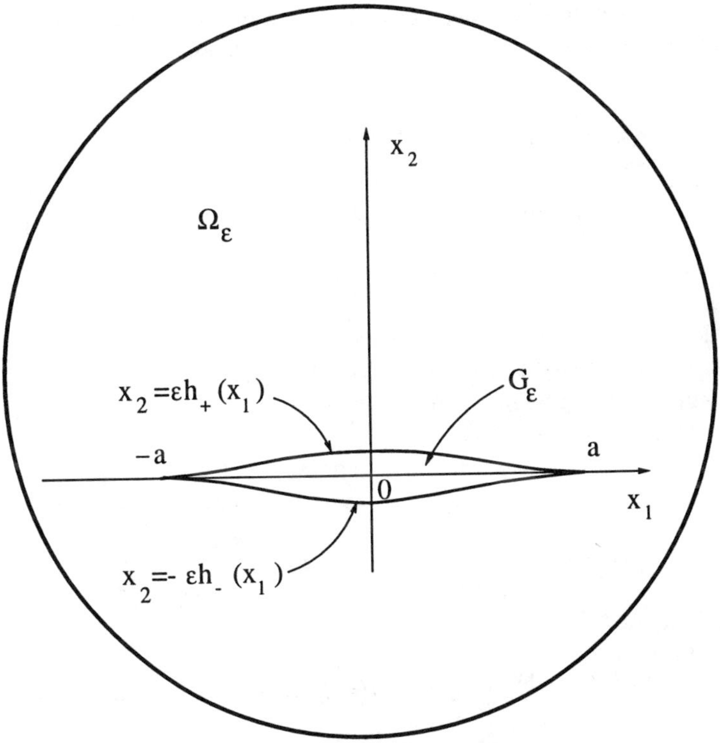

Fig. 2.1: Two–dimensional crack with smoothly closed edges.

The arc element ds on $\gamma_\varepsilon^\pm$ can be represented in the form

$$ds = \pm(1 + \varepsilon^2 h'_\pm(x_1)^2)^{1/2} dx_1.$$

It yields that

$$\int_{\partial\Omega} \boldsymbol{p}(\boldsymbol{x})ds = -\sum_{\pm} \int_{-a}^{a} \boldsymbol{p}^\pm(\varepsilon, x_1)(1 + \varepsilon^2 h'_\pm(x_1)^2)^{1/2} dx_1.$$

The following notation is used

$$p^{\pm}(\varepsilon, x_1) = (1 + \varepsilon^2 h'_{\pm}(x_1)^2)^{-1/2} q^{\pm}(x_1).$$

The vector $p^{\pm}$ can be expanded with respect to powers of ε

$$p^{\pm}(\varepsilon, x_1) = q^{\pm}(x_1) - \frac{1}{2}\varepsilon^2 h'_{\pm}(x_1)^2 q^{\pm}(x_1) + O(\varepsilon^4).$$

Consider the limiting case $\varepsilon = 0$. The domain Ω_ε becomes $\Omega_0 = \Omega \setminus M$ with the crack M, and the boundary value problem (2.1.2)–(2.1.4) takes the form

$$\mu\Delta u^{(0)}(x) + (\lambda + \mu)\nabla\nabla \cdot u^{(0)}(x) = 0, \quad x \in \Omega_0, \tag{2.1.5}$$

$$\sigma_{2j}(u^{(0)}; x_1, \pm 0) = \mp q_j^{\pm}, \quad |x_1| < a, \quad j = 1, 2, \tag{2.1.6}$$

$$\sigma^{(n)}(u^{(0)}; x) = p(x), \quad x \in \partial\Omega. \tag{2.1.7}$$

It is known (see, for example, Sedov [114]) that in a neighbourhood of the tips $O_{\pm} = (\pm a, 0)$ of the crack M the displacement vector $u^{(0)}$ has the asymptotic representation

$$u_r^{(0)\pm}(r_{\pm}, \varphi_{\pm}) = C_1^{\pm} \cos\varphi_{\pm} + C_2^{\pm} \sin\varphi_{\pm}$$

$$+\frac{1}{4\mu}\left(\frac{r_{\pm}}{2\pi}\right)^{1/2} \left\{ ((2\varkappa - 1)\cos\frac{\varphi_{\pm}}{2} - \cos\frac{3\varphi_{\pm}}{2}) K_I^{(0)\pm} \right.$$

$$\left. -((2\varkappa - 1)\sin\frac{\varphi_{\pm}}{2} - 3\sin\frac{3\varphi_{\pm}}{2}) K_{II}^{(0)\pm} \right\} + O(r_{\pm}),$$

$$u_\varphi^{(0)\pm}(r_{\pm}, \varphi_{\pm}) = -C_1^{\pm} \sin\varphi_{\pm} + C_2^{\pm} \cos\varphi_{\pm}$$

$$+\frac{1}{4\mu}\left(\frac{r_{\pm}}{2\pi}\right)^{1/2} \left\{ (-(2\varkappa + 1)\sin\frac{\varphi_{\pm}}{2} + \sin\frac{3\varphi_{\pm}}{2}) K_I^{(0)\pm} \right.$$

$$\left. -((2\varkappa + 1)\cos\frac{\varphi_{\pm}}{2} - 3\cos\frac{3\varphi_{\pm}}{2}) K_{II}^{(0)\pm} \right\} + O(r_{\pm}); \tag{2.1.8}$$

$$\varkappa = \frac{\lambda + 3\mu}{\lambda + \mu},$$

where $(r_\pm, \varphi_\pm)$ are polar coordinates with the centre $O_\pm$ and the polar axis directed along M; $C_j^\pm$ are rigid body displacements at the tips of the crack; $K_j^{(0)\pm}, j = I, II$, are the stress–intensity factors.

It is known (see, for example, [116]) that the coefficients $K_I^{(0)\pm}, K_{II}^{(0)\pm}$ are determined by formulae

$$K_j^{(0)\pm} = \frac{1}{2\sqrt{\pi a}}\left\{ \int_{\partial\Omega} \boldsymbol{p}(\boldsymbol{x}) \cdot \boldsymbol{\zeta}^{j,(\pm)}(\boldsymbol{x}) ds \right. \tag{2.1.9}$$

$$\left. + \sum_{\pm} \int_{-a}^{a} \boldsymbol{q}^\pm(x_1) \cdot \boldsymbol{\zeta}^{j,(\pm)}(x_1, \pm 0) dx_1 \right\}, \quad j = I, II,$$

where $\boldsymbol{\zeta}^{j,(\pm)}$ are the weight functions given by solutions (with the infinite elastic energy) of the homogeneous problem (2.1.5)–(2.1.7) which are bounded in $\Omega_0 \setminus O_\pm$ and have the following asymptotic representation near $O_\pm$

$$\boldsymbol{\zeta}^{I,(\pm)}(r_\pm, \varphi_\pm) = \frac{1}{2(1+\varkappa)\sqrt{2\pi r_\pm}} \left((2\varkappa + 1)\cos\frac{3\varphi_\pm}{2} - 3\cos\frac{\varphi_\pm}{2}, \right.$$

$$\left. 3\sin\frac{\varphi_\pm}{2} - (2\varkappa - 1)\sin\frac{3\varphi_\pm}{2} \right)^T + O(1), \tag{2.1.10}$$

$$\boldsymbol{\zeta}^{II,(\pm)}(r_\pm, \varphi_\pm) = \frac{1}{2(1+\varkappa)\sqrt{2\pi r_\pm}} \left((2\varkappa + 1)\sin\frac{3\varphi_\pm}{2} - \sin\frac{\varphi_\pm}{2}, \right.$$

$$\left. (2\varkappa - 1)\cos\frac{3\varphi_\pm}{2} - \cos\frac{\varphi_\pm}{2} \right)^T + O(1). \tag{2.1.11}$$

2.1.2 Asymptotic representation of the stress-intensity factors

It is important that the functions $h_\pm$, describing the geometry of the crack faces, are smooth and do vanish at the ends of the segment $[-a, a]$ together with their derivatives. Thus, the solution of the problem (2.1.2)-(2.1.4) satisfies the expansion (2.1.8) in the vicinity of the points $O_\pm$.

To calculate several terms in the asymptotic expansion of the stress-intensity factors, we seek the solution of (2.1.2)–(2.1.4) in the form of the asymptotic series

$$u(\varepsilon, x) \sim \sum_{k=0}^{\infty} \varepsilon^k u^{(k)}(x), \tag{2.1.12}$$

where $u^{(k)}$ are solutions of problems of the form (2.1.5)–(2.1.7) in the domain Ω_0. Due to the restrictions, prescribed for the functions $h_\pm$, the formulation (2.1.2)–(2.1.4) should be interpreted as a problem in a domain with a *regularly perturbed boundary.*

Consider the upper and lower parts of the boundary $\gamma_\varepsilon^\pm$ of the cavity G_ε. The unit inward (with respect to G_ε) normal vector $n^\pm$ on $\gamma_\varepsilon^\pm$ is

$$n^\pm(\varepsilon, x_1) = (1 + \varepsilon^2 h'_\pm(x_1)^2)^{-1/2}(\varepsilon h'_\pm(x_1), \mp 1)^T.$$

Consequently,

$$n_1^\pm(\varepsilon, x_1) = \varepsilon h'_\pm(x_1) + O(\varepsilon^3),$$

$$n_2^\pm(\varepsilon, x_1) = \mp(1 - \frac{1}{2}\varepsilon^2 h'_\pm(x_1)^2) + O(\varepsilon^4),$$

and, therefore, as $x_2 = \pm h_\pm(x_1)$, one has

$$\begin{aligned} \sigma_j^{(n)}(u; x) &= n_1^\pm(\varepsilon, x_1)\sigma_{1j}(u; x) + n_2^\pm(\varepsilon, x_1)\sigma_{2j}(u; x) \qquad (2.1.13) \\ &= \mp\sigma_{2j}(u; x) + \varepsilon h'_\pm(x_1)\sigma_{1j}(u; x) \\ &\pm \frac{1}{2}\varepsilon^2 h_\pm(x_1)^2 \sigma_{2j}(u; x) + O(\varepsilon^3). \end{aligned}$$

Consider the expansion

$$\begin{aligned} \sigma_{ij}(u; x_1, \pm\varepsilon h_\pm(x_1)) &= \sigma_{ij}(u; x_1, \pm 0) \\ &\pm \varepsilon h_\pm(x_1)\frac{\partial}{\partial x_2}\sigma_{ij}(u; x_1, \pm 0) \\ &+ \frac{1}{2}\varepsilon^2 h_\pm(x_1)^2 \frac{\partial^2}{\partial x_2^2}\sigma_{ij}(u; x_1, \pm 0) + O(\varepsilon^3), \end{aligned}$$

and represent the right–hand side of (2.1.13) in the form

$$\sigma_j^{(n)}(u; x_1, \pm\varepsilon h_\pm(x_1)) = \mp\sigma_{2j}(u; x_1, \pm 0)$$

$$+\varepsilon\{h'_{\pm}(x_1)\sigma_{1j}(\boldsymbol{u};x_1,\pm 0)-h_{\pm}(x_1)\sigma_{2j,2}(\boldsymbol{u};x_1,\pm 0)\} \tag{2.1.14}$$

$$\mp\frac{1}{2}\varepsilon^2\{h_{\pm}(x_1)^2\sigma_{2j,22}(\boldsymbol{u};x_1,\pm 0)-2h_{\pm}(x_1)h'_{\pm}(x_1)\sigma_{1j,2}(\boldsymbol{u};x_1,\pm 0)$$

$$-h'_{\pm}(x_1)^2\sigma_{2j}(\boldsymbol{u};x_1,\pm 0)\}+O(\varepsilon^3).$$

Combining expansion (2.1.12) of a solution to (2.1.2)–(2.1.4) and formulae (2.1.14), we derive that the vector $\boldsymbol{u}^{(0)}$ is a solution of the problem (2.1.5)–(2.1.7), and the vector $\boldsymbol{u}^{(1)}$ satisfies equation (2.1.5) and the boundary conditions

$$\boldsymbol{\sigma}^{(n)}(\boldsymbol{u}^{(1)};\boldsymbol{x})=\boldsymbol{0},\quad \boldsymbol{x}\in\partial\Omega, \tag{2.1.15}$$

$$\sigma_{2j}(\boldsymbol{u}^{(1)};x_1,\pm 0)=\pm h'_{\pm}(x_1)\sigma_{1j}(\boldsymbol{u}^{(0)};x_1,\pm 0) \tag{2.1.16}$$

$$\mp h_{\pm}(x_1)\sigma_{2j,2}(\boldsymbol{u}^{(0)};x_1,\pm 0),\quad |x_1|<a.$$

Due to the equilibrium equations, equality (2.1.16) can be reduced to the form

$$\sigma_{2j}(\boldsymbol{u}^{(1)};x_1,\pm 0)=\pm h_{\pm}(x_1)\sigma_{1j,1}(\boldsymbol{u}^{(0)};x_1,\pm 0)$$

$$\pm\, h'_{\pm}(x_1)\sigma_{1j}(\boldsymbol{u}^{(0)};x_1,\pm 0) \tag{2.1.17}$$

$$=\pm\frac{\partial}{\partial x_1}(h_{\pm}(x_1)\sigma_{1j}(\boldsymbol{u}^{(0)};x_1,\pm 0)),\quad |x_1|<a.$$

Since

$$h_{\pm}(x_1)=O(r_{\pm}^2),\quad \text{as } r_{\pm}\to 0,$$

(2.1.8) yields that the right–hand side of (2.1.17) is of order $O(r_{\pm}^{1/2})$. Thus, the displacement field $\boldsymbol{u}^{(1)}$, which satisfies (2.1.15), (2.1.17), admits the representation of the type (2.1.8), with the coefficients $K_I^{(1)\pm}, K_{II}^{(1)\pm}$.

Formulae (2.1.9) for the stress–intensity factors yields

$$K_j^{(1)+}=-\frac{1}{2\sqrt{\pi a}}\sum_{\pm}\int_{-a}^{a}\frac{\partial}{\partial x_1}(h_{\pm}(x_1)\boldsymbol{\sigma}^{(1)}(\boldsymbol{u}^{(0)};x_1,\pm 0))\cdot\boldsymbol{\zeta}^{j,(+)}(x_1,\pm 0)dx_1$$

$$=\frac{1}{2\sqrt{\pi a}}\sum_{\pm}\int_{-a}^{a}h_{\pm}(x_1)\boldsymbol{\sigma}^{(1)}(\boldsymbol{u}^{(0)};x_1,\pm 0)\cdot\frac{\partial\boldsymbol{\zeta}^{j,(+)}}{\partial x_1}(x_1,\pm 0)dx_1,\quad j=I,II;$$

$$\boldsymbol{\sigma}^{(i)}=(\sigma_{1i},\sigma_{2i}),\ i=1,2.$$

Similar formula is valid for $K_j^{(1)-}$.

Consider the next term $\boldsymbol{u}^{(2)}$ of the asymptotic approximation. From (2.1.12), (2.1.14) we derive that $\boldsymbol{u}^{(2)}$ is subjected to equations (2.1.5) with boundary conditions (2.1.15) and

$$\sigma_{2j}(\boldsymbol{u}^{(2)}; x_1, \pm 0) = \frac{\partial}{\partial x_1}(\frac{1}{2}h_\pm(x_1)^2 \sigma_{1j,2}(\boldsymbol{u}^{(0)}; x_1, \pm 0)$$

$$\pm h_\pm(x_1)\sigma_{1j}(\boldsymbol{u}^{(1)}; x_1, \pm 0)) + \frac{1}{2}h'_\pm(x_1)^2 \sigma_{2j}(\boldsymbol{u}^{(0)}; x_1, \pm 0)$$

$$\pm \frac{1}{2}h'_\pm(x_1)^2 q_j^\pm(x_1), \quad |x_1| < a, \quad j = 1, 2. \tag{2.1.18}$$

The right–hand side of the equality (2.1.18) is of order $O(r_\pm^{1/2})$, as $r_\pm \to 0$, and for the vector $\boldsymbol{u}^{(2)}$ the formulae (2.1.8), with the coefficients $K_j^{(2)\pm}$, are valid.

This asymptotic procedure can be also used for further coefficients of the asymptotic expansion.

As a consequence of the asymptotic representation for the displacement components we obtain the following formula for the stress–intensity factors

$$K_j^\pm(\varepsilon) = K_j^{(0)\pm} + \varepsilon K_j^{(1)\pm} + \varepsilon^2 K_j^{(2)\pm} + O(\varepsilon^2), \quad j = I, II. \tag{2.1.19}$$

2.1.3 Potential energy

Consider the potential energy functional

$$\mathcal{E}(\varepsilon) = -\frac{1}{2}\int_{\partial\Omega} \boldsymbol{p}(\boldsymbol{x}) \cdot \boldsymbol{u}(\varepsilon, \boldsymbol{x}) ds - \frac{1}{2}\sum_\pm \int_{\gamma_\varepsilon^\pm} \boldsymbol{p}^\pm(\varepsilon, x_1) \cdot \boldsymbol{u}(\varepsilon, \boldsymbol{x}) ds, \tag{2.1.20}$$

which corresponds to the boundary value problem (2.1.2)–(2.1.4), and find its asymptotic approximation, as $\varepsilon \to 0$.

From the representation of the external load $\boldsymbol{p}^\pm$ we obtain

$$\mathcal{E}(\varepsilon) = -\frac{1}{2}\int_{\partial\Omega} \boldsymbol{p}(\boldsymbol{x}) \cdot \boldsymbol{u}(\varepsilon, \boldsymbol{x}) ds$$

$$-\frac{1}{2}\sum_\pm \int_{-a}^{a} \boldsymbol{q}^\pm(x_1) \cdot \boldsymbol{u}(\varepsilon, x_1, \pm\varepsilon h_\pm(x_1)) dx_1. \tag{2.1.21}$$

Due to the asymptotic formula (2.1.12) for displacements

$$\int_{\partial\Omega} \boldsymbol{p}(\boldsymbol{x}) \cdot \boldsymbol{u}(\varepsilon, \boldsymbol{x}) ds = \sum_{j=0}^{2} \varepsilon^j \int_{\partial\Omega} \boldsymbol{p}(\boldsymbol{x}) \cdot \boldsymbol{u}^{(j)}(\boldsymbol{x}) ds + O(\varepsilon^3).$$

Since

$$\boldsymbol{p}(\boldsymbol{x}) = \boldsymbol{\sigma}^{(n)}(\boldsymbol{u}^{(0)}; \boldsymbol{x}), \ \boldsymbol{\sigma}^{(n)}(\boldsymbol{u}^{(j)}; \boldsymbol{x}) = \boldsymbol{0}, \ \ \boldsymbol{x} \in \partial\Omega, \ \ j = 1, 2,$$

the Betti formula yields

$$\begin{aligned}\int_{\partial\Omega} \boldsymbol{p}(\boldsymbol{x}) \cdot \boldsymbol{u}^{(j)}(\boldsymbol{x}) ds &= \int_{\partial\Omega} \{\boldsymbol{\sigma}^{(n)}(\boldsymbol{u}^{(0)}; \boldsymbol{x}) \cdot \boldsymbol{u}^{(j)}(\boldsymbol{x}) \\ &\quad -\boldsymbol{\sigma}^{(n)}(\boldsymbol{u}^{(j)}; \boldsymbol{x}) \cdot \boldsymbol{u}^{(0)}(\boldsymbol{x})\} ds \\ &= \sum_{\pm} \pm \int_{-a}^{a} \{\boldsymbol{\sigma}^{(2)}(\boldsymbol{u}^{(0)}; x_1, \pm 0) \cdot \boldsymbol{u}^{(j)}(x_1, \pm 0) \\ &\quad -\boldsymbol{\sigma}^{(2)}(\boldsymbol{u}^{(j)}; x_1, \pm 0) \cdot \boldsymbol{u}^{(0)}(x_1, \pm 0)\} dx_1, \\ &\quad j = 1, 2. \end{aligned} \tag{2.1.22}$$

Next, for the integrals over $\Gamma^{\pm}$ in the right–hand side of (2.1.21) we have

$$\begin{aligned}\int_{-a}^{a} \boldsymbol{q}^{\pm}(x_1) \cdot \boldsymbol{u}(\varepsilon, x_1, \pm\varepsilon h_{\pm}(x_1)) dx_1 &= \int_{-a}^{a} \boldsymbol{q}^{\pm}(x_1) \cdot \Bigg\{ \boldsymbol{u}^{(0)}(x_1, \pm 0) \\ &+\varepsilon \left(\boldsymbol{u}^{(1)}(x_1, \pm 0) \pm h_{\pm}(x_1) \frac{\partial \boldsymbol{u}^{(0)}}{\partial x_2}(x_1, \pm 0) \right) \\ &+\varepsilon^2 \Bigg(\boldsymbol{u}^{(2)}(x_1, \pm 0) \pm h_{\pm}(x_1) \frac{\partial \boldsymbol{u}^{(1)}}{\partial x_2}(x_1, \pm 0) \\ &+ \frac{1}{2} h_{\pm}(x_1)^2 \frac{\partial^2 \boldsymbol{u}^{(0)}}{\partial x_2^2}(x_1, \pm 0) \Bigg) \Bigg\} dx_1 + O(\varepsilon^3). \end{aligned} \tag{2.1.23}$$

Combining (2.1.22), (2.1.23) and using (2.1.21), we derive the following asymptotic formula

$$\mathcal{E}(\varepsilon) = \mathcal{E}_0 + \varepsilon\mathcal{E}_1 + \varepsilon^2\mathcal{E}_2 + O(\varepsilon^3), \tag{2.1.24}$$

where

$$\mathcal{E}_0 = -\frac{1}{2}\int_{\partial\Omega} \boldsymbol{p}(\boldsymbol{x}) \cdot \boldsymbol{u}^{(0)}(\boldsymbol{x})ds - \frac{1}{2}\sum_{\pm}\int_{-a}^{a} \boldsymbol{q}^{\pm}(x_1) \cdot \boldsymbol{u}^{(0)}(x_1, \pm 0)dx_1$$

is the potential energy corresponding to (2.1.5)–(2.1.7). Collecting the terms of order $O(\varepsilon)$ in the right-hand sides of (2.1.22) and (2.1.23), we obtain

$$\mathcal{E}_1 = -\frac{1}{2}\sum_{\pm}\int_{-a}^{a}\Big\{ \mp\boldsymbol{\sigma}^{(2)}(\boldsymbol{u}^{(1)}; x_1, \pm 0) \cdot \boldsymbol{u}^{(0)}(x_1, \pm 0)$$

$$\pm h_{\pm}(x_1)\boldsymbol{q}^{\pm}(x_1) \cdot \frac{\partial \boldsymbol{u}^{(0)}}{\partial x_2}(x_1, \pm 0)\Big\}dx_1. \tag{2.1.25}$$

Using (2.1.17), (2.1.6), one can reduce (2.1.25) to the form

$$\mathcal{E}_1 = \frac{1}{2}\sum_{\pm}\int_{-a}^{a}\left(\frac{\partial}{\partial x_1}(h_{\pm}(x_1)\boldsymbol{\sigma}^{(1)}(\boldsymbol{u}^{(0)}; x_1, \pm 0)) \cdot \boldsymbol{u}^{(0)}(x_1, \pm 0)\right.$$

$$\left. + h_{\pm}(x_1)\boldsymbol{\sigma}^{(2)}(\boldsymbol{u}^{(0)}; x_1, \pm 0) \cdot \frac{\partial \boldsymbol{u}^{(0)}}{\partial x_2}(x_1, \pm 0)\right)dx_1$$

$$= -\frac{1}{2}\sum_{\pm}\int_{-a}^{a} h_{\pm}(x_1)\Big\{\boldsymbol{\sigma}^{(1)}(\boldsymbol{u}^{(0)}; x_1, \pm 0) \cdot \frac{\partial \boldsymbol{u}^{(0)}}{\partial x_1}(x_1, \pm 0)$$

$$-\boldsymbol{\sigma}^{(2)}(\boldsymbol{u}^{(0)}; x_1, \pm 0) \cdot \frac{\partial \boldsymbol{u}^{(0)}}{\partial x_2}(x_1, \pm 0)\Big\}dx_1 \tag{2.1.26}$$

Calculate the expression $R_{\pm}(\boldsymbol{u}^{(0)}; x_1, \pm 0)$ from the curly brackets in (2.1.26)

$$R_{\pm}(\boldsymbol{u}^{(0)}; x_1, \pm 0) = (2\mu u^{(0)}_{1,1}(x_1, \pm 0)$$

$$+\lambda(u_{1,1}^{(0)}(x_1,\pm 0)+u_{2,2}^{(0)}(x_1,\pm 0)))u_{1,1}^{(0)}(x_1,\pm 0)$$

$$+\mu(u_{1,2}^{(0)}(x_1,\pm 0)+u_{2,1}^{(0)}(x_1,\pm 0))u_{2,1}^{(0)}(x_1,\pm 0)$$

$$-\mu(u_{1,2}^{(0)}(x_1,\pm 0)+u_{2,1}^{(0)}(x_1,\pm 0))u_{1,2}^{(0)}(x_1,\pm 0)$$

$$-(2\mu u_{2,2}^{(0)}(x_1,\pm 0)+\lambda(u_{1,1}^{(0)}(x_1,\pm 0)+u_{2,2}^{(0)}(x_1,\pm 0)))u_{2,2}^{(0)}(x_1,\pm 0)$$

$$=(2\mu+\lambda)(u_{1,1}^{(0)}(x_1,\pm 0)^2-u_{2,2}^{(0)}(x_1,\pm 0)^2)$$

$$+\mu(u_{2,1}^{(0)}(x_1,\pm 0)^2-u_{1,2}^{(0)}(x_1,\pm 0)^2).$$

Thus,

$$\mathcal{E}_1=-\frac{1}{2}\sum_{\pm}\int_{-a}^{a}h_{\pm}(x_1)\Bigg\{(2\mu+\lambda)(u_{1,1}^{(0)}(x_1,\pm 0)^2-u_{2,2}^{(0)}(x_1,\pm 0)^2)$$

$$+\mu(u_{2,1}^{(0)}(x_1,\pm 0)^2-u_{1,2}^{(0)}(x_1,\pm 0)^2)\Bigg\}dx_1. \tag{2.1.27}$$

Similarly, collecting terms of order $O(\varepsilon^2)$ in (2.1.22), (2.1.23) and taking into account (2.1.6), (2.1.18), we obtain

$$\mathcal{E}_2=-\frac{1}{2}\sum_{\pm}\int_{-a}^{a}\Bigg\{\mp\boldsymbol{\sigma}^{(2)}(\boldsymbol{u}^{(2)};x_1,\pm 0)\cdot\boldsymbol{u}^{(0)}(x_1,\pm 0)$$

$$\pm h_{\pm}(x_1)\boldsymbol{q}^{\pm}(x_1)\cdot\frac{\partial\boldsymbol{u}^{(1)}}{\partial x_2}(x_1,\pm 0)$$

$$+\frac{1}{2}h_{\pm}(x_1)^2\boldsymbol{q}^{\pm}(x_1)\cdot\frac{\partial^2\boldsymbol{u}^{(0)}}{\partial x_2^2}(x_1,\pm 0)\Bigg\}dx_1$$

$$=\sum_{\pm}\int_{-a}^{a}\Bigg\{\mp\frac{\partial}{\partial x_1}\Bigg(h_{\pm}(x_1)^2\frac{1}{2}\frac{\partial}{\partial x_2}\boldsymbol{\sigma}^{(1)}(\boldsymbol{u}^{(0)};x_1,\pm 0)$$

$$+h_{\pm}(x_1)\boldsymbol{\sigma}^{(1)}(\boldsymbol{u}^{(1)};x_1,\pm 0)\Bigg)\cdot\boldsymbol{u}^{(0)}(x_1,\pm 0)$$

$$-h_{\pm}(x_1)\boldsymbol{\sigma}^{(2)}(\boldsymbol{u}^{(0)};x_1,\pm 0)\cdot\frac{\partial\boldsymbol{u}^{(1)}}{\partial x_2}(x_1,\pm 0)$$

$$\mp\frac{1}{2}h_\pm(x_1)^2\boldsymbol{\sigma}^{(2)}(\boldsymbol{u}^{(0)};x_1,\pm 0)\cdot\frac{\partial^2\boldsymbol{u}^{(0)}}{\partial x_2^2}(x_1,\pm 0)\Bigg\}dx_1.$$

After the integration by parts it becomes

$$\mathcal{E}_2=-\frac{1}{2}\sum_\pm\int_{-a}^{a}\Bigg\{h_\pm(x_1)\Bigg(\boldsymbol{\sigma}^{(1)}(\boldsymbol{u}^{(1)};x_1,\pm 0)\cdot\frac{\partial\boldsymbol{u}^{(0)}}{\partial x_1}(x_1,\pm 0)$$

$$-\boldsymbol{\sigma}^{(2)}(\boldsymbol{u}^{(0)};x_1,\pm 0)\cdot\frac{\partial\boldsymbol{u}^{(1)}}{\partial x_2}(x_1,\pm 0)\Bigg)$$

$$\pm\frac{1}{2}h_\pm(x_1)^2\Bigg(\frac{\partial}{\partial x_2}\boldsymbol{\sigma}^{(1)}(\boldsymbol{u}^{(0)};x_1,\pm 0)\cdot\frac{\partial\boldsymbol{u}^{(0)}}{\partial x_1}(x_1,\pm 0)$$

$$-\boldsymbol{\sigma}^{(2)}(\boldsymbol{u}^{(0)};x_1,\pm 0)\cdot\frac{\partial^2\boldsymbol{u}^{(0)}}{\partial x_2^2}(x_1,\pm 0)\Bigg)\Bigg\}dx_1.$$

In particular case where the faces of the crack are free of tractions ($\boldsymbol{q}^\pm=\boldsymbol{0}$) from the boundary conditions (2.1.6) one can derive that

$$u_{1,2}^{(0)}(x_1,\pm 0)=-u_{2,1}^{(0)}(x_1,\pm 0),\quad u_{2,2}^{(0)}(x_1,\pm 0)=-\lambda(2\mu+\lambda)^{-1}u_{1,1}^{(0)}(x_1,\pm 0).$$

Thus,

$$\mathcal{E}_1=-2\mu\frac{\mu+\lambda}{2\mu+\lambda}\sum_\pm\int_{-a}^{a}h_\pm(x_1)u_{1,1}^{(0)}(x_1,\pm 0)^2dx_1,$$

and, in accordance with (2.1.24), we obtain

$$\mathcal{E}(\varepsilon)=\mathcal{E}_0-\frac{1-\nu^2}{2E}\,\varepsilon\sum_\pm\int_{-a}^{a}h_\pm(x_1)\sigma_{11}(\boldsymbol{u}^{(0)};x_1,\pm 0)^2dx_1$$

$$+O(\varepsilon^2). \tag{2.1.28}$$

2.1.4 Examples

Uniaxial tensile loading of a plane with a thin cavity $\mathbf{G}_\varepsilon$. Assume that the crack faces are free of tractions $\boldsymbol{q}^\pm=\boldsymbol{0}$. The boundary conditions (2.1.4) should be replaced by the following conditions at infinity

$$\sigma_{11}(\boldsymbol{u};\boldsymbol{x})\to p\cos^2\beta;\ \sigma_{12}(\boldsymbol{u};\boldsymbol{x})\to p\sin\beta\cos\beta;$$

$$\sigma_{22}(\boldsymbol{u};\boldsymbol{x}) \to p\sin^2\beta, \quad \text{as} \quad \|\boldsymbol{x}\| \to \infty,$$

where p is the load density, and β is the angle between the direction of loading and the Ox_1 axis.

Stresses, which correspond to the solution $\boldsymbol{u}^{(0)}$ of the problem (2.1.5)–(2.1.7), are specified by the formulae (see, for example, [114])

$$\sigma_{11}(\boldsymbol{u}^{(0)};\boldsymbol{x}) = p\Bigg\{\cos^2\beta + \sin^2\beta\Big(\mathrm{Re}Z(z) - x_2\mathrm{Im}Z'(z)\Big)$$

$$+\cos\beta\sin\beta\Big(2\mathrm{Im}Z(z) + x_2\mathrm{Re}Z'(z)\Big)\Bigg\},$$

$$\sigma_{22}(\boldsymbol{u}^{(0)};\boldsymbol{x}) = p\Bigg\{\sin^2\beta + \sin^2\beta\Big(\mathrm{Re}Z(z) + x_2\,\mathrm{Im}Z'(z)\Big)$$

$$-x_2\cos\beta\sin\beta\mathrm{Re}Z'(z)\Bigg\},$$

$$\sigma_{21}(\boldsymbol{u}^{(0)};\boldsymbol{x}) = p\Bigg\{\sin\beta\cos\beta - x_2\sin^2\beta\mathrm{Re}Z'(z)$$

$$+\cos\beta\sin\beta\Big(\mathrm{Re}Z(z) - x_2\mathrm{Im}Z'(z)\Big)\Bigg\}, \tag{2.1.29}$$

$$Z(z) = z(z^2 - a^2)^{-1/2} - 1, \quad z = x_1 + ix_2.$$

Equalities (2.1.29) and the boundary conditions (2.1.17) with respect to the vector $\boldsymbol{u}^{(1)}$ yield

$$\sigma_{21}(\boldsymbol{u}^{(1)};x_1,\pm 0) = \pm p\frac{\partial}{\partial x_1}\left(h_\pm(x_1)\left\{\cos 2\beta \mp \frac{\sin 2\beta}{\sqrt{a^2 - x_1^2}}x_1\right\}\right),$$

$$\sigma_{22}(\boldsymbol{u}^{(1)};x_1,\pm 0) = 0, \quad |x_1| < a. \tag{2.1.30}$$

The values of the vector functions $\boldsymbol{\zeta}^{j,(\pm)}$, subjected to (2.1.10), (2.1.11), are specified on the faces of the crack $\Gamma^\pm$ by the equalities (also see [38], [116])

$$\zeta_2^{I,(+)}(x_1,\pm 0) = \zeta_1^{II,(+)}(x_1,\pm 0) = \pm\sqrt{\frac{a + x_1}{a - x_1}},$$

$$\zeta_1^{I,(+)}(x_1, \pm 0) = \frac{\varkappa - 1}{\varkappa + 1}, \ \zeta_2^{II,(+)}(x_1, \pm 0) = -\frac{\varkappa - 1}{\varkappa + 1}.$$

Then equations (2.1.30) yield

$$K_I^{(1)+} = 0, \ K_{II}^{(1)+} = \frac{p}{\sqrt{\pi a}} \int_{-a}^{a} \left\{ (h_+(x_1) - h_-(x_1)) \cos 2\beta \right. \tag{2.1.31}$$

$$\left. -H(x_1) \frac{x_1 \sin 2\beta}{\sqrt{a^2 - x_1^2}} \right\} \frac{a \, dx_1}{(a - x_1)\sqrt{a^2 - x_1^2}} + O(\varepsilon^3).$$

where $H(x_1) = h_+(x_1) + h_-(x_1)$ is the normalized crack opening. Thus, formula (2.1.19) for $j = II$ takes the form

$$K_{II}^+(\varepsilon) = \frac{p}{\sqrt{\pi a}} \left\{ \pi a \sin\beta \cos\beta + \varepsilon \int_{-a}^{a} \left\{ [h_+(x_1) - h_-(x_1)] \cos 2\beta \right. \right.$$

$$\left. \left. -H(x_1) \frac{x_1 \sin 2\beta}{\sqrt{a^2 - x_1^2}} \right\} \frac{a \ dx_1}{(a - x_1)\sqrt{a^2 - x_1^2}} \right\} + O(\varepsilon^3). \tag{2.1.32}$$

Consider formula (2.1.19) and derive the representation for the term of order $O(\varepsilon^2)$ for $j = I$. Relations (2.1.18) yield that the average of $\sigma_{12}(\boldsymbol{u}^{(2)}; x_1, \pm 0)$ over the segment $[-a, a]$ is zero. Also,

$$\zeta_1^{I,(+)}(x_1, \pm 0) = \text{const}.$$

Thus, due to the formula (2.1.9), the boundary condition (2.1.18), with $j = 1$, does not contribute to the coefficient $K_I^{(2)+}$.

From the second relation (2.1.18) and the equilibrium equations we derive

$$\sigma_{22}(\boldsymbol{u}^{(2)}; x_1, \pm 0) = \frac{\partial}{\partial x_1} \left(\frac{1}{2} h_\pm(x_1)^2 \frac{\partial}{\partial x_2} \sigma_{12}(\boldsymbol{u}^{(0)}; x_1, \pm 0) \right.$$

$$\left. + h_\pm(x_1) \frac{\partial}{\partial x_1} \left(h_\pm(x_1) \sigma_{11}(\boldsymbol{u}^{(0)}; x_1, \pm 0) \right) \right)$$

$$= \frac{\partial^2}{\partial x_1^2} \left(\frac{1}{2} h_\pm(x_1)^2 \sigma_{11}(\boldsymbol{u}^{(0)}; x_1, \pm 0) \right).$$

Thus,

$$K_I^{(2)+} = \frac{1}{4\sqrt{\pi a}} \sum_{\pm} \int_{-a}^{a} \frac{\partial^2}{\partial x_1^2} \left(h_\pm(x_1)^2 \sigma_{11}(\boldsymbol{u}^{(0)}; x_1, \pm 0) \right) \sqrt{\frac{a+x_1}{a-x_1}} dx_1$$

$$= \frac{1}{4\sqrt{\pi a}} \sum_{\pm} \int_{-a}^{a} h_\pm(x_1)^2 \sigma_{11}(\boldsymbol{u}^{(0)}; x_1, \pm 0) \frac{\partial^2}{\partial x_1^2} \left(\sqrt{\frac{a+x_1}{a-x_1}} \right) dx_1$$

$$= -\frac{p}{4}\sqrt{\frac{a}{\pi}} \int_{-a}^{a} \Bigg\{ \sum_{\pm} h_\pm(x_1)^2 \cos 2\beta$$

$$- \left(h_+(x_1)^2 - h_-(x_1)^2 \right) \sin 2\beta \frac{x_1}{\sqrt{a^2 - x_1^2}} \Bigg\} \frac{(a+2x_1)\, dx_1}{(a-x_1)(a^2-x_1^2)^{3/2}},$$

and formula (2.1.19) for $j = I$ takes the form

$$K_I^+(\varepsilon) = p\sqrt{\frac{a}{\pi}} \Bigg\{ \pi(\sin\beta)^2 - \frac{\varepsilon^2}{4} \int_{-a}^{a} \Bigg\{ \sum_{\pm} h_\pm(x_1)^2 \cos 2\beta$$

$$- \left(h_+(x_1)^2 - h_-(x_1)^2 \right) \frac{x_1 \sin 2\beta}{\sqrt{a^2 - x_1^2}} \Bigg\} \Bigg\} \times$$

$$\times \frac{(a+2x_1)\, dx_1}{(a-x_1)(a^2-x_1^2)^{3/2}} + O(\varepsilon^3). \tag{2.1.33}$$

Consider the following particular cases: (a) the uniaxial tension with $\beta = \pi/2$ and $p = P > 0$; (b) the longitudinal compression with $\beta = 0$ and $p = -P$. Formulae (2.1.32) and (2.1.33) yield

$$K_I^+(\varepsilon) - K_I^{(0)+} = \frac{P\varepsilon^2}{4}\sqrt{\frac{a}{\pi}} \int_{-a}^{a} \sum_{\pm} h_\pm(x_1)^2 \frac{(a+2x_1)\, dx_1}{(a-x_1)(a^2-x_1^2)^{3/2}}$$

$$+ O(\varepsilon^3). \tag{2.1.34}$$

In the case of a tension along the Ox_2 axis

$$K_I^{(0)+} = P\sqrt{\pi a},$$

and in the second case of a longitudinal compression

$$K_I^{(0)+} = 0.$$

In both cases $K_{II}^+ = 0$. For a longitudinal load the coefficient K_I^+ may, in general, change the sign. However, for a symmetric contour, where the functions $h_\pm(x_1)$ are even, this stress–intensity factor is positive. Moreover, for a nonsymmetric contour the sum $K_I^+(\varepsilon) + K_I^-(\varepsilon)$ of the stress–intensity factors at the tips of the crack is positive. The same statement is valid for the second-order approximation (2.1.34) in the case of the axial tension across the crack.

Longitudinal compression along the crack axis. Consider an elastic domain Ω_ε and assume that a uniaxial compression is imposed on Ω_ε in the direction parallel to the crack axis. Thus, in boundary conditions (2.1.3) and (2.1.4) we have

$$\boldsymbol{p}^\pm = \boldsymbol{0} \text{ and } \boldsymbol{p}(\boldsymbol{x}) = -p(n_1, 0)^T,$$

where p is the intensity of the compressive load. Then the displacements are specified by

$$\boldsymbol{u}^{(0)} = \frac{p}{4\mu(\mu+\lambda)}(-(2\mu+\lambda)x_1, \lambda x_2)^T. \tag{2.1.35}$$

Substituting (2.1.35) into (2.1.28) we derive

$$\Delta\mathcal{E} = \mathcal{E}(\varepsilon) - \mathcal{E}_0 = -\frac{1-\nu^2}{2E}\,\varepsilon\, p^2 \sum_\pm \int_{-a}^{a} h_\pm(x_1)dx_1 + O(\varepsilon^2)$$

$$= -p^2\frac{(1-\nu^2)}{2E}S_\varepsilon + O(\varepsilon^2), \tag{2.1.36}$$

where S_ε is the area of the cavity G_ε. Thus, the leading term of the increment of the potential energy in a solid with a thin hole ("thin crack"), subjected to a longitudinal load, is proportional to the area of the crack.

Asymptotic estimates for the potential energy. Now we consider a general case of a uniaxial loading. The potential energy of an infinite solid is infinite. To calculate $\Delta\mathcal{E}$ we can use the approach discussed in [64].

Let D_R be a disk $\{\boldsymbol{x} \in \mathbb{R}^2 : \|\boldsymbol{x}\| < R\}$, with $R \gg 1$. Consider the displacement vector $\boldsymbol{u}_R(\varepsilon, \boldsymbol{x})$ corresponding to a deformation of $D_R \setminus \overline{G}_\varepsilon$;

the intensity of the load is p, and the angle between the direction of the loading and the crack line is β. Let $\boldsymbol{v}(\boldsymbol{x})$ be the solution of the corresponding boundary value problem in a homogeneous domain (without crack)

$$\boldsymbol{v}(\boldsymbol{x}) = \frac{p}{2\mu}\Big(((\cos\beta)^2 - \frac{\lambda}{2(\lambda+\mu)})x_1 + x_2\cos\beta\sin\beta,$$

$$x_1\cos\beta\sin\beta + ((\sin\beta)^2 - \frac{\lambda}{2(\lambda+\mu)})x_2\Big)^T, \qquad (2.1.37)$$

and let $\boldsymbol{w}(\varepsilon, \boldsymbol{x})$ denote the solution of the problem (2.1.2), (2.1.3) vanishing at infinity in $\mathbb{R}^2 \setminus \overline{G}_\varepsilon$. The load on the crack faces is assumed to be

$$\boldsymbol{p}^\pm(\varepsilon, \boldsymbol{x}) = \frac{\boldsymbol{q}^\pm(\varepsilon, \boldsymbol{x})}{\sqrt{1+\varepsilon^2 h_\pm'(x_1)^2}},$$

where

$$\boldsymbol{q}^\pm(\varepsilon, \boldsymbol{x}) = -p\{\varepsilon h_\pm'(x_1)\cos\beta \mp \sin\beta\}(\cos\beta, \sin\beta)^T.$$

The potential energy for the domains $D_R \setminus \overline{G}_\varepsilon$ and D_R will be denoted by $\mathcal{E}_R$ and $\mathcal{E}_R^0$, respectively. As $R \to \infty$, the following asymptotic approximation holds

$$\boldsymbol{u}_R(\varepsilon, \boldsymbol{x}) = \boldsymbol{v}(\boldsymbol{x}) + \boldsymbol{w}(\varepsilon, \boldsymbol{x}) + O(R^{-1}),$$

and

$$\mathcal{E}_R - \mathcal{E}_R^0 = -\frac{1}{2}\sum_{i,j=1}^{2}\Big\{\int_{G_\varepsilon}\sigma_{ij}(\boldsymbol{v}, \boldsymbol{x})\varepsilon_{ij}(\boldsymbol{v}; \boldsymbol{x})d\boldsymbol{x}$$

$$+\int_{\mathbb{R}^2\setminus\overline{G}_\varepsilon}\sigma_{ij}(\boldsymbol{w}; \varepsilon, \boldsymbol{x})\varepsilon_{ij}(\boldsymbol{w}; \varepsilon, \boldsymbol{x})d\boldsymbol{x}\Big\} + o(1),$$

where ε_{ij} are the components of the strain tensor $\boldsymbol{\varepsilon}$. Thus,

$$\Delta\mathcal{E} = \lim_{R\to\infty}(\mathcal{E}_R - \mathcal{E}_R^0) = -\frac{1}{2}\sum_{i,j=1}^{2}\Big\{\int_{G_\varepsilon}\sigma_{ij}(\boldsymbol{v}; \boldsymbol{x})\varepsilon_{ij}(\boldsymbol{v}; \boldsymbol{x})d\boldsymbol{x}$$

$$+\int_{\mathbb{R}^2\setminus\overline{G}_\varepsilon}\sigma_{ij}(\boldsymbol{w}; \varepsilon, \boldsymbol{x})\varepsilon_{ij}(\boldsymbol{w}; \varepsilon, \boldsymbol{x})d\boldsymbol{x}\Big\}. \qquad (2.1.38)$$

For the first integral in the right–hand side (2.1.38) from (2.1.36) we get

$$\sum_{i,j=1}^{2}\int_{G_\varepsilon}\sigma_{ij}(\boldsymbol{v};\boldsymbol{x})\varepsilon_{ij}(\boldsymbol{v};\boldsymbol{x})d\boldsymbol{x}=p^2\frac{2\mu+\lambda}{4\mu(\lambda+\mu)}S_\varepsilon. \tag{2.1.39}$$

To evaluate the second integral we use the formulae (2.1.24) and (2.1.27). The solution $\boldsymbol{w}^{(0)}$ of the limiting problem (2.1.5), (2.1.6) in $\mathbb{R}^2\setminus M$ can be represented in the form

$$\boldsymbol{w}^{(0)}(\varepsilon,\boldsymbol{x})=\boldsymbol{z}^{(0)}(\boldsymbol{x})+\varepsilon\boldsymbol{z}^{(1)}(\boldsymbol{x}),$$

where $\boldsymbol{z}^{(0)}$ and $\boldsymbol{z}^{(1)}$ are solutions of the same problem with the right–hand sides

$$\begin{aligned}\boldsymbol{q}^{\pm,0}(x_1)&=\pm p\sin\beta(\cos\beta,\sin\beta)^T,\\ \boldsymbol{q}^{\pm,1}(x_1)&=-ph_\pm'(x_1)\cos\beta(\cos\beta,\sin\beta)^T.\end{aligned}$$

For the second integral in (2.1.38) we derive

$$-\frac{1}{2}\sum_{i,j=1}^{2}\int_{\mathbb{R}^2\setminus\overline{G}_\varepsilon}\sigma_{ij}(\boldsymbol{w};\varepsilon,\boldsymbol{x})\varepsilon_{ij}(\boldsymbol{w};\varepsilon,\boldsymbol{x})d\boldsymbol{x} \tag{2.1.40}$$

$$=-\frac{1}{2}\int_{\partial G_\varepsilon}\boldsymbol{\sigma}^{(n)}(\boldsymbol{w};\varepsilon,\boldsymbol{x})\cdot\boldsymbol{w}(\varepsilon,\boldsymbol{x})ds$$

$$=-\frac{1}{2}\sum_{\pm}\int_{-a}^{a}\Big\{\mp\boldsymbol{\sigma}^{(2)}(\boldsymbol{w}^{(0)};\varepsilon,x_1,\pm0)\cdot\boldsymbol{w}^{(0)}(x_1,\pm0)$$

$$+\varepsilon h_\pm(x_1)\{(2\mu+\lambda)(w_{1,1}^{(0)}(\varepsilon,x_1,\pm0)^2-w_{2,2}^{(0)}(\varepsilon,x_1,\pm0)^2)$$

$$+\mu(w_{2,1}^{(0)}(\varepsilon,x_1,\pm0)^2-w_{1,2}^{(0)}(\varepsilon,x_1,\pm0)^2)\}\Big\}dx_1+O(\varepsilon^2)$$

$$=-\frac{\varepsilon}{2}\sum_{\pm}\int_{-a}^{a}h_\pm(x_1)\Big\{(2\mu+\lambda)(z_{1,1}^{(0)}(x_1,\pm0)^2-z_{2,2}^{(0)}(x_1,\pm0)^2)$$

$$+\mu(z_{2,1}^{(0)}(x_1,\pm0)^2-z_{1,2}^{(0)}(x_1,\pm0)^2)\Big\}dx_1$$

$$-\frac{1}{2}\sum_{i,j=1}^{2}\int_{\mathbb{R}^2\setminus M}\sigma_{ij}(z^{(0)}+\varepsilon z^{(1)};x)\varepsilon_{ij}(z^{(0)}+\varepsilon z^{(1)};x)dx+O(\varepsilon^2).$$

The last integral can be estimated by

$$-\frac{1}{2}\sum_{i,j=1}^{2}\int_{\mathbb{R}^2\setminus M}(\sigma_{ij}(z^{(0)};x)\varepsilon_{ij}(z^{(0)};x)$$

$$+2\varepsilon\sigma_{ij}(z^{(0)};x)\varepsilon_{ij}(z^{(1)};x))dx+O(\varepsilon^2) \tag{2.1.41}$$

$$=-\frac{1}{2}\sum_{\pm}\mp\int_{-a}^{a}(\sigma^{(2)}(z^{(0)};x_1,\pm 0)$$

$$+2\varepsilon\sigma^{(2)}(z^{(1)};x_1,\pm 0))\cdot z^{(0)}(x_1,\pm 0)dx_1+O(\varepsilon^2).$$

The values of the displacement $z^{(0)}$ on the faces $\Gamma^{\pm}$ of the crack M are

$$z^{(0)}(x_1,\pm 0)=\pm\frac{1-\nu}{\mu}\,p\,\sqrt{a^2-x_1^2}\,\sin\beta\,(\cos\beta,\sin\beta)^T,$$

and,
therefore, formulae (2.1.38)–(2.1.41) yield the following representation for the increment of the potential energy

$$\Delta\mathcal{E}=-\frac{\pi a^2p^2(1-\nu)(\sin\beta)^2}{2\mu}-\frac{p^2(2\mu+\lambda)}{8\mu(\lambda+\mu)}\Big\{S_\varepsilon(1+(\sin\beta\sin 2\beta)^2)$$

$$-\varepsilon(a\sin\beta\sin 2\beta)^2\int_{-a}^{a}\frac{H(x_1)dx_1}{a^2-x_1^2}\Big\}+O(\varepsilon^2).$$

In particular, for the case of a longitudinal load the energy increment for an elastic plane with a "thin crack " G_ε equals

$$-\frac{p^2(2\mu+\lambda)}{8\mu(\lambda+\mu)}S_\varepsilon+O(\varepsilon^2).$$

As expected, this result is the same as in (2.1.36).

2.1.5 Three-dimensional crack with smoothly closed edges

Here we consider an elastic space $\mathbb{R}^3$ with a plane cut

$$M=\{x\in\mathbb{R}^3:x_3=0,\ x'=(x_1,x_2)\in g\},$$

where g is a two–dimensional domain with a smooth boundary ∂g.

Consider a "thin set"

$$G_\varepsilon = \{\boldsymbol{x} \in \mathbb{R}^3 : \ \boldsymbol{x}' \in g, \ -\varepsilon h_-(\boldsymbol{x}') < x_3 < \varepsilon h_+(\boldsymbol{x}')\}.$$

Here ε is a small nondimensional positive parameter which can be defined as a maximum thickness of G_ε normalized by diam g.

We also assume that at the crack edge

$$h_\pm(\boldsymbol{x}') = 0, \ \nabla h_\pm(\boldsymbol{x}') = \boldsymbol{0}, \ \ \boldsymbol{x}' \in \partial g.$$

Consider the deformation of an elastic solid $\Omega_\varepsilon = \Omega \setminus \overline{G}_\varepsilon$ subjected to the external self–balanced load applied to the boundary $\partial\Omega_\varepsilon$; the body force density is assumed to be zero. Then the displacement vector satisfies the boundary value problem

$$\mu\Delta\boldsymbol{u}(\varepsilon, \boldsymbol{x}) + (\lambda + \mu)\nabla\nabla \cdot \boldsymbol{u}(\varepsilon, \boldsymbol{x}) = \boldsymbol{0}, \ \ \boldsymbol{x} \in \Omega_\varepsilon, \tag{2.1.42}$$

$$\boldsymbol{\sigma}^{(n)}(\boldsymbol{u}; \varepsilon, \boldsymbol{x}) = \boldsymbol{p}(\boldsymbol{x}), \ \ \boldsymbol{x} \in \partial\Omega, \tag{2.1.43}$$

$$\boldsymbol{\sigma}^{(n)}(\boldsymbol{u}; \varepsilon, \boldsymbol{x}) = \boldsymbol{p}^\pm(\varepsilon, \boldsymbol{x}), \ \ \boldsymbol{x} \in \gamma_\varepsilon^\pm. \tag{2.1.44}$$

Here $\boldsymbol{n}$ is the unit outward normal vector with respect to $\partial\Omega_\varepsilon$; $\boldsymbol{p}^\pm$, $\boldsymbol{p}$ are smooth vector–valued functions; $\boldsymbol{\sigma}^{(n)} = \boldsymbol{\sigma n}$; $\gamma_\varepsilon^\pm = \{\boldsymbol{x} \in \mathbb{R}^3 : x_3 = \pm\varepsilon h_\pm(\boldsymbol{x}'), \ \boldsymbol{x}' \in g\}$.

Introduce the following notation

$$\boldsymbol{q}^\pm(\varepsilon, \boldsymbol{x}) = \sqrt{1 + \varepsilon^2 ||\nabla h_\pm(\boldsymbol{x}')||^2} \, \boldsymbol{p}^\pm(\varepsilon, \boldsymbol{x}).$$

As $\varepsilon \to 0$, the domain Ω_ε is transformed into Ω_0, and the problem (2.1.42)–(2.1.44) describes a deformation of the solid Ω with a plane crack M which faces are subjected to the load $\boldsymbol{q}^\pm(\varepsilon, \boldsymbol{x})$. The solution of this boundary value problem will be denoted by $\boldsymbol{u}^{(0)}$.

We seek the displacement field $\boldsymbol{u}(\varepsilon, \boldsymbol{x})$ which admits the asymptotic representation

$$\boldsymbol{u}(\varepsilon, \boldsymbol{x}) \sim \boldsymbol{u}^{(0)}(\boldsymbol{x}) + \varepsilon\boldsymbol{u}^{(1)}(\boldsymbol{x}) + \varepsilon^2\boldsymbol{u}^{(2)}(\boldsymbol{x}). \tag{2.1.45}$$

The vector functions $\boldsymbol{u}^{(1)}$ and $\boldsymbol{u}^{(2)}$ should be subjected to the homogeneous equations (2.1.42) in Ω_0 and the homogeneous boundary conditions (2.1.43) on $\partial\Omega$. Since

$$n(\varepsilon, x) = \frac{(\varepsilon \nabla h_\pm(x'),\ \pm 1)^T}{\sqrt{1 + \varepsilon^2 \|\nabla h_\pm(x')\|^2}}, \quad x \in \gamma_\varepsilon^\pm,$$

the boundary conditions for the functions $u^{(1)}$, $u^{(2)}$ on the crack faces $\Gamma^\pm$ are

$$\sigma_{3k}(u^{(j)}; x) = \pm \nabla \cdot (F_1^{j,k,\pm}(x), F_2^{j,k,\pm}(x)), \quad x \in \Gamma^\pm,$$

$$k = 1, 2, 3; \tag{2.1.46}$$

where

$$F_p^{j,k,\pm}(x) = h_\pm(x')\sigma_{pk}(u^{(j-1)}; x)$$

$$\mp \frac{\delta_{j2}}{2} h_\pm(x')^2 \frac{\partial}{\partial x_3} \sigma_{pk}(u^{(j-2)}; x), \ j = 1, 2,$$

δ_{ij} is Kronecker delta. The right–hand sides (2.1.46) correspond to a self–balanced load which yields the existence of static solutions $u^{(1)}$ and $u^{(2)}$.

The potential energy functional is defined by

$$\mathcal{E}(\varepsilon) = -\frac{1}{2} \int_{\partial\Omega} p(x) \cdot u(\varepsilon, x) dx - \frac{1}{2} \sum_\pm \int_{\gamma_\varepsilon^\pm} p^\pm(\varepsilon, x) \cdot u(\varepsilon, x) dx, \tag{2.1.47}$$

and we shall use the following asymptotic representation

$$\mathcal{E}(\varepsilon) = \mathcal{E}_0 + \varepsilon \mathcal{E}_1 + O(\varepsilon^2). \tag{2.1.48}$$

Here, $\mathcal{E}_0$ is the potential energy corresponding to a deformation of an elastic solid with a plane crack.

In particular, if the crack faces are free of tractions then the asymptotic formula (2.1.48) becomes

$$\mathcal{E}(\varepsilon) = \mathcal{E}_0 - \frac{\varepsilon}{2\mu} \sum_\pm \int_g h_\pm(x') \Big\{ \sigma_{12}(u^{(0)}, x', \pm 0)^2 \tag{2.1.49}$$

$$+ \frac{1}{(2\mu + 3\lambda)^2} \Big(-2\lambda(\mu + \lambda)(\sigma_{11}(u^{(0)}; x', \pm 0) - \sigma_{22}(u^{(0)}; x', \pm 0))^2$$

$$+(2\mu+\lambda)\Big((\mu+\lambda)(\sigma_{22}(\boldsymbol{u}^{(0)};\boldsymbol{x}',\pm 0)^2+\sigma_{11}(\boldsymbol{u}^{(0)};\boldsymbol{x}',\pm 0)^2)$$

$$+\lambda\sigma_{11}(\boldsymbol{u}^{(0)};\boldsymbol{x}',\pm 0)\sigma_{22}(\boldsymbol{u}^{(0)};\boldsymbol{x}',\pm 0)\Big)\Big)\Big\}d\boldsymbol{x}'+O(\varepsilon^2).$$

For a circular crack subjected to a longitudinal axisymmetric compression, the asymptotic equality (2.1.49) yields

$$\Delta\mathcal{E}=\mathcal{E}(\varepsilon)-\mathcal{E}_0=-\frac{2\mu+\lambda}{2\mu(2\mu+3\lambda)}p^2\mathrm{mes}G_\varepsilon+O(\varepsilon^2), \tag{2.1.50}$$

where $\mathrm{mes}G_\varepsilon$ is the volume of a "thin" cavity G_ε. Thus, the leading term of the potential energy increment in an elastic solid with a crack, parallel to the direction of the axisymmetric loading, is proportional to the volume of a "thin" cavity.

Let $K_i(\varepsilon,s)$, $i=I,II,III$, denote the stress-intensity factors at the point s of the contour ∂g. Expansion (2.1.45) yields

$$K_i(\varepsilon,s)=K_i^{(0)}(s)+\varepsilon K_i^{(1)}(s)+\varepsilon^2K_i^{(2)}(s)+O(\varepsilon^3),$$

where $K_i^{(j)}, i=I,II,III$, are the stress-intensity factors for the displacement fileds $\boldsymbol{u}^{(j)}$.

Consider a penny-shaped cavity G_ε with $h_+(\boldsymbol{x}')=h_-(\boldsymbol{x}')=h(\|\boldsymbol{x}'\|)$. It is known (see, for example [19]) that for the axisymmetric loading of the crack M the following equalities hold

$$K_I(\varepsilon)=-\frac{2}{\sqrt{\pi a}}\int_0^a\frac{\sigma_{33}(\boldsymbol{u};\varepsilon,r,0)}{\sqrt{a^2-r^2}}\,r\,dr,$$

$$K_{II}(\varepsilon)=K_{III}(\varepsilon)=0, \tag{2.1.51}$$

where a is the radius of the circle, $r=\|\boldsymbol{x}'\|$.

Assume that the elastic body $\mathbb{R}^3\setminus\overline{G_\varepsilon}$ is subjected to an axisymmetric compression of intensity P. In this case

$$K_{II}=K_{III}=0,$$

and the formula (2.1.51) yields

$$K_I(\varepsilon)=\varepsilon^2\frac{P}{\sqrt{\pi a}}\int_0^a\frac{h(r)^2(2a^2+r^2)}{(a^2-r^2)^{5/2}}\,r\,dr+O(\varepsilon^3). \tag{2.1.52}$$

Now, consider the equilibrium of a crack with a smoothly closed faces within restrictions formulated above.

Two cases are considered.

First, assume that the crack does not change its shape as it grows (it is subjected to scaling only). The stress–intensity factor is calculated as follows

$$K_I(\varepsilon) = \varepsilon^2 P\sqrt{\frac{a}{\pi}}\int_0^1 \frac{h_1(t)^2(2+t^2)}{(1-t^2)^{5/2}}\, t\, dt + O(\varepsilon^3), \tag{2.1.53}$$

where

$$h(\|\boldsymbol{x}'\|) = ah_1(t),\ \|\boldsymbol{x}'\| = at,\ t \in [0,1].$$

Evidently, as the radius a of the crack increases, the stress–intensity factor K_I grows as well, and, therefore, for the crack G_ε this equilibrium state is unstable.

The second idealized example that we are going to consider is related to a crack that grows without opening. In this case the function h satisfies the condition

$$h(\|\boldsymbol{x}'\|) = 0, \quad \text{as } 1 < \|\boldsymbol{x}'\| \leq a.$$

Then the stress–intensity factor is defined by

$$K_I(\varepsilon) = \varepsilon^2 \frac{P}{\sqrt{\pi a}}\int_0^1 \frac{h(r)^2(2a^2+r^2)}{(a^2-r^2)^{5/2}}\, r\, dr + O(\varepsilon^3). \tag{2.1.54}$$

It can be checked that

$$\frac{\partial K_I}{\partial a} < 0,$$

which yields that with increasing of the radius of the crack the stress-intensity factor increases. It means that in the case of an axisymmetric compression the circular crack G_ε, which grows without opening of its faces, is stable.

2.2 Thin rectangular holes

One important question in fracture mechanics concerns the mathematical description of cracks in an elastic solid. In well–known papers [29], [30] the energy criterion was formulated, and for a crack, represented by a thin ellipse, it was derived that the fracture occurs on the plane where the normal tensile stress takes the critical value (which depends, of course, on the particular elastic material). Another way is to represent a crack by a cut [114]. This model is very simple, and the analytic methods can be applied for evaluation of the stress–intensity factors which are further used in the fracture criteria. The asymptotic behaviour of the stress–field near the tips of a crack is described by Sneddon's formulae.

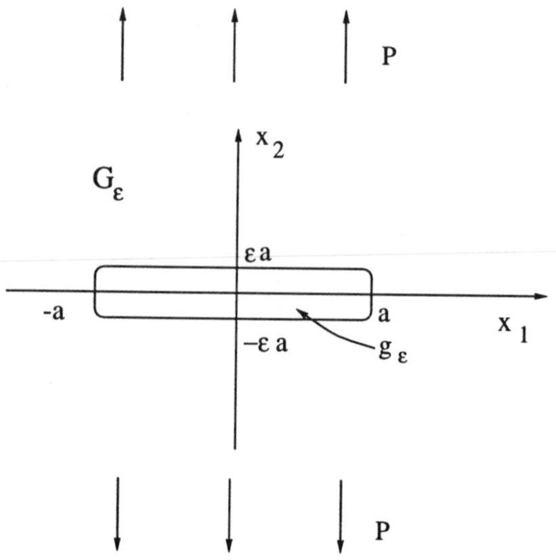

Fig. 2.2: Symmetric normal load of a plane with a thin rectangular cavity.

In the previous section the crack was modelled by a thin cavity with smoothly closed edges which corresponds to a regular perturbation of the boundary as the crack opening tends to zero. Also, in the paper [128] the example of a singular perturbation was considered. The model followed the idea [29], [30], and the crack was approximated by a "thin" cavity (but not necessarily elliptic) with a smooth contour. The last model needs the auxiliary field of the boundary layer type in a neighbourhood of the crack tips.

In the present section we consider an elastic plane containing a thin rectangular cavity. In fact the approximation that we use here allows the corners to be smoothed. The general case of a uni–axial load, applied at infinity, is considered. For the case of a longitudinal compression we shall need two terms of the asymptotic expansion of the displacement field in a neighbourhood of the tips of a crack, and it enables us to explain the effect of a rock fracture where the region of failure is characterized by a triangular shape (this effect is described, for example, in [19]).

2.2.1 Formulation of the problem

Let

$$G_\varepsilon = \{\boldsymbol{x} \in \mathbb{R}^2 : -\varepsilon h(x_1) \leq x_2 \leq \varepsilon h(x_1),\ x_1 \in (-a, a); |x_2| < b,\ |x_1| = a\}$$

be a "thin" cavity. Here, $h \in C^\infty(-a, a)$, $\varepsilon = b/a$ is a nondimensional small parameter. Assume that ∂G_ε is free of tractions, and the uniform normal load or shear stresses are prescribed at infinity (see Fig. 2.2 and Fig. 2.3).

We consider the plane–strain problem in a domain $\Omega_\varepsilon = \mathbb{R}^2 \setminus \overline{G}_\varepsilon$. The stress components satisfy the equilibrium equations

$$\frac{\partial}{\partial x_1}\sigma_{11}(\boldsymbol{u};\varepsilon,\boldsymbol{x}) + \frac{\partial}{\partial x_2}\sigma_{12}(\boldsymbol{u};\varepsilon,\boldsymbol{x}) = 0,$$

$$\frac{\partial}{\partial x_1}\sigma_{12}(\boldsymbol{u};\varepsilon,\boldsymbol{x}) + \frac{\partial}{\partial x_2}\sigma_{22}(\boldsymbol{u};\varepsilon,\boldsymbol{x}) = 0,\ \ \boldsymbol{x} \in \Omega_\varepsilon, \tag{2.2.1}$$

the compatibility condition

$$\Delta(\sigma_{11}(\boldsymbol{u};\varepsilon,\boldsymbol{x}) + \sigma_{22}(\boldsymbol{u};\varepsilon,\boldsymbol{x})) = 0,\ \boldsymbol{x} \in \Omega_\varepsilon, \tag{2.2.2}$$

and the homogeneous traction boundary conditions on the crack faces

$$\sigma_{ij}(\boldsymbol{u};\varepsilon,\boldsymbol{x})n_j(\boldsymbol{x}) = 0,\ \ \boldsymbol{x} \in \partial G_\varepsilon;\ \ i,j = 1,2, \tag{2.2.3}$$

where n_j are components of the unit outward normal vector $\boldsymbol{n}$ with respect to ∂G_ε.

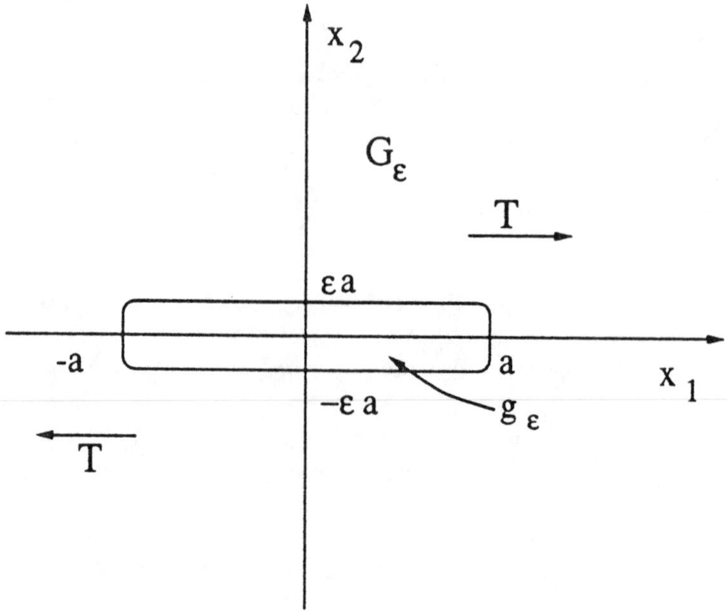

Fig. 2.3: A shear load of a plane with a thin rectangular cavity.

The following conditions are prescribed for the stress components at infinity

$$\sigma_{22}(\boldsymbol{u};\varepsilon,\boldsymbol{x}) = P,\ \sigma_{12}(\boldsymbol{u};\varepsilon,\boldsymbol{x}) = \sigma_{11}(\boldsymbol{u};\varepsilon,\boldsymbol{x}) = 0,\ ||\boldsymbol{x}|| \to \infty, \quad (2.2.4)$$

for the case of symmetric loading; and

$$\sigma_{12}(\boldsymbol{u};\varepsilon,\boldsymbol{x}) = T,\ \sigma_{22}(\boldsymbol{u};\varepsilon,\boldsymbol{x}) = \sigma_{11}(\boldsymbol{u};\varepsilon,\boldsymbol{x}) = 0,\ ||\boldsymbol{x}|| \to \infty, \quad (2.2.5)$$

for the case of shear.

2.2.2 Asymptotic approximation of the solution

Outside a neighbourhood of the cavity G_ε the stress state, denoted by σ_{ij}^0 say, corresponds to the limit problem ($\varepsilon = 0$) on an elastic plane with a cut $[-a, a]$ along the axis Ox_1. This solution is well-known (see, for example, [116], [114]). The Kolosov–Muskhelishvili complex potentials can be determined with the use of the Keldysh–Sedov function, and for the components of the stress tensor the asymptotic formulae by Sneddon are valid.

As we approach G_ε, the discrepancy in the boundary conditions should be compensated. To do this we need to construct the boundary layer type solution described in stretched coordinates in the vicinity of the crack tips. Introduce the following scaled coordinates in the vicinity of the right edge (the consideration is similar for the left edge)

$$\xi_1 = \frac{x_1 - a}{2\varepsilon a}, \quad \xi_2 = \frac{x_2}{2\varepsilon a}.$$

As $\varepsilon \to 0$, we obtain the boundary layer limit problem in the exterior of a semi-infinite strip $\Pi = \{\boldsymbol{\xi} \in \mathbb{R}^2 : |\xi_2| < 1/2, \xi_1 < 0\}$ (see Fig. 2.4).

For the case of a homogeneous traction boundary conditions the Kolosov–Muskhelishvili complex potentials ϕ and ψ should satisfy the equation (see [85])

$$\phi(t) + t\overline{\phi'(t)} + \overline{\psi(t)} = 0, \ t \in \partial\Pi.$$

To satisfy the asymptotic matching principle we require the asymptotic representation of the elastic field in a plane with a semi-infinite cut to hold at infinity.

We use the conformal mapping of the domain $\Omega = \mathbb{R}^2 \setminus \overline{\Pi}$ into the unit circle. This mapping will be approximated by the rational function

$$\omega(\zeta) = \frac{A}{(1+\zeta)^2} \sum_{k=0}^{N} T_k \zeta^k, \tag{2.2.6}$$

where A and T_k are constant coefficients.

We use the modification of the classic Kolosov–Muskhelishvili method [85]. Consider the relation

$$\overline{\phi(t)} + \bar{t}\phi'(t) + \psi(t) = 0, \ t \in \partial\Pi.$$

Introduce the complex potentials $\phi_*(\zeta) = \phi(\omega(\zeta))$, $\psi_*(\zeta) = \psi(\omega(\zeta))$ in the form

$$\phi_*(\zeta) = \frac{D}{1+\zeta} \sum_{k=0}^{N-1} a_k \zeta^k; \tag{2.2.7}$$

$$\psi_*(\zeta) = -\frac{\overline{D}}{1+\zeta}[a_0\zeta + a_1]$$

$$-\frac{\omega(1/\zeta)}{\omega'(\zeta)}\phi_*'(\zeta) + \frac{D}{\zeta^{N-2}(1+\zeta)} \sum_{k=0}^{N-3} f_k \zeta^k,$$

where

$$D = \begin{cases} 1, & \text{Mode-I load} \\ -i, & \text{Mode-II load} \end{cases}$$

$$f_l = \sum_{k=0}^{l} d_k t_{l-k},\ l = 0, 1, \ldots, N-3;$$

$$t_k = a_k(k-1) + a_{k+1}(k+1),\ k = 0, 1, \ldots, N-2;$$

$$\mathbf{d} = \mathbf{P}^{-1}\mathbf{g};\ \mathbf{d} = (d_0, d_1, \ldots, d_N)^T,\ \mathbf{g} = (g_0, g_1, \ldots, g_N)^T;$$

$$b_k = (k+1)T_{k+1} + (k-2)T_k,\quad g_k = T_{N-k},$$

$$k = 0, 1, \ldots, N;\ T_j = 0,\ j > N.$$

$$\mathbf{P} = \begin{pmatrix} b_0 & 0 & 0 & \ldots & 0 & 0 \\ b_1 & b_0 & 0 & \ldots & 0 & 0 \\ \cdot & \cdot & \cdot & & \cdot & \cdot \\ \cdot & \cdot & & \cdot & \cdot & \cdot \\ \cdot & \cdot & & & \cdot & \cdot \\ b_{N-1} & b_{N-2} & \ldots & \ldots & b_0 & 0 \\ b_N & b_{N-1} & \ldots & \ldots & b_1 & b_0 \end{pmatrix}. \tag{2.2.8}$$

Thus, the coefficients a_k satisfy the following system of linear algebraic equations

$$\sum_{l=0}^{N-1} a_l \left[\delta_{lk} + (-1)^{m_0} \sum_{m=k+1}^{N} T_m [l A_{(m-k-1),(l-1)\operatorname{sign} l}\right.$$

$$\left. + (l-1)A_{(m-k-1),l}]\right] = 0, \tag{2.2.9}$$

$$\sum_{k=0}^{N-1}(-1)^k a_k = \frac{\sqrt{2}(1+\sqrt{2})}{\pi} S_{I(II)}; \; m_0 = D^2,$$

where S_I, S_{II} are constant coefficients determined by the matching principle, and

$$A_{kp} = C_{ij}, \; i = k+1, \; j = p+1; \; (C_{ij})_{i,j=1}^{N+1} = \mathbf{P}^{-1}. \qquad (2.2.10)$$

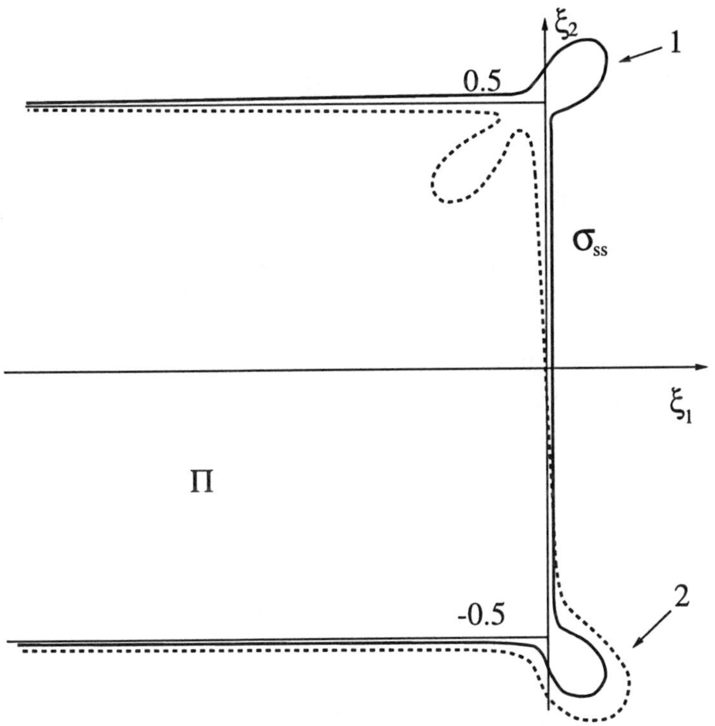

Fig. 2.4: Distribution of the tensile stress on the contour: 1 – normal load, 2 – shear.

The use of (2.2.7)–(2.2.9) and the Kolosov–Muskhelishvili formulae [85] yield the stress components σ_{ij}^{+}.

The constants S_I and S_{II} are determined by the matching of solutions $\boldsymbol{\sigma}^0$ and $\boldsymbol{\sigma}^+$ at the distance $O(\sqrt{a\varepsilon})$ from a crack tip. In our case

$$S_{I(II)} = \frac{1}{\sqrt{2a\varepsilon}} K_{I(II)},$$

where K_I, K_{II} are Mode–I and Mode–II stress–intensity factors.

A smooth cut–off function χ is introduced in such a way that $\chi(t) = 1$ for $t < 1$, and $\chi(t) = 0$ for $t > 2$.

The following asymptotic approximation of the stress tensor is used for numerical computation

$$\sigma_{ij}(\varepsilon, z) = \chi\left(\frac{z+a}{\sqrt{a\varepsilon}}\right)\chi\left(\frac{z-a}{\sqrt{a\varepsilon}}\right)\sigma^0_{ij}(z)$$

$$+\left[1-\chi\left(\frac{z-a}{\sqrt{a\varepsilon}}\right)\right]\sigma^+_{ij}(h_1(\varepsilon,z))+\left[1-\chi\left(\frac{z+a}{\sqrt{a\varepsilon}}\right)\right]\sigma^-_{ij}(h_2(\varepsilon,z)). \quad (2.2.11)$$

Here $z = x_1+ix_2$, $h_i, i = 1, 2$, are approximations of the function inverse to $\omega(\zeta)$ in the vicinity of the tips of the crack, $\sigma^\pm_{ij}$ corresponds to the solution of the boundary layer type near the ends of the crack.

In Fig. 2.4 we show the distribution of the tensile stress σ_{ss} on the contour of the rectangular cavity. The calculation has been done for $N = 20$ (see formula (2.2.7)). As expected, σ_{ss} takes its maximum value at the corner of the rectangular cavity.

2.2.3 Critical stress. Fracture

We use the fracture criterion established by Novozhilov in [98]. The intensity P of the critical external load, which causes fracture, satisfies the relation

$$\frac{2a\varepsilon}{d}\,P\int_0^{d/2a\varepsilon}\sigma_{ss}(\rho,\alpha)d\rho = \sigma_c, \quad (2.2.12)$$

where d is a structural parameter of the material, σ_c is the theoretical limit of elastic strength, (ρ, α) are local polar coordinates with the centre at a point of the contour where σ_{ss}, corresponding to the unit intensity of external load, takes its maximum value. The value α in (2.2.12) corresponds to the direction of the normal vector to the contour. Thus, to evaluate the critical value of external load we take the average of σ_{ss} along a line inside the domain. The stress decreases as we go away from the contour and, consequently, the critical load decreases as the structural parameter d increases.

Numerical calculations for a symmetric load can be compared with [128]. For $\varepsilon = 10^{-3}$, $N = 20$, $a = 1$ the value P agrees with P^0 obtained in [128]: as $d = 0.4\varepsilon a$, $P = 1.44 \cdot 10^{-2}\sigma_c$, and $P^0 = 1.42 \cdot 10^{-2}\sigma_c$. Increasing of d gives the critical intensity $P = 3.03 \cdot 10^{-2}\sigma_c$ which approaches the fracture limit obtained by Griffith.

For relatively big values of the structural parameter the values of the critical external load are nearly the same for different shapes of a crack tip. However, the direction of the further propagation of the crack is different for different geometries. For example, for an elliptical crack $\alpha = 0$; for a thin cavity with a rectangular shape of the edge $\alpha = 45°$.

2.2.4 Longitudinal load of a plane with a rectangular cavity

In comparison with the first part of the section we take into account next-order terms of the asymptotic expansion of the stress field in the vicinity of a thin rectangular cavity. This section is based on the results [78].

For the same geometry and elastic parameters of the material we assume that equations (2.2.1)–(2.2.3) hold, and the loading conditions at infinity have the form

$$\sigma_{12}(\boldsymbol{u};\varepsilon,\boldsymbol{x}) = \sigma_{22}(\boldsymbol{u};\varepsilon,\boldsymbol{x}) = 0, \ \sigma_{11}(\boldsymbol{u};\varepsilon,\boldsymbol{x}) = P, \ \|\boldsymbol{x}\| \to \infty. \quad (2.2.13)$$

The leading term $\boldsymbol{u}^{(0)}$ of the displacement field corresponds to the uniaxial load (along the Ox_1 axis) of the homogeneous elastic plane. The next order term $\varepsilon\boldsymbol{u}^{(1)}(\boldsymbol{x})$ is specified by

$$\boldsymbol{u}^{(1)}(\boldsymbol{x}) = aP\mathbf{V}(\boldsymbol{x}),$$

where $\mathbf{V}(\boldsymbol{x})$ is the displacement which corresponds to a presence of a concentrated load at the tips of the crack.

In the exterior of a neighbourhood of a thin cut the following asymptotic representation for displacement holds

$$\boldsymbol{u}(\varepsilon,\boldsymbol{x}) \sim \boldsymbol{u}^{(0)}(\boldsymbol{x}) + \varepsilon\boldsymbol{u}^{(1)}(\boldsymbol{x}).$$

The first term $\boldsymbol{u}^{(0)}(\boldsymbol{x})$ introduces a discrepancy in the traction boundary conditions on the faces of the thin cavity. As before, this discrepancy is compensated by the field of the boundary layer type.

Consider the boundary layer constructed in the vicinity of the right end of a thin cut (the algorithm for the left end is similar). Introduce nondimensional scaled variables

$$(\xi_1,\xi_2) = (2a\varepsilon)^{-1}(x_1 - a, x_2).$$

The solution of the boundary layer type in the domain $\mathbb{R}^2 \setminus \overline{\Pi}$ satisfies the boundary value problem

$$\mu\Delta\mathbf{U}(\boldsymbol{\xi}) + (\lambda+\mu)\nabla\nabla\cdot\mathbf{U}(\boldsymbol{\xi}) = \mathbf{0}, \boldsymbol{\xi} \in \mathbb{R}^2 \setminus \overline{\Pi}; \tag{2.2.14}$$

$$\boldsymbol{\sigma}^{(n)}(\mathbf{U};\boldsymbol{\xi}) = \begin{cases} \mathbf{0}, & |\xi_2| = 1/2,\ \xi_1 < 0, \\ P\mathbf{e}^{(1)}, & \xi_1 = 0,\ |\xi_2| < 1/2, \end{cases} \tag{2.2.15}$$

with the following conditions at infinity

$$\begin{aligned}(U_r, U_\varphi)(\boldsymbol{\xi}) &= -P[8\pi\mu]^{-1}\{(\varkappa+1)\ln r(\cos\varphi, -\sin\varphi) \\ &\quad +((\varkappa-1)\varphi\sin\varphi, \\ &\quad (\varkappa-1)\varphi\cos\varphi - 2\sin\varphi)\} + O(r^{-1}),\ r\to\infty.\end{aligned} \tag{2.2.16}$$

Here $\boldsymbol{\xi} = re^{i\varphi}$, $\varkappa = (\lambda+3\mu)(\lambda+\mu)^{-1}$. The solution of the problem (2.2.14)–(2.2.16) is constructed by the modification of the Kolosov–Muskhelishvili method [85]. As a result we obtain that the complex potentials $\phi^*(\zeta) = \phi(\omega(\zeta))$, $\psi^*(\zeta) = \psi(\omega(\zeta))$ admit the representation

$$\phi^*(\zeta) = \phi^{**}(\zeta) - \frac{c_*}{(1+\zeta)^2},\ \phi^{**}(\zeta) = \frac{1}{(1+\zeta)^2}\sum_{k=0}^{\infty} a_k\zeta^k;$$

$$\psi^*(\zeta) = \psi^{**}(\zeta) + \frac{2c_*}{(1+\zeta)^2},\ \psi^{**}(\zeta) = \frac{1}{(1+\zeta)^2}\sum_{k=1}^{\infty} c_k\zeta^k;$$

$$c_* = P(1+\sqrt{2})^2/\pi. \tag{2.2.17}$$

Here $\omega(\zeta)$ is the complex–valued function which provides the conformal mapping of a unit disk on the exterior of a semi–infinite rectangular cavity (see [43]).

As above, we keep a finite number of terms in the expansion (2.2.6), and, therefore, deal with a smooth contour which is close to a semi–infinite rectangular one. As $\|\boldsymbol{x}\| \to \infty$, the domain does not exactly coincide with a semi–infinite strip. However, due to multiplying by the cut–off function (see (2.2.11)) the solution of the boundary layer type is used in a large but finite domain, and, therefore, the small deflection of the contour for large values of $\|\boldsymbol{x}\|$ does not contribute into the leading term of the asymptotic expansion of the displacement field.

The coefficients a_k and c_k from (2.2.17) are determined due to the analysis of the following integral equations

$$\frac{1}{2\pi i}\int_{C_1}\frac{\phi^*(\sigma)(\sigma+1)^2}{\sigma-\zeta}d\sigma+\frac{1}{2\pi i}\int_{C_1}\frac{\omega(\sigma)\overline{\phi^{*\prime}(\sigma)}(\sigma+1)^2}{\overline{\omega'(\sigma)}(\sigma-\zeta)}d\sigma$$

$$+\frac{1}{2\pi i}\int_{C_1}\frac{\overline{\psi^*(\sigma)}(\sigma+1)^2}{\sigma-\zeta}d\sigma=0;$$

$$\frac{1}{2\pi i}\int_{C_1}\frac{\overline{\phi^*(\sigma)}(\sigma+1)^2}{\sigma-\zeta}d\sigma+\frac{1}{2\pi i}\int_{C_1}\frac{\overline{\omega(\sigma)}\phi^{*\prime}(\sigma)(\sigma+1)^2}{\omega'(\sigma)(\sigma-\zeta)}d\sigma$$

$$+\frac{1}{2\pi i}\int_{C_1}\frac{\psi^*(\sigma)(\sigma+1)^2}{\sigma-\zeta}d\sigma=0,\ C_1=\{\sigma:|\sigma|=1\}. \tag{2.2.18}$$

Denote by I_i, $i=1,2,\ldots,6$, the integrals from (2.2.18), and also define the quantities $a_i^*=a_i-\delta_{i0}c_*$, $c_0^*=c_0+2c_*$. To calculate the integrals from (2.2.18) the well-known statements on the Cauchy integrals [85] is used. Since the functions $\phi^*(\zeta)(\zeta+1)^2$, $\psi^*(\zeta)(\zeta+1)^2$ are analytic in $S^+=\{\zeta:|\zeta|<1\}$ and continuous in $\bar{S}^+$, one has

$$I_1=\frac{1}{2\pi i}\int_{C_1}\frac{\phi^*(\sigma)(\sigma+1)^2}{\sigma-\zeta}d\sigma=\phi^*(\zeta)(\zeta+1)^2=\sum_{k=0}^{N}a_k^*\zeta^k,\ \zeta\in S^+,$$

$$I_6=\frac{1}{2\pi i}\int_{C_1}\frac{\psi^*(\sigma)(\sigma+1)^2}{\sigma-\zeta}d\sigma=\psi^*(\zeta)(\zeta+1)^2,\zeta\in S^+.$$

Continuous on C_1 functions $\overline{\psi^*(\sigma)}(\sigma+1)^2$, $\overline{\phi^*(\sigma)}(\sigma+1)^2$ are boundary values of functions $\psi^*(1/\zeta)(\zeta+1)^2$, $\phi^*(1/\zeta)(\zeta+1)^2$ in $S^-=\{\zeta:|\zeta|>1\}$, with the second-order pole at infinity. In this case

$$I_3=\frac{1}{2\pi i}\int_{C_1}\frac{\overline{\phi^*(\sigma)}(\sigma+1)^2}{\sigma-\zeta}d\sigma=c_0^*\zeta^2+c_1\zeta+c_2,\ \zeta\in S^+;$$

$$I_4=\frac{1}{2\pi i}\int_{C_1}\frac{\overline{\phi^*(\sigma)}(\sigma+1)^2}{\sigma-\zeta}d\sigma=a_0^*\zeta^2+a_1^*\zeta+a_2^*,\ \zeta\in S^+.$$

The quantity

$$\frac{\omega(\sigma)}{\overline{\omega'(\sigma)}}\overline{\phi^{*\prime}(\sigma)}(\sigma+1)^2,\ \sigma\in C_1,$$

is the boundary value of the function

$$\frac{\omega(\zeta)}{\omega'(1/\zeta)}\phi^{*\prime}(1/\zeta)(1+\zeta)^2,$$

which is regular in S^- and has a pole of order N at infinity.

As a result, we have that

$$I_2 = \frac{1}{2\pi i}\int_{C_1} \frac{\omega(\sigma)\overline{\phi^{*\prime}(\sigma)}(\sigma+1)^2}{\overline{\omega'(\sigma)}(\sigma-\zeta)}d\sigma = \sum_{k=0}^{N} M_k\zeta^k, \ \zeta \in S^+;$$

$$M_k = -\sum_{m=k}^{N}\sum_{l=0}^{N} a_l^*\{T_m[lA_{m-k,(l-1)\text{sign}\, l} + (l-2)A_{m-k,l}]\};$$

here $A_{kp} = C_{k+1,p+1}$ and the components of the matrix C were specified in the previous section (see (2.2.10)).

Similarly,

$$\frac{\overline{\omega(\sigma)}}{\omega'(\sigma)}\phi^{*\prime}(\sigma)(\sigma+1)^2, \ \sigma \in C_1,$$

is the boundary value of the function

$$\frac{\omega(1/\zeta)}{\omega'(\zeta)}\phi^{*\prime}(\zeta)(1+\zeta)^2,$$

which is regular in $S^+ \setminus 0$, continuous in $\overline{S}^+ \setminus 0$ and has a pole of order $N-2$ at the point $\zeta = 0$.

Thus, the integral I_5 from the second equation of system (2.2.18) has the form

$$I_5 = \frac{1}{2\pi i}\int_{C_1} \frac{\overline{\omega(\sigma)}\phi^{*\prime}(\sigma)(\sigma+1)^2}{\omega'(\sigma)(\sigma-\zeta)}d\sigma = \frac{\omega(1/\zeta)\phi^{*\prime}(\zeta)}{\omega'(\zeta)}(\zeta+1)^2$$

$$-\sum_{k=0}^{N-3} f_k\zeta^{k-N+2}, \ \zeta \in S^+,$$

where the coefficients f_k are specified by (2.2.8), with $t_k = a_k^*(k-2) + a_{k+1}^*(k+1)$.

Next, the integrals I_j, $j = 1, \ldots, 6$ can be substituted into (2.2.18) which, consequently, yields

$$\sum_{k=0}^{N} a_k^* \zeta^2 + \sum_{k=0}^{N} M_k \zeta^k + c_0^* \zeta^2 + c_1 \zeta + c_2 = 0; \qquad (2.2.19)$$

$$a_0^* \zeta^2 + a_1^* \zeta + a_2^* + \frac{\omega(1/\zeta)}{\omega'(\zeta)} \phi^{*\prime}(\zeta)(\zeta+1)^2$$

$$-\zeta^{2-N} \sum_{k=0}^{N-3} f_k \zeta^k + \psi^*(\zeta)(\zeta+1)^2 = 0. \qquad (2.2.20)$$

Equating in (2.2.19) the coefficients near like powers of ζ and using condition (2.2.16) at infinity, we obtain that coefficients a_k (see (2.2.17)) satisfy the following system of linear algebraic equations:

$$\begin{cases} \sum_{l=0}^{N} a_l \Big\{ \delta_{l,k+1} - \sum_{m=k+1}^{N} T_m [l A_{m-k-1,(l-1)\text{sign} l} \\ +(l-2) A_{(m-k-l),l}] \Big\} = 0, \; k = 1, \ldots, N-1; \\ \sum_{k=0}^{N} (-1)^k a_k = \pi^{-1}(1+\sqrt{2})^2 p; \\ \sum_{k=0}^{N} a_k k (-1)^{k-1} = 0. \end{cases}$$

The functions $\phi^{**}(\zeta)$, $\psi^{**}(\zeta)$ are given by the equalities

$$\phi^{**}(\zeta) = \frac{1}{(1+\zeta)^2} \sum_{k=0}^{N} a_k \zeta^k;$$

$$\psi^{**}(\zeta) = -\frac{1}{(1+\zeta)^2}(a_0 \zeta^2 + a_1 \zeta + a_2) - \frac{\omega(1/\zeta)}{\omega'(\zeta)} \phi^{**\prime}(\zeta)$$

$$+ \frac{1}{\zeta^{N-2}(1+\zeta)^2} \sum_{k=0}^{N-3} f_k \zeta^k. \qquad (2.2.21)$$

Then the Kolosov–Muskhelishvili formulae can be used to obtain the stress–strain state in a neighbourhood of the end region of a thin rectangular cavity.

Introduce a cut–off function $\chi(t) \in C^\infty(\mathbb{R}^1_+)$, such that $\chi(t) = 1$ for $t > 1$ and $\chi(t) = 0$ for $t < 1/2$. The asymptotic representation of the stress tensor is

$$\sigma(\boldsymbol{u};\varepsilon,\boldsymbol{x}) = \sigma(\boldsymbol{u}^{(0)};\boldsymbol{x})+\varepsilon\left\{\sigma(\boldsymbol{u}^{(1)};\boldsymbol{x})\chi\left(\frac{\|\boldsymbol{x}-(a,0)\|}{a\varepsilon}\right)\chi\left(\frac{\|\boldsymbol{x}+(a,0)\|}{a\varepsilon}\right)\right.$$

$$+\varepsilon^{-1}\left(1-\chi\left(\frac{\|\boldsymbol{x}-(a,0)\|}{a\varepsilon}\right)\right)\sigma(\mathbf{U}^+;\boldsymbol{\xi})$$

$$\left.+\varepsilon^{-1}\left(1-\chi\left(\frac{\|\boldsymbol{x}+(a,0)\|}{a\varepsilon}\right)\right)\sigma(\mathbf{U}^-;\boldsymbol{\xi})\right\}+O(\varepsilon^2).$$

Here $\mathbf{U}^+, \mathbf{U}^-$ are fields of the boundary layer type which represent displacements in the vicinity of the ends $(\pm a, 0)$ of the thin rectangular cavity.

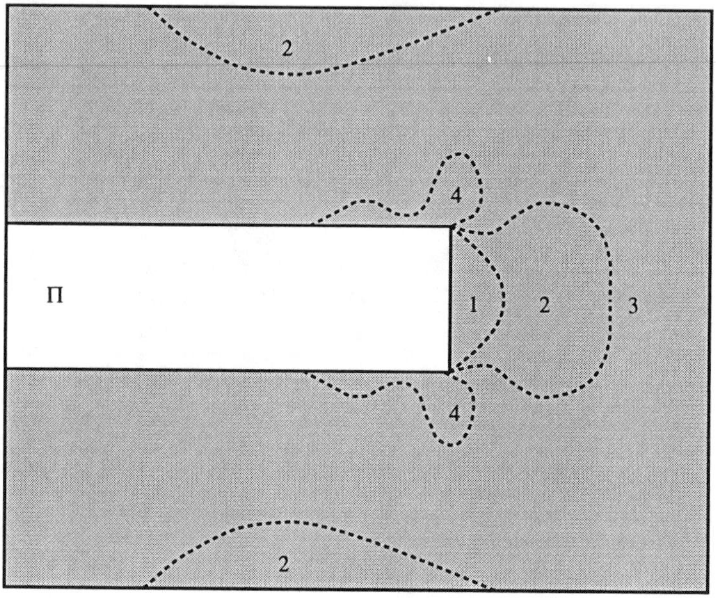

Fig. 2.5: Longitudinal compression of a plane with a rectangular cavity.

Numerical analysis. Taking into account the boundary layer effect we plot in Fig. 2.5 the regions of values of the function $f(x_1, x_2) = -\min(\sigma_1, \sigma_2)\sqrt{2\pi}P^{-1}(1+\sqrt{2})^{-1}$ for the case of a longitudinal symmetric load along the crack. Here σ_1, σ_2 are principal components of stress. The marked regions correspond to the following intervals of values of the quantity f:

$$\mathbf{1}: (0.41, 1.03), \quad \mathbf{2}: (-0.22, 0.41),$$

$$\mathbf{3}: (-0.85, -0.22), \quad \mathbf{4}: (-4.6, -0.85).$$

We can see that the longitudinal compressive loading gives a wedge–shaped region of positive normal stresses in the vicinity of the end of the rectangular cavity. This effect was mentioned in [19] with regard to the rock fracture phenomena.

Let us note that the symmetric transversal load and the shear load do not give similar effects near the crack ends. Also, it is important that the solution corresponding to the longitudinal compression has the leading order with respect to a distance from the crack edge in comparison with the terms related to the normal and to the shear loading.

Illustrative example. Consider an axisymmetric problem of linear elasticity in a lower half-space containing a thin penny-shaped cavity with a rectangular radial cross-section, and assume that the gravity field, directed downwards, is applied to the medium. This problem can be formulated in two ways (also, see Section 3.5) corresponding to the following conditions at infinity: *(i)* in a layer of a finite thickness $u_r = o(1)$, as $r \to \infty$, or *(ii)* $\sigma_{rr} = o(1)$, $\sigma_{\varphi\varphi} = o(1)$, $\sigma_{rz} = o(1)$; here (r, φ, z) are cylindrical coordinates. Due to the solution of the boundary layer type, the first case provides a wedge-shaped region of positive values of principal components of stress which is preceding to the rock fracture phenomena (see [19]) that was observed in mining engineering. The second formulation does not cause a longitudinal compression, and, therefore, the above mentioned rock fracture effect does not occur in this case.

2.3 Formation of a Griffith's crack in a nonuniform stress field

In papers [29], [30] Griffith studied the energy balance for a crack in a uniform stress field, and it was shown that

$$\Delta\mathcal{E}_G + \mathfrak{S} < 0, \tag{2.3.1}$$

where

$$\Delta\mathcal{E}_G = -\frac{\pi(1-\nu^2)p^2 l^2}{E}$$

is the increment of the strain energy, and

$$\mathfrak{S} = 4\gamma l$$

is the surface energy. Here γ is the surface energy density, $2l$ is the crack length, p is the intensity of a tensile load at infinity, E is the Young modulus and ν is the Poisson ratio. Relation (2.3.1) does not hold for infinitesimal cracks, since

$$\Delta\mathcal{E}_G = O(l^2), \quad \mathfrak{S} = O(l), \quad \text{as } l \to 0,$$

and, therefore, equation (2.3.1) cannot be used as a criterion of the crack formation.

Real elastic solids are not homogeneous, and the objective of the present section is to study the crack formation in a stress field with the local extremum point.

Example. Consider a region

$$\Pi = \{\boldsymbol{x} = (x_1, x_2) \in \mathbb{R}^2 : |x_2| \le \delta h_0 + \frac{x_1^2}{h}\}$$

located between two rigid inclusions of a parabolic shape and subjected to the plane strain load; the elastic medium is in perfect contact with the rigid contours. The quantities h_0 and h are assumed to have the same order of magnitude; δ is a small positive parameter. The rigid bodies are subjected to displacements $(0, \rho), (0, -\rho)$. Thus, in the region Π the leading parts of displacement components and components of the stress tensor are specified by

$$\boldsymbol{u}(\boldsymbol{x}) \sim \rho\{(t - \frac{2h_0}{3h}\frac{\delta\nu}{1-2\nu}t(t^2-1))\mathrm{e}^{(2)}$$

$$+\frac{x_1(t^2-1)}{h(1-2\nu)}\mathrm{e}^{(1)}\}, \qquad (2.3.2)$$

$$t = x_2(\delta h_0 + x_1^2/h)^{-1};$$

$$\sigma_{11}(\boldsymbol{u};\boldsymbol{x}) \sim p_1 + [-q_1 x_2^2 + q_2(x_1^2 - 2x_2^2)],$$

$$\sigma_{12}(\boldsymbol{u};\boldsymbol{x}) \sim -2q_2 x_1 x_2,$$

$$\sigma_{22}(\boldsymbol{u};\boldsymbol{x}) \sim p_2 + [q_1 x_1^2 + q_2 x_2^2]; \qquad (2.3.3)$$

where

$$q_1 = -\frac{E(1-\nu)\rho}{h(h_0\delta)^2(1+\nu)(1-2\nu)},$$

$$q_2 = -\frac{E\nu\rho}{h(h_0\delta)^2(1+\nu)(1-2\nu)},$$

$$p_1 = \frac{E\rho}{(1+\nu)(1-2\nu)}\left\{-\frac{1-\nu}{h(1-2\nu)} + \nu(\frac{1}{\delta h_0} + \frac{2\nu}{3h(1-2\nu)})\right\},$$

$$p_2 = \frac{E\rho}{(1+\nu)(1-2\nu)}\left\{-\frac{\nu}{h(1-2\nu)} + (1-\nu)(\frac{1}{\delta h_0} + \frac{2\nu}{3h(1-2\nu)})\right\}.$$

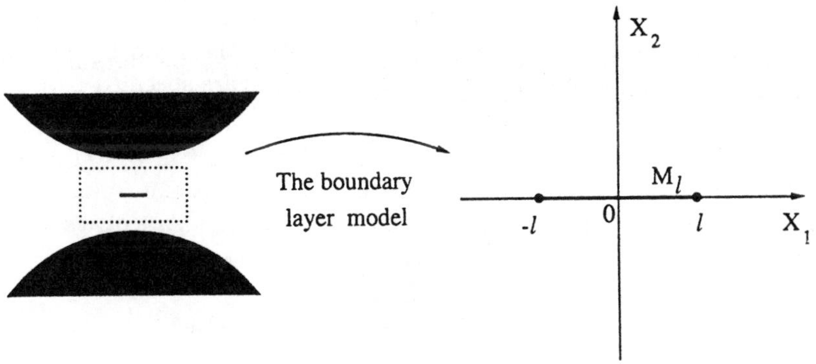

Fig. 2.6: A crack in a thin region between two rigid inclusions.

From (2.3.2) one can see that in the region $\{\boldsymbol{x} \in \Pi : |x_1| < h_0\sqrt{\delta}\}$ the leading part of the displacement field is represented by a cubic vector polynomial. For the case of the axial tension ($\rho > 0$) along the Ox_2 axis stress components σ_{11} on the axis Ox_2 and σ_{22} on the axis Ox_1 have local minimum and maximum at the origin, respectively.

2.3.1 Elastic plane with a crack in a nonhomogeneous stress field

Suppose that a small crack occurs in a thin region between two rigid contours described above.

Then the boundary layer (see Fig. 2.6) should be constructed in the vicinity of the defect; it is described in the scaled variables $\mathbf{X}$ and is related to an elastic plane with a finite crack and the conditions at infinity that provide matching of inner and outer asymptotic expansions.

Consider an elastic plane and assume that the stress state is specified by expressions of the form (2.3.3). We shall analyze the formation of the Mode–I crack $M_l = \{\mathbf{X} \in \mathbb{R}^2 : X_2 = 0, \ |X_1| \le l\}$ on the OX_1 axis. Formulae (2.3.3) yield that, as $p_2 > 0$, the stress component σ_{22} is positive in the vicinity of the origin, and, as $q_1 < 0$, it has a local maximum at $X_1 = 0$. The Mode–I stress–intensity factor has the form

$$K_I(l) = \sqrt{\pi l}(p_2 + q_1 l^2/2). \tag{2.3.4}$$

The increment of the potential strain energy can be determined in a conventional way suggested by Griffith [29], [30] (see also [51]). Namely, we calculate the energy increment for a large domain $\Omega_R = \{\mathbf{X} : R^{-1}\mathbf{X} \in \Omega\}$, where Ω is a simply connected domain with the smooth boundary $\partial\Omega$ containing the origin. The loading conditions are specified by

$$\boldsymbol{\sigma}^{(n)}(\boldsymbol{u}^{(0)}; \mathbf{X}) = \boldsymbol{\sigma}(\boldsymbol{u}^{(0)}; \mathbf{X})\boldsymbol{n}(\mathbf{X}), \quad \mathbf{X} \in \partial\Omega_R, \tag{2.3.5}$$

where $\boldsymbol{n}$ is the unit outward normal vector with respect to $\partial\Omega_R$. A polynomial solution of the homogeneous Lamé system, which satisfies (2.3.5), can be represented in the form

$$\boldsymbol{u}^{(0)}(\mathbf{X}) = \sum_{j=1}^{4} a_j \mathbf{V}^{(j)}(\mathbf{X}), \tag{2.3.6}$$

where

$$\mathbf{V}^{(j)}(\mathbf{X}) = X_j \boldsymbol{e}^{(j)}, \quad j = 1, 2,$$

$$\mathbf{V}^{(3)}(\mathbf{X}) = \frac{2}{1-2\nu}(-\nu X_1^3 - 3(1-\nu)X_2^2 X_1, 3(1-\nu)X_2 X_1^2 + \nu X_2^3)^T,$$

$$\mathbf{V}^{(4)}(\mathbf{X}) = \frac{2}{1-2\nu}((1-\nu)X_1^3 - 3(2-\nu)X_2^2 X_1,$$

$$-3\nu X_2 X_1^2 + (1+\nu) X_2^3)^T; \qquad (2.3.7)$$

The coefficients a_i, $i = 1, 2, 3, 4$, are given by

$$a_1 = \frac{1+\nu}{E}((1-\nu)p_1 - \nu p_2), \quad a_2 = \frac{1+\nu}{E}((1-\nu)p_2 - \nu p_1),$$

$$a_3 = \frac{(1+\nu)(1-2\nu)}{6E} q_1, \quad a_4 = \frac{(1+\nu)(1-2\nu)}{6E} q_2. \qquad (2.3.8)$$

The presence of a small crack of the length $2l/R$ in a domain Ω_R corresponds to a singular perturbation of the boundary. Thus, the boundary layer occurs in a neighbourhood of the origin, and the leading order term of the asymptotic expansion of the displacement field near M_l (inner expansion) corresponds to the solution $\boldsymbol{w}$ of a boundary value problem for a finite crack in an infinite elastic plane. The following asymptotic approximation holds:

$$\boldsymbol{w}(\mathbf{X}) = \boldsymbol{u}^{(0)}(\mathbf{X}) + \sum_{j=1}^{4} d_j (\mathbf{V}_1^{(j)}(\frac{\partial}{\partial \mathbf{X}})\mathbf{T}^{(1)}(\mathbf{X})$$

$$+\mathbf{V}_2^{(j)}(\frac{\partial}{\partial \mathbf{X}})\mathbf{T}^{(2)}(\mathbf{X})) + O(||\mathbf{X}||^{-2}); \qquad (2.3.9)$$

the vector differential operators $\mathbf{V}_i^{(j)}(\partial/\partial \mathbf{X})$ are specified by the vector–valued functions (2.3.7), where X_i should be replaced by $\partial/\partial X_i$; $\mathbf{T}^{(j)}$ are the columns of the Somigliana tensor. The coefficients d_j have the form

$$d_1 = \frac{\nu}{1-\nu} d_2, \quad d_2 = -\frac{1}{2(1-2\nu)} \pi (1-\nu)^2 (4p_2 + q_1 l^2) l^2,$$

$$d_3 = -d_4(1 + 4(3 + 4\nu^2 - 6\nu)^{-1}),$$

$$d_4 = \pi/192(1-\nu)(1-2\nu)(3p_2 + q_1 l^2) l^4. \qquad (2.3.10)$$

Using the matching principle, we can see that the leading term of the inner expansion is specified by

$$\boldsymbol{u}^{(0)} + \sum_{j=1}^{4} d_j \zeta^{(j)}(\mathbf{X}),$$

where the vectors $\zeta^{(j)}$ are solutions of the boundary value problems

$$\mathbf{L}(\partial/\partial\mathbf{X})\zeta^{(j)}(\mathbf{X}) + \mathbf{V}^{(j)}(\partial/\partial\mathbf{X})\delta(\mathbf{X}) = \mathbf{0}, \quad \mathbf{X} \in \Omega_R, \tag{2.3.11}$$

$$\boldsymbol{\sigma}^{(n)}(\zeta^{(j)}; \mathbf{X}) = \mathbf{0}, \quad \mathbf{X} \in \partial\Omega_R. \tag{2.3.12}$$

The potential strain energy can be evaluated as follows:

$$\mathcal{E}(\Omega_R \setminus M) = -\frac{1}{2}\int_{\partial\Omega_R} (\boldsymbol{u}^{(0)} + \sum_{j=1}^{4} d_j \zeta^{(j)}) \cdot \boldsymbol{\sigma}(\mathbf{X})\boldsymbol{n}(\mathbf{X})ds + O(R^{-1})$$

$$= \mathcal{E}(\Omega_R) - 1/2I + O(R^{-1}), \quad R \to \infty, \tag{2.3.13}$$

where

$$I = \sum_{j=1}^{4} d_j \int_{\partial\Omega_R} \zeta^{(j)}(\mathbf{X}) \cdot \boldsymbol{\sigma}^{(n)}(\boldsymbol{u}^{(0)}; \mathbf{X})ds$$

$$= -\sum_{j=1}^{4} d_j \int_{\Omega_R} \boldsymbol{u}^{(0)} \cdot \mathbf{L}(\frac{\partial}{\partial\mathbf{X}})\zeta^{(j)}(\mathbf{X})d\mathbf{X}$$

$$= \sum_{j=1}^{4} d_j \int_{\mathbb{R}^2} \boldsymbol{u}^{(0)}(\mathbf{X}) \cdot \mathbf{V}^{(j)}(\partial/\partial\mathbf{X})\delta(\mathbf{X})d\mathbf{X}$$

$$= -\sum_{k=1}^{4}\sum_{j=1}^{4} d_j a_k \mathbf{V}^{(j)}(\partial/\partial\mathbf{X})\mathbf{V}^{(k)}(\mathbf{X})|_{\mathbf{X}=\mathbf{0}}.$$

Taking the limit $R \to \infty$, we derive the representation for the increment of the potential strain energy due to the presence of a small crack in an elastic domain

$$\Delta\mathcal{E}(l) = -\frac{l^2\pi(1-\nu^2)}{4E}(4p_2^2 + 2l^2 p_2 q_1 + \frac{1}{3}q_1^2 l^4). \tag{2.3.14}$$

On the basis of relations (2.3.4) and (2.3.14) we introduce the loading parameter α (it has the dimension of length)

$$\alpha = \sqrt{p_2/|q_1|}. \tag{2.3.15}$$

It yields

$$p_2 = p, \quad q_1 = -p/\alpha^2,$$

(here we assume that the quantity α does not depend on the normalized intensity of the load p).

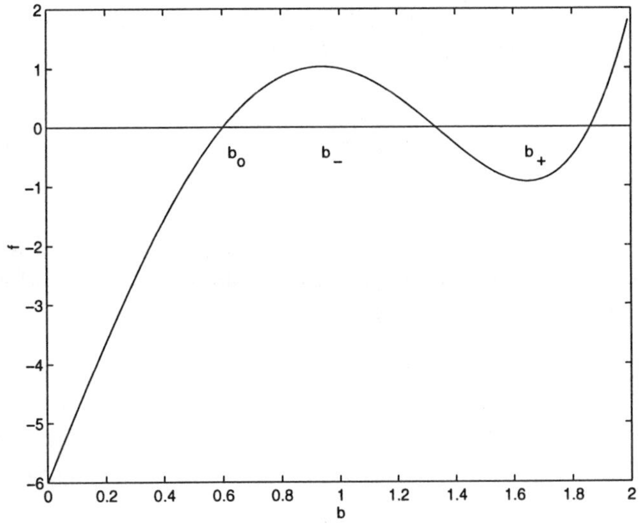

Fig. 2.7: The plot of f versus b; $B = 6$.

The length l of the crack is subjected to the following inequalities

$$\Delta\mathcal{E}(l) + \mathfrak{S}(l) \leq 0, \tag{2.3.16}$$

$$0 \leq K_I(l) \leq K_c := \sqrt{E\gamma/(1-\nu)^2}. \tag{2.3.17}$$

For a crack in a state of stable equilibrium we assume that, if the stress–intensity factor takes its critical value $K_I(l) = K_c$, then the local stability condition

$$K'_I(l) < 0 \tag{2.3.18}$$

is satisfied.

2.3.2 Analysis of the crack formation in an elastic domain

The equality corresponding to (2.3.16) can be written as follows

$$f(b) := b(b^4 - 6b^2 + 12) - B = 0. \tag{2.3.19}$$

Here

$$b = l/a, \quad B = 16K_c^2/(\pi \alpha p^2).$$

As $b > 0$, the function f has two extremum points

$$b_\pm = \left(\frac{9 \pm \sqrt{21}}{5} \right)^{1/2}, \tag{2.3.20}$$

(see Fig. 2.7). It should be mentioned that $b_-^2 < 2 < b_+^2$, and b_- is the point of local maximum. In order to satisfy the relation (2.3.16) we have to make sure that $f(b_-) \geq 0$, or, equivalently,

$$p > p_c = cK_c/\sqrt{\alpha}, \tag{2.3.21}$$

$$c = \left(50(9 + 21^{1/2})^{1/2}/[3^{3/2}\pi(11 + 21^{1/2})] \right)^{1/2} \approx 0.851.$$

If the normalized intensity of the load p is less than the critical value p_c, then the crack formation is impossible. The quantity p_c is defined by the critical value K_c and by the loading parameter α.

For the root b_0 of equation (2.3.19), such that $b_0 \in (0, \sqrt{2})$, the following estimates hold

$$\frac{4K_c^2}{3\alpha\pi p^2} < b_0 < \frac{100K_c^2}{3\alpha\pi(11 + 21^{1/2}p^2)}. \tag{2.3.22}$$

It seems that the upper bound yields the decreasing of the crack length for greater values of p. However, we shall show that for large enough values of p the crack $M_l = \{\mathbf{X} : \ |X_1| \leq l, \ X_2 = 0\}$, with $l = \alpha b_0$, is not stable, and, consequently, the governing relations for the crack length are represented by inequalities (2.3.17) and (2.3.18).

Suppose that $K_I(\alpha b_0) \leq K_c$. Taking the square of both sides and using (2.3.4) we derive

$$g(b) := 4b(b^4 - 4b^2 + 4) - B \leq 0. \tag{2.3.23}$$

Clearly (see (2.3.19)),

$$0 = f(b_0) \le f(b_0) - g(b_0) = -b_0(3b_0^4 - 10b_0^2 + 4). \qquad (2.3.24)$$

As $b_0 \in (0, b_*)$, where $b_*^2 = (5-\sqrt{13})/3$, the quantity (2.3.23) is negative, which is unphysical, and, therefore, $K_I(\alpha b_0) > K_c$ for $b_0 < b_*$. The plots of functions $f + B$ and $g + B$ are presented in Fig. 2.8.

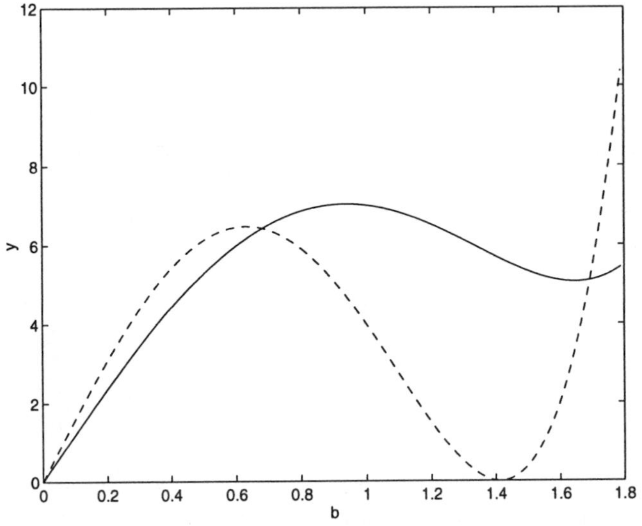

Fig. 2.8: The plots $y = f(b) + B$ (solid line) and $y = g(b) + B$ (dashed line).

The stress intensity factor $K_I(b\alpha)$ has maximum at the point $b = b_1 := \sqrt{2/5}$, and it vanishes at $b = b_2 := \sqrt{2}$; in this case $b_1 < b_* < b_2$. Let p_c^* denote such intensity p that provides the quantity b_* satisfies the equation (2.3.19), and, therefore,

$$p_c^* = c^* K_c/\sqrt{\alpha} \approx 1.045 p_c, \qquad (2.3.25)$$

$$c^* = 4(\pi b_*(b_*^4 - 6b_*^2 + 12))^{-1/2} \approx 0.89.$$

If $p > p_c^*$, then the root b_0 is less than b_*, and the crack of the length $2ab_0$ is not stable. Thus, the crack should grow until its length reaches the value $2\alpha b'$, where b' is the root of the equation $K_I(\alpha b') = K_c$, or, in equivalent form, of the equation

$$g(b') = 0. \tag{2.3.26}$$

The root of equation (2.3.26) should be sought on the interval (b_1, b_2) that corresponds to the monotonically decreasing stress–intensity factor and, therefore, provides the local stability condition (2.3.18).

At this stage qualitative effects related to formation of a small crack in a nonhomogeneous stress field can be discussed. Clearly, if $p < p_c$, then the crack formation is impossible. When $p_c < p < p_c^*$, the crack of the length $2\alpha b_0$, where $b_0 \in (b_*, b_-)$, can occur. As $p \geq p_c^*$, the governing equation for the crack length $2l' = 2\alpha b'$ is given by (2.3.26), and in this case $b' \in [b_*, \sqrt{2})$. For such cracks the energy dissipation occurs. It corresponds to the strict inequality in the energy balance condition (2.3.16).

For large values of p formulae (2.3.23), (2.3.26) yield the following asymptotic equality

$$b' = \sqrt{2} - \frac{K_c}{p\sqrt{\pi\alpha}8^{1/2}} + O(K_c^2/(\alpha p^2)). \tag{2.3.27}$$

The crack with this length is locally stable and its length cannot exceed $\alpha\sqrt{2}$.

As a final remark, we would like to calculate the loading parameter α for a simple example on a crack formation between a couple of rigid circular inclusions. For the case of stretching of a "thin layer" along the Ox_2 axis and the formation of a crack on the Ox_1 axis one has

$$\alpha = h_0\delta\left\{\frac{h}{\delta h_0} - \frac{\nu(1+2\nu)}{3(1-2\nu)(1-\nu)}\right\}^{1/2}.$$

For the case of compression along the Ox_2 axis and a small crack on the same axis the parameter α is given by

$$\alpha = h_0\delta\left\{\frac{\nu h}{(1+\nu)\delta h_0} + \frac{2\nu^2+3\nu-3}{3(1+\nu)(1-2\nu)}\right\}^{1/2}.$$

If a uniform tension is applied at infinity the parameter α is constant, which can be treated as a structural parameter of the material.

2.4 Generalization of the Novozhilov problem for stationary propagating cracks

The present section is based on the results of Novozhilov related to the equilibrium theory of cracks in an elastic medium. The bonding forces between the crack faces are introduced in the vicinity of the crack tip. We discuss the generalization of the Novozhilov criterion for the case of stationary propagating cracks.

2.4.1 Novozhilov's criterion of fracture

In papers [97], [98] Novozhilov proposed the concept of the fracture phenomenon regarding cracks in elastic solids as nontrivial modes of equilibrium deformation. The formation of a crack is interpreted as a loss of stability on the macro scale of the trivial modes of equilibrium.

Novozhilov's approach involves parameters of the discrete atomic structure ("interatomic distance" d) and the continuum model developed by Leonov and Panasyuk [48]. The latter model deals with a crack where the interaction between surfaces occurs near the end points. Let the crack be given by

$$M_l = \{\boldsymbol{x} : x_2 = 0, |x_1| < l\}.$$

The parts of the boundary

$$\{\boldsymbol{x} : x_2 = \pm l, |x_1| < l_0\},\ l_0 < l,$$

are assumed to be free of tractions, and the tractions of the constant intensity are applied near the ends, so one has

$$\sigma_{22} = \sigma_c, \quad \text{as } l_0 < |x_1| < l,\ x_2 = \pm 0.$$

Here the quantity σ_c is regarded as the rupture yield strength.

The body force density is assumed to be zero, and the components of the stress tensor satisfy the equilibrium equations. The normal load of constant intensity is set at infinity:

$$\sigma_{11}^{\infty} = \sigma_{12}^{\infty} = 0,\ \sigma_{22}^{\infty} = \sigma.$$

The quantities σ and $\delta = l - l_0$ are assumed to be small in comparison with σ_c and l, respectively.

$$|\frac{\sigma}{\sigma_c}| \ll 1, \ \frac{\delta}{l} \ll 1. \tag{2.4.1}$$

The solution, obtained by Leonov and Panasyuk [48], yields the following distribution of the normal stress σ_{22} and displacement u_2:

$$\sigma_{22}(x_1, 0) = \sigma_c + \frac{x_1}{\sqrt{x_1^2 - l^2}}\left(\sigma - \frac{2\sigma_c}{\pi}\cos^{-1}\frac{l_0}{l}\right)$$

$$+\frac{\sigma_c}{\pi}\left(\sin^{-1}\frac{l^2 - x_1 l_0}{l(x_1 - l_0)} - \sin^{-1}\frac{l^2 + x_1 l_0}{l(x_1 + l_0)}\right), \ |x_1| \geq l, \tag{2.4.2}$$

$$u_2(x_1, +0) = \frac{2}{E}\left(\sigma - \frac{2}{\pi}\sigma_c \cos^{-1}\frac{l_0}{l}\right)\sqrt{l^2 - x_1^2}$$

$$+\frac{\sigma_c}{\pi E}\left((x_1 - l_0)G(l, x_1, l_0) - (x_1 + l_0)G(l, x_1, -l_0)\right), \ |x_1| \leq l, \tag{2.4.3}$$

where E is the Young modulus of the material and

$$G(l, x_1, k) = \ln \frac{l^2 - x_1 k - \sqrt{(l^2 - x_1^2)(l^2 - k^2)}}{l^2 - x_1 k + \sqrt{(l^2 - x_1^2)(l^2 - k^2)}}.$$

The following condition of the energy balance is set in order to specify the length of the region of interaction between the crack faces:

$$u_2(l_0, +0) = \frac{4\sigma_c \delta \gamma}{\pi E \sigma_c}, \tag{2.4.4}$$

with γ being the density of the surface energy of the elastic solid.

Using the assumption (2.4.1) Novozhilov performed a second-order asymptotic analysis of the system (2.4.2), (2.4.3), (2.4.4) and introduced one more condition,

$$\frac{1}{d}\int_0^d \sigma_{22}(l + s, 0)ds \leq \sigma_c, \tag{2.4.5}$$

where the sign "=" corresponds to the critical state of equilibrium and further increase of load will cause fracture.

In comparison with the Griffith criterion of fracture Novozhilov's approach allows one to observe a *trapping effect* and obtain a certain interval for the length $L = 2l$ of an equilibrium crack. Namely,

$$L_G < L < L_c,$$

where

$$L_G = \frac{4}{\pi}\frac{E\gamma}{\sigma^2}$$

is the value obtained by Griffith, and

$$L_c = \left(\frac{2}{\pi}\beta + \frac{4}{\pi^2}\alpha + \frac{\beta^2}{4\alpha}\right)\frac{d\sigma_c^2}{\sigma^2},$$

with given $\beta = \gamma E/(d\sigma_c^2)$, and $\alpha = \delta/d$ being a solution of the transcendental equation

$$\beta = \sqrt{\alpha}(1+\alpha)\left(1 - \frac{2}{\pi}\sin^{-1}\frac{\alpha-1}{\alpha+1}\right), \tag{2.4.6}$$

which has been solved by Novozhilov numerically.

One can notice that Novozhilov's model does not give direct methods of evaluation of the structural parameter d, and, in a certain sense, this is a disadvantage of the approach. However, this model provides a qualitative description of the trapping effect for a crack in an elastic solid.

2.4.2 A crack propagating with a constant speed

Consider an elastic plane with a semi-infinite crack $M_t = \{(x_1, x_2) : x_2 = 0, x_1 < vt\}$ which moves with a constant velocity v along the Ox_1 axis. Assume that in a neighbourhood of the crack tip the crack faces are subjected to a symmetric load such that

$$\sigma_{22} = -P, \sigma_{12} = 0, \quad -a < x_1 - vt < -b, \quad a > b > 0, \quad x_2 = \pm 0,$$

$$\sigma_{22} = \sigma_{12} = 0, \quad x_1 - vt \in \mathbb{R}^1_- \backslash (-a, -b), \quad x_2 = \pm 0. \tag{2.4.7}$$

The displacement vector $\boldsymbol{u}$ is assumed to satisfy the system of equations

$$L\left(\frac{\partial}{\partial x_1}, \frac{\partial}{\partial x_2}\right)\boldsymbol{u} - \rho\ddot{\boldsymbol{u}} = \boldsymbol{0}, \quad (x_1, x_2) \in \mathbb{R}^2 \backslash M_t, \tag{2.4.8}$$

and the following conditions at infinity

$$\sigma_{ij}(\boldsymbol{u}; x_1, x_2) = O(\tau^{-2}), \quad \text{as } \tau = ((x_1 - vt)^2 + x_2^2)^{\frac{1}{2}} \to \infty. \tag{2.4.9}$$

The displacement vector can be expressed in terms of scalar and vector potentials Φ and $(0, \Psi)$

$$u_1 = \partial_1\Phi + \partial_2\Psi, \quad u_2 = \partial_2\Phi - \partial_1\Psi, \tag{2.4.10}$$

$$\Box_1\Phi(x_1, x_2) = 0, \quad \Box_2\Psi(x_1, x_2) = 0, \tag{2.4.11}$$

where

$$\Box_i = \nabla^2 - \frac{1}{c_i^2}\frac{\partial^2}{\partial t^2},$$

and

$$c_1^2 = \frac{\lambda + 2\mu}{\rho}, \quad c_2^2 = \frac{\mu}{\rho}$$

are the values of the speed of dilatational and shear waves; λ and μ are the Lamé constants. Here we assume that $v < c_R$, where c_R is the speed of propagation of the Rayleigh waves.

The problem on the stationary motion of a crack, whose faces are subjected to a distributed load, was solved in [21]. In the system of coordinates (X, Y_j) such that

$$X = x_1 - vt, \quad Y_j = \left(1 - \frac{v^2}{c_j^2}\right)^{1/2} x_2,$$

equations (2.4.11) can be written in the form

$$\Delta_{XY_1}\Phi = 0, \quad \Delta_{XY_2}\Psi = 0. \tag{2.4.12}$$

The boundary conditions (2.4.7) are represented by

$$\sigma_{22} = -P, \quad \sigma_{12} = 0, \quad -b < X < -a. \tag{2.4.13}$$

One can also use the complex potentials $w_1(z_1), w_2(z_2)$, which are the regular functions of the variables $z_j = X + iY_j,\ j = 1,2;\ \ Re\ w_1(z_1) = \Phi,\ Re w_2(z_2) = \Psi$. The following notations will be used:

$$\phi = Im\ w_1(z_1), \quad \psi = Im\ w_2(z_2).$$

Consequently, the problem is reduced to determining the regular functions $w_1(z_1),\ w_2(z_2)$ in the regions

$$-\pi < arg\ z_j < \pi, \quad j = 1,2, \tag{2.4.14}$$

with the conditions at infinity

$$\tau \frac{d^2 w_1}{dz_1^2} \to 0, \quad \text{as } |z_1| \to \infty, \tag{2.4.15}$$

$$\tau \frac{d^2 w_2}{dz_2^2} \to 0, \quad \text{as } |z_2| \to \infty. \tag{2.4.16}$$

The boundary conditions (2.4.13) take the form

$$2\left(1 - \frac{v^2}{c_1^2}\right)^{1/2} \frac{\partial^2 \psi}{\partial X^2} + \left(2 - \frac{v^2}{c_2^2}\right) \frac{\partial^2 \Psi}{\partial X^2} = 0, \quad x_2 = \pm 0, \tag{2.4.17}$$

$$\left(2 - \frac{v^2}{c_2^2}\right) \frac{\partial^2 \Phi}{\partial X^2} - 2\left(1 - \frac{v^2}{c_2^2}\right)^{1/2} \frac{\partial^2 \phi}{\partial X^2} = \mu^{-1} f, \quad x_2 = \pm 0,$$

$$f = \begin{cases} P, -a < X < -b, \\ 0, X \in \mathbb{R}^1_- \backslash (-a,-b). \end{cases} \tag{2.4.18}$$

The functions $W_j(\zeta_j) = d^2 w_j / dz_j^2,\ \zeta_j = i z_j^{1/2},\ j = 1,2$, analytic in the upper half-plane $0 < \arg \zeta_j < \pi$, were determined in [21]:

$$C(v) W_1(\zeta_1) = \frac{1}{\pi\mu} \int_{-\infty}^{\infty} \frac{i(2 - v^2 c_2^{-2})}{t - \zeta_1} f(t^2) dt,$$

$$C(v) W_2(\zeta_2) = \frac{1}{\pi\mu} \int_{-\infty}^{\infty} \frac{2(1 - v^2 c_1^{-2})^{1/2}}{t - \zeta_1} f(t^2) dt,$$

$$C(v) = 4(1 - v^2 c_2^{-2})^{1/2} (1 - v^2 c_1^{-2})^{1/2} - (2 - v^2 c_2^{-2})^2. \tag{2.4.19}$$

For the case of load, applied on the interval $-a < X < -b$, the following equalities hold

$$\pi\mu C(v)W_1 = iP\left(2-\frac{v^2}{c_2^2}\right)\left\{\ln\frac{a^{1/2}-iz_1^{1/2}}{b^{1/2}-iz_1^{1/2}}\right.$$

$$\left.-\ln\frac{a^{1/2}+iz_1^{1/2}}{b^{1/2}+iz_1^{1/2}}-2i\frac{a^{1/2}-b^{1/2}}{z_1^{1/2}}\right\},$$

$$\pi\mu C(v)W_2 = -2P\left(1-\frac{v^2}{c_1^2}\right)^{1/2}\left\{\ln\frac{a^{1/2}-iz_2^{1/2}}{b^{1/2}-iz_2^{1/2}}\right.$$

$$\left.-\ln\frac{a^{1/2}+iz_2^{1/2}}{b^{1/2}+iz_2^{1/2}}-2i\frac{a^{1/2}-b^{1/2}}{z_2^{1/2}}\right\}. \tag{2.4.20}$$

If a concentrated force P is applied on the crack face at the distance b from the crack tip, then

$$\pi\mu C(v)W_1 = \left(2-\frac{v^2}{c_2^2}\right)\frac{P}{b^{1/2}}\left\{\frac{1}{z_1^{1/2}}-\frac{z_1^{1/2}}{b+z_1}\right\},$$

$$\pi\mu C(v)W_2 = \frac{2Pi}{b^{1/2}}\left(1-\frac{v^2}{c_1^2}\right)^{1/2}\left\{\frac{1}{z_2^{1/2}}-\frac{z_2^{1/2}}{b+z_2}\right\}. \tag{2.4.21}$$

We calculate the stress σ_{22} ahead of the crack front and the component u_2 of the displacement vector at $X < 0,\ x_2 = \pm 0$:

$$\sigma_{22} = -\mu\left(2-\frac{v^2}{c_2^2}\right)Re\frac{d^2w_1}{dz_1^2}$$

$$+2\mu\left(1-\frac{v^2}{c_2^2}\right)^{1/2}Im\frac{d^2w_2}{dz_2^2}, \tag{2.4.22}$$

$$u_2 = -\left\{\left(1-\frac{v^2}{c_1^2}\right)^{1/2}Im\frac{dw_1}{dz_1}+Re\frac{dw_2}{dz_2}\right\}. \tag{2.4.23}$$

For a pair of concentrated forces, applied to the faces of the crack at

$X = -b$, we have

$$\frac{dw_1}{dz_1} = \frac{2P(2 - v^2 c_2^{-2})}{\pi \mu C(v)} \tan^{-1}\left(\frac{z_1}{b}\right)^{1/2},$$

$$\frac{dw_2}{dz_2} = \frac{4Pi(1 - v^2 c_1^{-2})^{1/2}}{\pi \mu C(v)} \tan^{-1}\left(\frac{z_2}{b}\right)^{1/2}. \qquad (2.4.24)$$

Equations (2.4.22)–(2.4.24) yield

$$u_2(-X, +0) = \frac{P(1 - v^2 c_1^{-2})^{1/2}}{\pi \mu C(v)} \frac{v^2}{c_2^2} \ln\left|\frac{X^{1/2} + b^{1/2}}{X^{1/2} - b^{1/2}}\right|; \qquad (2.4.25)$$

$$\sigma_{22}(\boldsymbol{u}; X, 0) = \frac{P}{\pi b^{1/2}}\left(\frac{1}{X^{1/2}} - \frac{X^{1/2}}{b + X}\right), \quad X > 0. \qquad (2.4.26)$$

If the bonding forces with the intensity σ_c are distributed in the vicinity of the crack tip in the region $\{(X, Y) : -\delta < X < 0,\ Y = \pm 0\}$ and the opening by concentrated forces of intensity P are applied on the crack faces at $X = -b$, then

$$u_2(-X, +0) = \frac{(1 - v^2 c_1^{-2})^{1/2}}{\pi \mu C(v)} \frac{v^2}{c_2^2} \left\{ P \ln\left|\frac{X^{1/2} + b^{1/2}}{X^{1/2} - b^{1/2}}\right| \right.$$

$$-\sigma_c(\delta - X) \ln \frac{X^{1/2} + \delta^{1/2}}{X^{1/2} - \delta^{1/2}}$$

$$\left. -2\sigma_c X^{1/2} \delta^{1/2} \right\}, \qquad (2.4.27)$$

$$\sigma_{22}(\boldsymbol{u}; X, 0) = \frac{1}{\pi}\left\{ \frac{P}{b^{1/2}}\left(\frac{1}{X^{1/2}} - \frac{X^{1/2}}{b + X}\right) - \right.$$

$$\left. -2\sigma_c\left(\left(\frac{\delta}{X}\right)^{1/2} - \tan^{-1}\left(\frac{\delta}{X}\right)^{1/2}\right) \right\}. \qquad (2.4.28)$$

One can notice that relations (2.4.27), (2.4.28) are similar to the representations derived in [48] for the static problem.

Following the approach [97], [98], we assume that as the crack is growing the conditions

$$\frac{1}{d}\int_0^d \sigma_{22}(X,0)dX \geq \sigma_c, \tag{2.4.29}$$

$$u_2(-\delta,+0) = \gamma/\sigma_c \tag{2.4.30}$$

are satisfied.

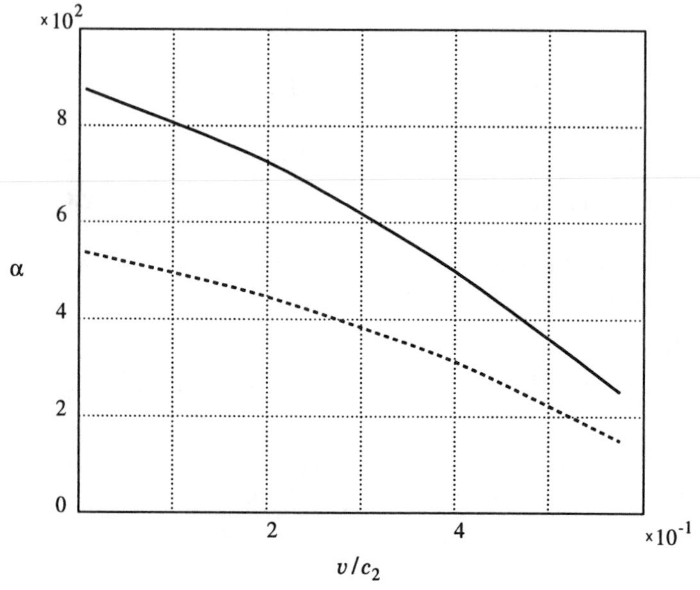

Fig. 2.9: The plot for the quantity α; $\mu = 0.3$, $\beta = 1$, $A = 1.4$ (solid line), $A = 1.6$ (dashed line).

Here d is the structural parameter, σ_c is the theoretical strength limit of the material and γ is the surface energy density. Direct calculations show that

$$\int_{-\delta}^{0} \sigma_c \frac{du_2}{dt}(-\xi,+0)d\xi = \gamma v.$$

Using (2.4.27), (2.4.28), we can transform the conditions (2.4.29),

(2.4.30) to the form

$$A\tan^{-1}\left(\frac{1}{B^{1/2}}\right) - \alpha^{1/2} - (1+\alpha)\tan^{-1}\left(\frac{1}{\alpha^{1/2}}\right) = 0, \qquad (2.4.31)$$

$$\frac{1}{2}A\ln\frac{(\alpha/B)^{1/2}+1}{1-(\alpha/B)^{1/2}} - \alpha = M, \qquad (2.4.32)$$

with

$$M = \beta\pi C(v)\{4(1-v^2c_1^{-2})^{1/2}(1+\nu)v^2c_2^{-2}\}^{-1},$$

$$A = \frac{P}{d\sigma_c}, \quad B = \frac{b}{d}, \quad \alpha = \frac{\delta}{d}, \quad \beta = \frac{\gamma E}{d\sigma_c^2}.$$

If $\delta/b \ll 1$, then equalities (2.4.31), (2.4.32) yield

$$AB^{-1/2} = (1+\alpha)\tan^{-1}(\alpha^{-1/2}) + \alpha^{1/2} + O(\delta/b), \qquad (2.4.33)$$

$$AB^{-1/2}\alpha^{1/2} - \alpha = M + O(\delta/b). \qquad (2.4.34)$$

The results of calculations show that at fixed A the value of α decreases with the increasing of the speed. This corresponds to reduction of the length of interaction region where the bonding forces occur (see Fig. 2.9).

The solution (2.4.27) decreases at infinity. Following the approach [48] one can determine the length l at which the distance between the crack faces is equal to $\Delta = 2u_2(-\delta, +0)$. One can assume that during further closure of the edges at $X < -l$ the crack "closes up" as a result of the presence of bonding forces between the crack faces. The quantity $L = ld^{-1}$ satisfies the equation

$$\frac{\pi\beta C(v)}{2(1+\nu)(1-v^2c_1^{-2})^{1/2}}\frac{c_2^2}{v^2} = A\ln\frac{L^{1/2}+B^{1/2}}{L^{1/2}-B^{1/2}}$$

$$-(\alpha - L)\ln\frac{L^{1/2}+\alpha^{1/2}}{L^{1/2}-\alpha^{1/2}} - 2L^{1/2}\alpha^{1/2}.$$

Of course, this is the simplified model. However, it gives a clear way of calculation for the length of the region of opening in the case of a crack propagating steadily in an elastic continuum.

2.5 State of stress in a plane with a thin elastic inclusion

In the present section we consider an example of singularly perturbed boundary problem (plane strain) for the elasticity equations in a composite region containing an elastic inclusion with the width which is much smaller than its length. We also assume that Young's moduli of materials of the inclusion and of the matrix have the same order of magnitude. The main attention is paid to the asymptotic representation of the stress and displacement components near the edges of the inclusion.

2.5.1 Specification of the problem

Let $\Omega \subset \mathbb{R}^2$ contain a segment $M = \{\boldsymbol{x} : x_2 = 0, |x_1| \leq a\}$, and $h_\pm(t) := (a^2 - t^2)^{1/2} h^0_\pm(t)$, where $h^0_\pm \in \mathbb{C}^\infty[-a, a]$. Assume that $h^0_+(\pm a) = h^0_-(\pm a) = 0$. Introduce a thin domain $G_\varepsilon = \{\boldsymbol{x} : |x_1| < a, -\varepsilon h_-(x_1) < x_2 < \varepsilon h_+(x_1)\}$ and $\Omega_\varepsilon = \Omega \setminus \overline{G}_\varepsilon$ (see Fig. 2.10), where ε is a positive small nondimensional parameter. Note that the interface boundary is smooth at the end points $O_\pm = (\pm a,\ 0)$. Assume that the region Ω_ε is occupied by a material with the Lamé constants λ and μ, and the material of the inclusion G_ε has the Lamé constants λ° and μ°. Young's moduli E and E° are assumed to have the same order of magnitude.

The external load, applied at the outer boundary $\partial\Omega$, is supposed to be self-balanced. The displacement vectors $\boldsymbol{u}$, $\boldsymbol{u}^\circ$ satisfy the Lamé system of equations

$$\mu \Delta \boldsymbol{u}(\varepsilon, \boldsymbol{x}) + (\lambda + \mu)\nabla\nabla \cdot \boldsymbol{u}(\varepsilon, \boldsymbol{x}) = \boldsymbol{0}, \boldsymbol{x} \in \Omega_\varepsilon, \tag{2.5.1}$$

$$\mu^\circ \Delta \boldsymbol{u}^\circ(\varepsilon, \boldsymbol{x}) + (\lambda^\circ + \mu^\circ)\nabla\nabla \cdot \boldsymbol{u}^\circ(\varepsilon, \boldsymbol{x}) = \boldsymbol{0}, \boldsymbol{x} \in G_\varepsilon, \tag{2.5.2}$$

the continuity conditions on the interface boundary

$$\boldsymbol{u}(\varepsilon, \boldsymbol{x}) = \boldsymbol{u}^\circ(\varepsilon, \boldsymbol{x}), \boldsymbol{\sigma}^{\circ,(n)}(\boldsymbol{u}^\circ; \varepsilon, \boldsymbol{x}) = \boldsymbol{\sigma}^{(n)}(\boldsymbol{u}; \varepsilon, \boldsymbol{x}), \boldsymbol{x} \in \partial G_\varepsilon, \tag{2.5.3}$$

and the traction boundary conditions

$$\boldsymbol{\sigma}^{(n)}(\boldsymbol{u}; \varepsilon, \boldsymbol{x}) = \boldsymbol{p}(\boldsymbol{x}), \boldsymbol{x} \in \partial\Omega, \tag{2.5.4}$$

with $\boldsymbol{p}(\boldsymbol{x})$ such that

$$\int_{\partial\Omega} \boldsymbol{p}(\boldsymbol{x})ds = \mathbf{0}, \int_{\partial\Omega} \boldsymbol{x} \times \boldsymbol{p}(\boldsymbol{x})ds = \mathbf{0}. \tag{2.5.5}$$

As usual, $\boldsymbol{\sigma}^{(n)}$ denotes the traction vector with components $\sigma_i^{(n)} = \sigma_{ij}n_j$, where $\boldsymbol{n}$ is the unit outward normal vector with respect to $\partial\Omega_\varepsilon$.

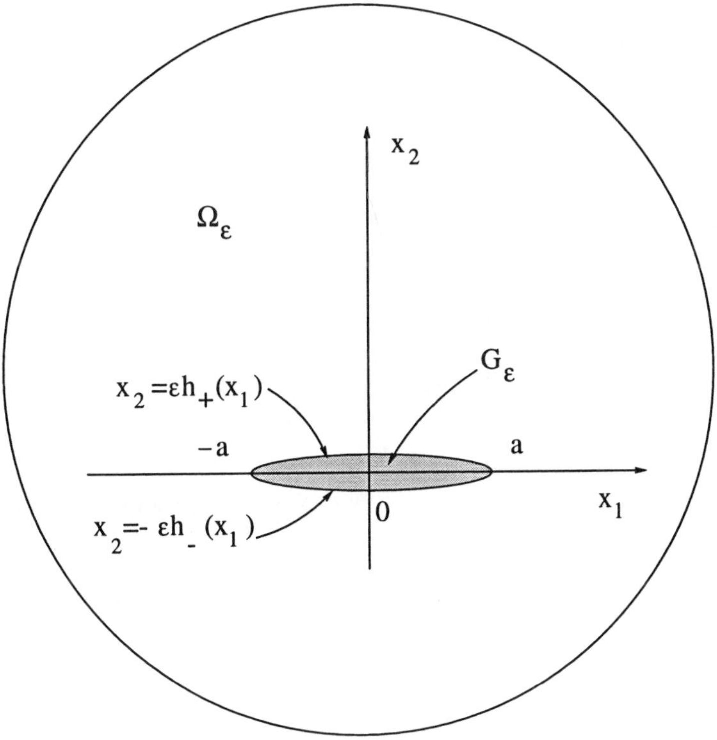

Fig. 2.10: A thin elastic inclusion.

2.5.2 Leading order terms of the asymptotic expansion

Introduce the scaled variable

$$\zeta = \frac{2x_2 + \varepsilon(h_-(x_1) - h_+(x_1))}{2\varepsilon H(x_1)},$$

where $H(x_1) = h_+(x_1) + h_-(x_1)$ is the normalized thickness of the inclusion.

The asymptotic approximation of the displacement $\boldsymbol{u}^\circ$ in a thin region G_ε outside neighbourhoods of the points $O_\pm$ has the form

$$\boldsymbol{u}^\circ(\varepsilon, \boldsymbol{x}) \sim \sum_{k=0}^{\infty} \varepsilon^k \boldsymbol{w}^{(k)}(x_1, \zeta). \tag{2.5.6}$$

For an arbitrary smooth function $U(x_1, x_2) = W(x_1, \zeta)$ the following equalities hold:

$$\partial_1 U(x_1, x_2) = \partial_1 W(x_1, \zeta)$$

$$+ \left\{ \frac{h'_-(x_1) - h'_+(x_1)}{2H(x_1)} - \zeta \frac{H'(x_1)}{H(x_1)} \right\} \partial_\zeta W(x_1, \zeta),$$

$$\partial_2 U(x_1, x_2) = \frac{1}{\varepsilon H(x_1)} \partial_\zeta W(x_1, \zeta). \tag{2.5.7}$$

Here we use the notations $\partial_j W = \partial W / \partial x_j$, $\partial_\zeta W = \partial W / \partial \zeta$.

Using (2.5.7) one can derive

$$\boldsymbol{L}^\circ \left(\frac{\partial}{\partial \boldsymbol{x}} \right) \boldsymbol{w}^{(0)}(x_1, \zeta) = [\varepsilon H(x_1)]^{-2} \boldsymbol{Q}^\circ \partial_\zeta^2 \boldsymbol{w}^{(0)}(x_1, \zeta) + O(\varepsilon^{-1}), \tag{2.5.8}$$

where $\boldsymbol{L}^\circ$ is the matrix differential operator of the Lamé system with the elastic moduli λ°, μ°, and

$$\boldsymbol{Q}^\circ = \mathrm{diag}\{\mu^\circ, 2\mu^\circ + \lambda^\circ\}.$$

Equation (2.5.8) yields that $\boldsymbol{w}^{(0)}$ depends linearly on ζ.

The unit outward normal vector with respect to ∂G_ε is represented by

$$\boldsymbol{n}^\pm = [1 + \varepsilon^2 h'_\pm(x_1)^2]^{-1/2} (-\varepsilon h'_\pm(x_1), \pm 1)^T,$$

and, therefore, at $x_2 = \pm\varepsilon h_\pm(x_1)$,

$$\sigma^{(n)}(\boldsymbol{u};\varepsilon,\boldsymbol{x}) = \pm\,\sigma^{(2)}(\boldsymbol{u};\varepsilon,\boldsymbol{x}) - \varepsilon h'_\pm(x_1)\sigma^{(1)}(\boldsymbol{u};\varepsilon,\boldsymbol{x})$$

$$\mp\,\frac{1}{2}\varepsilon^2 h'_\pm(x_1)^2\sigma^{(2)}(\boldsymbol{u};\varepsilon,\boldsymbol{x}) + O(\varepsilon^3). \qquad (2.5.9)$$

Substitute (2.5.6) into the interface boundary conditions. To leading order approximation one can derive

$$\boldsymbol{u}(\varepsilon, x_1, \pm\varepsilon h_\pm(x_1)) = \boldsymbol{w}^{(0)}(x_1, \pm 1/2) + O(\varepsilon), \qquad (2.5.10)$$

$$\sigma^{(2)}(\boldsymbol{u};\varepsilon, x_1, \pm\varepsilon h_\pm(x_1)) = \frac{1}{\varepsilon H(x_1)}\boldsymbol{Q}^\circ\partial_\zeta\boldsymbol{w}^{(0)}(x_1,\pm 1/2) + O(1). \qquad (2.5.11)$$

From (2.5.11) one can see that

$$\partial_\zeta\boldsymbol{w}^{(0)} = \boldsymbol{0},$$

and, therefore, the vector $\boldsymbol{w}^{(0)}$ depends on x_1 only.

Consider the second term of the asymptotic expansion (2.5.6). Using (2.5.8) and (2.5.11) we obtain

$$\boldsymbol{L}^\circ\left(\frac{\partial}{\partial\boldsymbol{x}}\right)\boldsymbol{u}^\circ(\varepsilon,\boldsymbol{x}) = [\varepsilon H(x_1)]^{-2}\boldsymbol{Q}^\circ\partial_\zeta^2\boldsymbol{w}^{(0)}(x_1,\zeta)$$

$$+\ \varepsilon^{-1}\Big\{H(x_1)^{-2}\boldsymbol{Q}^\circ\partial_\zeta^2\boldsymbol{w}^{(1)}(x_1,\zeta) + (\lambda^\circ+\mu^\circ)(\partial_1$$

$$+\ b(x_1,\zeta)\partial_\zeta)H(x_1)^{-1}\boldsymbol{J}\partial_\zeta\boldsymbol{w}^{(0)}(x_1,\zeta)\Big\} + O(1), x\in G_\varepsilon, \quad (2.5.12)$$

$$\sigma^{(2)}(\boldsymbol{u};\varepsilon,\pm\varepsilon h_\pm(x_1)) = [\varepsilon H(x_1)]^{-1}\boldsymbol{Q}^\circ\partial_\zeta\boldsymbol{w}^{(0)}(x_1,\pm 1/2)$$

$$+\ H(x_1)^{-1}\boldsymbol{Q}^\circ\partial_\zeta\boldsymbol{w}^{(1)}(x_1,\pm 1/2)$$

$$+\ \boldsymbol{B}^\circ(\partial_1 + b(x_1,\zeta)\partial_\zeta)\boldsymbol{w}^{(0)}(x_1,\pm 1/2) + O(\varepsilon),\ |x_1| < a, \quad (2.5.13)$$

where

$$b(x_1,\zeta) = \frac{h'_-(x_1) - h'_+(x_1) - 2H'(x_1)\zeta}{2H(x_1)},$$

$$\boldsymbol{B}^\circ = \begin{pmatrix} 0 & \mu^\circ \\ \lambda^\circ & 0 \end{pmatrix}, \quad \boldsymbol{J} = \begin{pmatrix} 0 & 1 \\ 1 & 0 \end{pmatrix}.$$

Thus, the vector function $\boldsymbol{w}^{(1)}$ satisfies the following equations:

$$\boldsymbol{Q}^\circ \partial_\zeta^2 \boldsymbol{w}^{(1)}(x_1, \zeta) = O(\varepsilon), \ \zeta \in (-1/2, 1/2),$$

$$H(x_1)^{-1} \boldsymbol{Q}^\circ \partial_\zeta \boldsymbol{w}^{(1)}(x_1, \pm 1/2) = -\boldsymbol{B}^\circ \partial_1 \boldsymbol{w}^{(0)}(x_1, \pm 1/2)$$

$$+\boldsymbol{\sigma}^{(2)}(\boldsymbol{u}; \varepsilon, x_1, \pm \varepsilon h_\pm(x_1)) + O(\varepsilon), \tag{2.5.14}$$

and, consequently,

$$\boldsymbol{\sigma}^{(2)}(\boldsymbol{u}; \varepsilon, x_1, \varepsilon h_+(x_1)) = \boldsymbol{\sigma}^{(2)}(\boldsymbol{u}; \varepsilon, x_1, -\varepsilon h_-(x_1)) + O(\varepsilon). \tag{2.5.15}$$

We seek the leading order term of the asymptotic expansion of the matrix displacement $\boldsymbol{u}$ in the form

$$\boldsymbol{u}(\varepsilon, \boldsymbol{x}) = \boldsymbol{v}^{(0)}(\boldsymbol{x}) + O(\varepsilon), \ \boldsymbol{x} \in \Omega_\varepsilon. \tag{2.5.16}$$

As $\varepsilon \to 0$, the region Ω_ε reduces to $\Omega_0 = \Omega \backslash M$, and the vector function $\boldsymbol{v}^{(0)}$ satisfies the homogeneous Lamé system in Ω_0. The following conditions on the sides $\Gamma^\pm$ of the segment M hold

$$\boldsymbol{v}^{(0)}(x_1, +0) = \boldsymbol{v}^{(0)}(x_1, -0),$$

$$\boldsymbol{\sigma}^{(2)}(\boldsymbol{v}^{(0)}; x_1, +0) = \boldsymbol{\sigma}^{(2)}(\boldsymbol{v}^{(0)}; x_1, -0), \ |x_1| < a. \tag{2.5.17}$$

The first two terms of the asymptotic expansion of the displacement field within the inclusion are given by

$$\boldsymbol{w}^{(0)}(x_1, \zeta) = \boldsymbol{v}^{(0)}(x_1, 0), \tag{2.5.18}$$

$$\boldsymbol{w}^{(1)}(x_1, \zeta) = \boldsymbol{W}^{(1)}(x_1)$$

$$+H(x_1)\zeta(\boldsymbol{Q}^\circ)^{-1} \left\{ \boldsymbol{\sigma}^{(2)}(\boldsymbol{v}^{(0)}; x_1, 0) - \boldsymbol{B}^\circ \partial_1 \boldsymbol{v}^{(0)}(x_1, 0) \right\}$$

$$= \boldsymbol{W}^{(1)}(x_1) + H(x_1)\zeta \left\{ (\boldsymbol{Q}^\circ)^{-1} \boldsymbol{Q} \partial_2 \boldsymbol{v}^{(0)}(x_1, 0) \right.$$

$$\left. + \boldsymbol{Q}^{-1}(\boldsymbol{B} - \boldsymbol{B}^\circ) \partial_1 \boldsymbol{v}^{(0)}(x_1, 0) \right\}. \tag{2.5.19}$$

The vector function $\boldsymbol{W}^{(1)}$ will be specified below. The matrices $\boldsymbol{Q}$ and $\boldsymbol{B}$ are defined in the same way as $\boldsymbol{Q}^\circ, \boldsymbol{B}^\circ$ with λ°, μ° replaced by λ, μ.

Thus, the leading order term $\boldsymbol{v}^{(0)}$ of the displacement field in the matrix is represented by a solution of the problem on the deformation of the homogeneous region Ω.

2.5.3 Second-order approximation of the matrix displacement field

Within the matrix Ω_ε the vector function $\boldsymbol{u}$ is approximated by

$$\boldsymbol{u}(\varepsilon, \boldsymbol{x}) = \boldsymbol{v}^{(0)}(\boldsymbol{x}) + \varepsilon \boldsymbol{v}^{(1)}(\boldsymbol{x}) + O(\varepsilon^2). \tag{2.5.20}$$

As before, $\boldsymbol{v}^{(1)}$ satisfies the homogeneous Lamé system in Ω_0. Using the interface boundary conditions, we derive

$$\begin{aligned} \boldsymbol{v}^{(0)} & (x_1, 0) + \varepsilon(\pm\frac{1}{2}H(x_1)(\boldsymbol{Q}^\circ)^{-1}(\boldsymbol{Q}\partial_2 \boldsymbol{v}^{(0)}(x_1, 0) \\ & +(\boldsymbol{B} - \boldsymbol{B}^\circ)\partial_1 \boldsymbol{v}^{(0)}(x_1, 0)) + \boldsymbol{W}^{(1)}(x_1)) \\ = \; & \boldsymbol{v}^{(0)}(x_1, \pm\varepsilon h_\pm(x_1)) + \varepsilon \boldsymbol{v}^{(1)}(x_1, \pm\varepsilon h_\pm(x_1)). \end{aligned} \tag{2.5.21}$$

Hence, we can specify the vector function $\mathbf{W}^{(1)}$ and the jump $[\boldsymbol{v}^{(1)}]$ across M:

$$\begin{aligned} \boldsymbol{W}^{(1)}(x_1) = \frac{1}{2}(\boldsymbol{v}^{(1)}(x_1, +0) + \boldsymbol{v}^{(1)}(x_1, -0) \\ + (h_+(x_1) - h_-(x_1))\partial_2 \boldsymbol{v}^{(0)}(x_1, 0)), \end{aligned} \tag{2.5.22}$$

$$\begin{aligned} [\boldsymbol{v}^{(1)}](x_1) = H(x_1)\{(\boldsymbol{Q}^\circ)^{-1}(\boldsymbol{B} - \boldsymbol{B}^\circ)\partial_1 \boldsymbol{v}^{(0)}(x_1, 0) \\ + ((\boldsymbol{Q}^\circ)^{-1}\boldsymbol{Q} - \mathbf{I})\partial_2 \boldsymbol{v}^{(0)}(x_1, 0)\}. \end{aligned} \tag{2.5.23}$$

In order to evaluate the jump in stresses across M, we shall need the term $\boldsymbol{w}^{(2)}$ from the expansion (2.5.6). Relations (2.5.12), (2.5.18), (2.5.19) yield

$$\begin{aligned} \boldsymbol{L}^\circ\left(\frac{\partial}{\partial \boldsymbol{x}}\right)\boldsymbol{u}^\circ(\varepsilon, \boldsymbol{x}) \; = \; & \{(H(x_1))^{-2}\boldsymbol{Q}^\circ \partial_\zeta^2 \boldsymbol{w}^{(2)}(x_1, \zeta) \\ & +\boldsymbol{Q}^{\circ,(1)}\partial_1^2 \boldsymbol{v}^{(0)}(x_1, 0) \end{aligned}$$

$$+ (\lambda^\circ + \mu^\circ)\boldsymbol{J}(\boldsymbol{Q}^\circ)^{-1}\boldsymbol{Q}^\circ \partial_{12}^2 \boldsymbol{v}^{(0)}(x_1, 0)\}$$

$$+ O(\varepsilon), \tag{2.5.24}$$

$$\boldsymbol{Q}^{\circ,(1)} = \text{diag}\{2\mu^\circ + \lambda - \mu^\circ(\lambda - \lambda^\circ)(2\mu^\circ + \lambda^\circ)^{-1},\ \lambda^\circ(\mu^\circ)^{-1}\mu - \lambda^\circ + \mu\}.$$

By virtue of (2.5.9) we derive

$$\boldsymbol{\sigma}^{(n)}(\boldsymbol{v}^{(0)} + \varepsilon \boldsymbol{v}^{(1)}; \varepsilon, x_1, \pm\varepsilon h_\pm(x_1)) = \pm\boldsymbol{\sigma}^{(2)}(\boldsymbol{v}^{(0)} + \varepsilon \boldsymbol{v}^{(1)}; x_1, \pm 0)$$

$$+\varepsilon h_\pm(x_1)\partial_2 \boldsymbol{\sigma}^{(2)}(\boldsymbol{v}^{(0)}; x_1, 0) - \varepsilon h'_\pm(x_1)\boldsymbol{\sigma}^{(1)}(\boldsymbol{v}^{(0)}; x_1, \pm 0) + O(\varepsilon^2)$$

$$= \pm\boldsymbol{\sigma}^{(2)}(\boldsymbol{v}^{(0)}; x_1, 0) + \varepsilon\{\pm\boldsymbol{\sigma}^{(2)}(\boldsymbol{v}^{(1)}; x_1, \pm 0)$$

$$-\partial_1(h_\pm(x_1)\boldsymbol{\sigma}^{(1)}(\boldsymbol{v}^{(0)}; x_1, 0))\} + O(\varepsilon^2). \tag{2.5.25}$$

Similar calculations for the displacement field within the inclusion give

$$\boldsymbol{\sigma}^{\circ,(n)}(\boldsymbol{w}^{(0)} + \varepsilon \boldsymbol{w}^{(1)} + \varepsilon^2 \boldsymbol{w}^{(2)}; \varepsilon, x_1, \pm\varepsilon h_\pm(x_1))$$

$$= \pm\frac{1}{H(x_1)}\boldsymbol{Q}^\circ \partial_\zeta \boldsymbol{w}^{(1)}(x_1, \pm 1/2) + \boldsymbol{B}^\circ \partial_1 \boldsymbol{w}^{(0)}(x_1)$$

$$+\varepsilon\left\{\pm\left[\frac{1}{H(x_1)}\boldsymbol{Q}^\circ \partial_\zeta \boldsymbol{w}^{(2)}(x_1, \pm 1/2)\right.\right.$$

$$\left.+ \boldsymbol{B}^\circ(\partial_1 + b(x_1, \pm 1/2)\partial_\zeta)\boldsymbol{w}^{(1)}(x_1, \pm 1/2)\right]$$

$$\left.- h'_\pm(x_1)(\boldsymbol{P}^\circ \partial_1 \boldsymbol{w}^{(0)}(x_1) + \frac{1}{H(x_1)}\mathbf{D}^\circ \partial_\zeta \boldsymbol{w}^{(1)}(x_1, \pm 1/2))\right\}$$

$$+O(\varepsilon^2), \tag{2.5.26}$$

where

$$\boldsymbol{P}^\circ = \begin{pmatrix} 2\mu^\circ + \lambda^\circ & 0 \\ 0 & \mu^\circ \end{pmatrix}, \quad \mathbf{D}^\circ = \begin{pmatrix} 0 & \lambda^\circ \\ \mu^\circ & 0 \end{pmatrix}.$$

Using (2.5.24)–(2.5.26) and the interface boundary conditions at ∂G_ε, we obtain the following relations for the jump of tractions across M :

$$[\sigma_{21}(\boldsymbol{v}^{(1)})](x_1) = -2\partial_1\{H(x_1)[\alpha_1\partial_1 v_1^{(0)}(x_1,0) + \alpha_2\partial_2 v_2^{(0)}(x_1,0)]\},$$

$$[\sigma_{22}(\boldsymbol{v}^{(1)})](x_1) = 0, \quad |x_1| < a, \tag{2.5.27}$$

where $\alpha_j = (2\mu^\circ + \lambda^\circ)^{-1}\{(\lambda^\circ - \lambda)\mu^\circ + (2\delta_{j2}\mu^\circ + \lambda^\circ)(\mu^\circ - \mu)\}$. Thus, the displacement field $\boldsymbol{v}^{(1)}$ satisfies the homogeneous Lamé system in Ω_0, the homogeneous traction boundary condition on $\partial\Omega$, and the jump conditions (2.5.23), (2.5.27) across M.

2.5.4 Asymptotic representation of $\boldsymbol{v}^{(1)}$ near the ends of the inclusion

Since $H(x_1) = O(r_\pm^{1/2})$, as $r_\pm \to 0$, and components of $\boldsymbol{v}^{(0)}$ are smooth, the jump conditions for components of $\boldsymbol{v}^{(1)}$ and $\boldsymbol{\sigma}^{(2)}(\boldsymbol{v}^{(1)};\boldsymbol{x})$ in neighbourhoods of the points $(\pm a, 0)$ can be written in the form

$$[\boldsymbol{v}^{(1)}](x_1) = r_\pm^{1/2}\sqrt{2a}\, b_\pm \Big\{(\boldsymbol{Q}^\circ)^{-1}(\boldsymbol{B} - \boldsymbol{B}^\circ)\partial_1\boldsymbol{v}^{(0)}(\pm a, 0)$$

$$+ \ ((\boldsymbol{Q}^\circ)^{-1}\boldsymbol{Q} - \boldsymbol{I})\partial_2\boldsymbol{v}^{(0)}(\pm a, 0)\Big\} + O(r_\pm^{3/2}), \tag{2.5.28}$$

$$[\sigma_{21}(\boldsymbol{v}^{(1)})](x_1) = -r_\pm^{-1/2}\sqrt{2a}\, b_\pm \Big\{\alpha_1\partial_1 v_1^{(0)}(\pm a, 0)$$

$$+ \alpha_2\partial_2 v_2^{(0)}(\pm a, 0)\Big\} + O(r_\pm^{1/2}), \tag{2.5.29}$$

$$[\sigma_{22}(\boldsymbol{v}^{(1)})](x_1) = 0, \tag{2.5.30}$$

where $(r_\pm, \varphi_\pm)$ are polar coordinates with centres $(\pm a, 0)$, and the sides Γ^+, Γ^- of M are specified by the relations $\varphi_+ = \pi, \varphi_+ = -\pi$ and $r_+ \leq 2a$.

The presence of the square root singularity in the stress jump across M yields a singularity in components of the tensor $\boldsymbol{\sigma}(\boldsymbol{v}^{(1)};\boldsymbol{x})$. It follows from [38] that

$$\boldsymbol{v}^{(1)}(\boldsymbol{x}) = r_\pm^{1/2}\boldsymbol{\xi}^\pm(\varphi_\pm) + \boldsymbol{V}(\boldsymbol{x}) + O(r_\pm^{3/2}), \tag{2.5.31}$$

and

$$\boldsymbol{\sigma}(\boldsymbol{v}^{(1)};\boldsymbol{x}) = r_\pm^{-1/2}\Xi^\pm(\varphi_\pm) + \boldsymbol{\sigma}(\mathbf{V};\boldsymbol{x}) + O(r_\pm^{1/2}). \tag{2.5.32}$$

Here $\mathbf{V}$ is a vector valued function with smooth components, and the angular parts $\boldsymbol{\xi}^\pm(\varphi_\pm)$ and $\Xi^\pm(\varphi_\pm)$ are defined by the equalities

$$(\xi_r^\pm(\varphi_\pm), \xi_\varphi^\pm(\varphi_\pm)) = c_1\left(\cos\frac{3\varphi_\pm}{2}, -\sin\frac{3\varphi_\pm}{2}\right)$$

$$+c_2\left(\sin\frac{3\varphi_\pm}{2}, \cos\frac{3\varphi_\pm}{2}\right) + c_3\left(\cos\frac{\varphi_\pm}{2}, \frac{2\varkappa+1}{2\varkappa-1}\sin\frac{\varphi_\pm}{2}\right)$$

$$+ c_4\left(\sin\frac{\varphi_\pm}{2}, \frac{2\varkappa+1}{2\varkappa-1}\cos\frac{\varphi_\pm}{2}\right), \tag{2.5.33}$$

$$(\Xi_{rr}^\pm(\varphi_\pm),\ \Xi_{r\varphi}^\pm(\varphi_\pm), \Xi_{\varphi\varphi}^\pm(\varphi_\pm))$$

$$= \mu\left[c_1\left(\cos\frac{3\varphi_\pm}{2}, -\sin\frac{3\varphi_\pm}{2}, -\cos\frac{3\varphi_\pm}{2}\right)\right.$$

$$+c_2\left(\sin\frac{3\varphi_\pm}{2}, \cos\frac{3\varphi_\pm}{2}, -\sin\frac{3\varphi_\pm}{2}\right)$$

$$+c_3\left(\frac{2\mu}{\lambda+5\mu}\cos\frac{\varphi_\pm}{2}, \frac{\lambda+\mu}{\lambda+5\mu}\sin\frac{\varphi_\pm}{2}, -\frac{3(\lambda+\mu)}{\lambda+5\mu}\cos\frac{\varphi_\pm}{2}\right)$$

$$\left.+ c_4\left(\frac{2\mu}{\lambda+5\mu}\sin\frac{\varphi_\pm}{2}, -\frac{\lambda+\mu}{\lambda+5\mu}\cos\frac{\varphi_\pm}{2}, -\frac{3(\lambda+\mu)}{\lambda+5\mu}\sin\frac{\varphi_\pm}{2}\right)\right], \tag{2.5.34}$$

where $\varkappa = (\lambda+3\mu)(\lambda+\mu)^{-1}$, and the constants c_1, c_2, c_3, c_4 are specified by

$$c_1 = \frac{\sqrt{2a}\, b_\pm}{4(\lambda+2\mu)}\left\{\left[\frac{\lambda^\circ-\lambda}{2\mu^\circ+\lambda^\circ}\left(\frac{3\lambda+7\mu}{\mu}\mu^\circ - \lambda - \mu\right)\right.\right.$$

$$\left. + \frac{3\lambda+7\mu}{\mu}(\mu-\mu^\circ)\right]\partial_1 v_1^{(0)}(\pm a, 0)$$

$$+ \left[\frac{2(\lambda+\mu)(\mu-\mu^\circ+\lambda-\lambda^\circ)}{2\mu^\circ+\lambda^\circ} \right.$$

$$\left. - \frac{(3\lambda+7\mu)(\mu\lambda^\circ-\lambda\mu^\circ)}{\mu(2\mu^\circ+\lambda^\circ)} \right] \partial_2 v_2^{(0)}(\pm a, 0) \Big\},$$

$$c_2 = -\frac{3\sqrt{2a}\, b_\pm(\lambda+\mu)}{4(\lambda+2\mu)} \, \frac{\mu-\mu^\circ}{\mu} \left\{ \partial_1 v_2^{(0)}(\pm a, 0) + \partial_2 v_1^{(0)}(\pm a, 0) \right\},$$

$$c_3 = -\frac{(\lambda+5\mu)\sqrt{2a}\, b_\pm}{4(\lambda+2\mu)} \left\{ \left[\frac{\mu+\mu^\circ}{2\mu^\circ+\lambda^\circ} \, \frac{\lambda-\lambda^\circ}{\mu} - \frac{\mu-\mu^\circ}{\mu} \right] \partial_1 v_1^{(0)}(\pm a, 0) \right.$$

$$\left. + \frac{2(\mu-\mu^\circ)+2(\lambda-\lambda^\circ)+\lambda^\circ-\lambda\mu^\circ\mu^{-1}}{2\mu^\circ+\lambda^\circ} \partial_2 v_2^{(0)}(\pm a, 0) \right\},$$

$$c_4 = \frac{\lambda+5\mu}{4(\lambda+2\mu)} \sqrt{2a}\, b_\pm \frac{\mu-\mu^\circ}{\mu^\circ} \left[\partial_1 v_2^{(0)}(\pm a, 0) + \partial_2 v_1^{(0)}(\pm a, 0) \right]. \quad (2.5.35)$$

2.5.5 Boundary layer

The expansions (2.5.6), (2.5.16) work outside neighbourhoods of the ends of the elastic inclusion.

Consider, for example, the right end of the inclusion, and introduce scaled coordinates

$$\boldsymbol{x} \mapsto \boldsymbol{y} = \varepsilon^{-2}(x_1 - a, x_2). \quad (2.5.36)$$

It follows from the equality

$$y_2 = \pm\varepsilon \left[a^2 - (a+\varepsilon^2 y_1)^2 \right]^{1/2} h_\pm^0(a+\varepsilon^2 y_1)$$

$$= \pm b_+[(-2ay_1)^{1/2} + O(\varepsilon^2 |y_1|)], \quad (2.5.37)$$

that in the limit $\varepsilon \to 0$ the contour ∂G_ε is transformed into a parabola $\partial\Pi$.

Introduce the following notation,

$$\Pi = \{ \boldsymbol{y} : y_1 < 0, \ |y_2| \le b_+(-2ay_1)^{1/2} \},$$

and assume that the boundary layer fields $\boldsymbol{Z}, \boldsymbol{Z}^\circ$ satisfy the homogeneous

Lamé equations

$$\mu\Delta \boldsymbol{Z}(\boldsymbol{y}) + (\lambda+\mu)\nabla\nabla\cdot \boldsymbol{Z}(\boldsymbol{y}) = \boldsymbol{0}, \quad \boldsymbol{y}\in\mathbb{R}^2\backslash\bar{\Pi}, \tag{2.5.38}$$

$$\mu^\circ\Delta \boldsymbol{Z}^\circ(\boldsymbol{y}) + (\lambda^\circ+\mu^\circ)\nabla\nabla\cdot \boldsymbol{Z}^\circ(\boldsymbol{y}) = \boldsymbol{0}, \quad \boldsymbol{y}\in\Pi, \tag{2.5.39}$$

and the interface boundary conditions

$$\boldsymbol{Z}(\boldsymbol{y}) = \boldsymbol{Z}^\circ(\boldsymbol{y}), \boldsymbol{\sigma}^{(n)}(\boldsymbol{Z};\boldsymbol{y}) = \boldsymbol{\sigma}^{\circ,(n)}(\boldsymbol{Z}^\circ;\boldsymbol{y}), \quad \boldsymbol{y}\in\partial\Pi. \tag{2.5.40}$$

We also need the conditions at infinity that should be specified for the vector $\boldsymbol{Z}$. The following expansion holds in a neighbourhood of the right end of the inclusion

$$\boldsymbol{v}^{(0)}(\boldsymbol{x}) + \varepsilon\boldsymbol{v}^{(1)}(\boldsymbol{x}) = \boldsymbol{v}^{(0)}(a,0) + \varepsilon\mathbf{V}(a,0) \tag{2.5.41}$$

$$+ \varepsilon^2\Big(\sum_{i=1}^{2}\boldsymbol{y}\cdot\nabla(v_i^{(0)}(a,0))\boldsymbol{e}^{(i)} + \rho_+^{1/2}\boldsymbol{\xi}^+(\varphi_+)\Big) + O(\varepsilon^{2+\delta}),$$

as $\|\boldsymbol{x}-O_+\| = O(\varepsilon^{1+\delta})$, where $\rho_+ = \varepsilon^{-2}r_+$. The components of the stress tensor are specified by

$$\boldsymbol{\sigma}(\boldsymbol{v}^{(0)}+\varepsilon\boldsymbol{v}^{(1)};\varepsilon,\boldsymbol{x}) = \boldsymbol{\sigma}(\boldsymbol{v}^{(0)};a,0) + \rho_+^{-1/2}\boldsymbol{\Xi}^+(\varphi_+) + O(\varepsilon). \tag{2.5.42}$$

For the displacement vector inside G_ε we have

$$\boldsymbol{w}^{(0)}(x_1) + \varepsilon\boldsymbol{w}^{(1)}(x_1,\zeta) \sim \boldsymbol{w}^{(0)}(a) + (x_1-a)\partial_1\boldsymbol{w}^{(0)}(a)$$

$$+ \varepsilon[\boldsymbol{W}^{(1)}(x_1) + H(x_1)\zeta\{(\boldsymbol{Q}^\circ)^{-1}\boldsymbol{Q}\partial_2\boldsymbol{v}^{(0)}(x_1,0)$$

$$+ (\boldsymbol{Q}^\circ)^{-1}(\boldsymbol{B}-\boldsymbol{B}^\circ)\partial_1\boldsymbol{v}^{(0)}(x_1,0)\}] + O(\varepsilon(a-x_1)^{3/2}). \tag{2.5.43}$$

Since $\zeta\sim y_2/[2b_+(-2ay_1)^{1/2}]$, the interface boundary conditions yield

$$\boldsymbol{w}^{(0)}(x_1) + \varepsilon\boldsymbol{w}^{(1)}(x_1,\zeta) = \boldsymbol{v}^{(0)}(a,0) + \varepsilon\boldsymbol{V}(a,0)$$

$$+ \varepsilon^2\{y_1\partial_1\boldsymbol{v}^{(0)}(a,0) + y_2(\boldsymbol{Q}^\circ)^{-1}[\boldsymbol{Q}\partial_2\boldsymbol{v}^{(0)}(a,0)$$

$$+ (\boldsymbol{B}-\boldsymbol{B}^\circ)\partial_1\boldsymbol{v}^{(0)}(a,0)]$$

$$+\ \tfrac{1}{2}(-y_1)^{1/2}[\boldsymbol{\xi}^+(0)+\boldsymbol{\xi}^+(2\pi)]\}+O(\varepsilon^{2+\delta}), \tag{2.5.44}$$

as $a - x_1 \sim \varepsilon^{1+\delta}$.

The components of the stress tensor are specified by

$$\boldsymbol{\sigma}^\circ(\boldsymbol{w}^{(0)}+\varepsilon\boldsymbol{w}^{(1)};\varepsilon,\boldsymbol{x}) = \boldsymbol{\Sigma}(\boldsymbol{v}^{(0)})+O(\varepsilon^\delta),$$

$$\boldsymbol{\Sigma}^{(2)}(\boldsymbol{v}^{(0)}) = (\Sigma_{21}(\boldsymbol{v}^{(0)}),\Sigma_{22}(\boldsymbol{v}^{(0)}))^T := \boldsymbol{\sigma}^{(2)}(\boldsymbol{v}^{(0)};a,0),$$

$$\boldsymbol{\Sigma}_{11}(\boldsymbol{v}^{(0)}) = \frac{\lambda^\circ}{2\mu^\circ+\lambda^\circ}\sigma_{22}(\boldsymbol{v}^{(0)};a,0)$$

$$+\frac{4\mu^\circ(\lambda^\circ+\mu^\circ)}{2\mu^\circ+\lambda^\circ}\partial_1 v_1^{(0)}(a,0). \tag{2.5.45}$$

Following the method of matched asymptotic expansions we obtain that the displacement field $\boldsymbol{u}$ admits the representation

$$\boldsymbol{u}(\varepsilon,\boldsymbol{x}) \sim \boldsymbol{v}^{(0)}(a,0)+\varepsilon\boldsymbol{V}(a,0)+\varepsilon^2\boldsymbol{Z}(\boldsymbol{y}), \tag{2.5.46}$$

where $\boldsymbol{Z}$ satisfies (2.5.38)–(2.5.40) and the following conditions at infinity:

$$\boldsymbol{\sigma}(\boldsymbol{Z};\boldsymbol{y}) = \boldsymbol{\sigma}(\boldsymbol{v}^{(0)};a,0)$$

$$+\,\rho_+^{-1/2}\boldsymbol{\Xi}(\varphi_+)+o(\rho_+^{-1/2}),\quad \boldsymbol{y}\in\mathbb{R}^2\backslash\Pi,\ \rho_+\to\infty, \tag{2.5.47}$$

$$\boldsymbol{\sigma}^\circ(\boldsymbol{Z};\boldsymbol{y}) = \boldsymbol{\Sigma}(\boldsymbol{v}^{(0)})+o(1),\quad \boldsymbol{y}\in\Pi,\quad \rho_+\to\infty. \tag{2.5.48}$$

Here, the components of $\boldsymbol{\Sigma}$ are defined by (2.5.45).

It was shown in [19], [20] that the solution of (2.5.38)–(2.5.40), (2.5.47), (2.5.48) can be written in terms of complex potentials

$$\phi^\circ(z) = \mu^\circ(1+\varkappa)(A+B)z,\ \psi^\circ(z) = 2\mu^\circ(1+\varkappa)(A-B+2iC)z,$$

$$\phi(z) = z(\mu^\circ+\mu\varkappa)(A+B)+z(\mu^\circ-\mu)(A-B)$$

$$+2(\mu^\circ-\mu)(A-B+2iC)[i2ab_\pm^2-z+(-8iab_\pm^2 z)^{1/2}],$$

$$\psi(z) = 2\{(\mu^\circ\varkappa+\mu)(A-B+2iC)z$$

$$+[2iab_\pm^2-z+(-8iab_\pm^2 z)^{1/2}]$$

$$\times[(\mu^\circ(\varkappa-1)-\mu(\varkappa^\circ-1))(A+B)$$

$$+(\mu^\circ-\mu)(A-B-2iC)(1+\varkappa-(-2iab_\pm^2/z)^{1/2})]\}, \quad (2.5.49)$$

where A, B and C are certain constants. Let (d,τ) be polar coordinates with the centre at the focus $(-1/2ab_\pm^2, 0)$ of the parabola Π such that $\tau \in (-\pi, \pi)$ and the axis Oy_1 is specified by $\tau = 0$. Using (2.5.49) we can represent the stress tensor components in the form

$$(\sigma_{11}^\circ, \sigma_{12}^\circ, \sigma_{22}^\circ) = 4\mu^\circ(1+\varkappa)(B, C, A), \quad (2.5.50)$$

$$(\sigma_{dd}, \sigma_{d\tau}, \sigma_{\tau\tau}) = (\varkappa-1)(\lambda+\mu)\left\{ B\left[\frac{\mu^\circ}{\mu} + (\varkappa^\circ - 3)\right]\right.$$

$$\left. + \; A(\varkappa^\circ+1)\right\}(1,0,1) + 4\mu^\circ(1+\varkappa)C(-\sin(2\tau),$$

$$- \; \cos(2\tau), \sin(2\tau)) + \{B[2\mu^\circ(\varkappa-1)+\mu(3-\varkappa^\circ)]$$

$$- \; \mu A(\varkappa^\circ+1)\}\,(-\cos(2\tau), \sin(2\tau), \cos(2\tau))$$

$$+ \; b_\pm\left(\frac{2a}{d}\right)^{1/2}\left\{[3(\mu^\circ-\mu)(A-B)+2(\mu^\circ(\varkappa-1)\right.$$

$$- \; \mu(\varkappa^\circ-1))(A+B)]\left(\cos\frac{3\tau}{2}, -\sin\frac{3\tau}{2}, \cos\frac{3\tau}{2}\right)$$

$$+ \; 6C(\mu^\circ-\mu)\left(\sin\frac{3\tau}{2}, \cos\frac{3\tau}{2}, -\sin\frac{3\tau}{2}\right)$$

$$+ \; (A-B)(2\varkappa-1)\frac{(\mu^\circ-\mu)(\lambda+\mu)}{\lambda+5\mu}\left(5\cos\frac{\tau}{2}, \sin\frac{\tau}{2}, 3\cos\frac{\tau}{2}\right)$$

$$\left. - \; 2C(\mu^\circ-\mu)(2\varkappa-1)\frac{\lambda+\mu}{\lambda+5\mu}\left(5\cos\frac{\tau}{2}, -\cos\frac{\tau}{2}, 3\sin\frac{\tau}{2}\right)\right\}$$

$$+ 2\left(\frac{ab_{\pm}^2}{d}\right)^{3/2} (\mu - \mu^\circ)\left\{[2C + B - A]\left(\sin\frac{\tau}{2}, \cos\frac{\tau}{2}, -\sin\frac{\tau}{2}\right)\right.$$

$$\left. + \ [2C + A - B]\left(\cos\frac{\tau}{2}, -\sin\frac{\tau}{2}, -\cos\frac{\tau}{2}\right)\right\}. \tag{2.5.51}$$

Using (2.5.47), (2.5.48) and (2.5.50), (2.5.51) we obtain

$$(B, C, A) = [4\mu^\circ(1 + \varkappa)]^{-1}(\Sigma_{11}(\boldsymbol{v}^{(0)}), \Sigma_{12}(\boldsymbol{v}^{(0)}), \Sigma_{22}(\boldsymbol{v}^{(0)})).$$

Thus, the boundary layer fields $\boldsymbol{Z}, \boldsymbol{Z}^\circ$ are specified by the complex potentials (2.5.49), where the constants A, B, C are given by (2.5.53).

2.5.6 Justification of the asymptotic expansion

It can be noticed that in original coordinates $\boldsymbol{x} = (\varepsilon^2 y_1 \pm a, \varepsilon^2 y_2)$ the contour, corresponding to $\partial\Pi$, does not necessarily coincide with ∂G_ε. In order to take into account this deflection we shall introduce the coordinates $\mathbf{Y}^\pm(\varepsilon, \boldsymbol{x})$ in the following way. Take the point O_+ and assume that G_ε and Π are defined by

$$\varphi_+ \in (-A_-(\varepsilon, r_+),\ A_+(\varepsilon, r_+)),$$

and

$$\varphi \in (-\alpha(\rho),\ \alpha(\rho)),$$

where $A_\pm, \alpha$ describe smooth contours. Let $\mathbf{Y}^+$ be defined by the equality

$$\mathbf{Y}^+(\varepsilon, \boldsymbol{x}) = \varepsilon^{-2}(r_+ \cos\Theta(\varepsilon, r_+, \varphi_+), r_+ \sin\Theta(\varepsilon, r_+, \varphi_+)).$$

Here

$$\Theta(\varepsilon, r_+, \varphi_+) = \frac{\alpha(\varepsilon^{-2} r_+)(2\varphi_+ - A_+(\varepsilon, r_+) + A_-(\varepsilon, r_+))}{A_+(\varepsilon, r_+) + A_-(\varepsilon, r_+)}.$$

Coordinates $\boldsymbol{Y}^-$ are defined in a similar manner. Inside the inclusion G_ε one has to introduce the coordinates $(\varepsilon^{-2}x_1, \eta^\pm(\varepsilon, \boldsymbol{x}))$, where

$$\eta^\pm(\varepsilon, \boldsymbol{x}) = \varepsilon^{-2} b_\pm(-2ax_1)^{1/2}\,\frac{2x_2 + \varepsilon(h_-(x_1) - h_+(x_1))}{h_-(x_1) + h_+(x_1)}.$$

Then, we shall also need a cut-off function $\chi \in \mathbb{C}^\infty(\mathbb{R}^1)$ such that

$$\chi(t) = 0, \text{ as } t < d/4; \ \chi(t) = 1, \text{ as } t > d/2. \tag{2.5.52}$$

Here $d = \min(a, \mathrm{dist}(O_\pm, \partial\Omega))$. The second cut-off function is introduced as follows:

$$\mathfrak{X}(\varepsilon, \boldsymbol{x}) = (1 - \chi(\varepsilon^{-1} r_+))(1 - \chi(\varepsilon^{-1} r_-)).$$

The displacement fields $\boldsymbol{u}, \boldsymbol{u}^\circ$ are approximated by $\boldsymbol{U}, \boldsymbol{U}^\circ$. The asymptotic approximation is uniformly valid in $\Omega_\varepsilon \cup G_\varepsilon$, and

$$\begin{aligned} \boldsymbol{U}(\varepsilon, \boldsymbol{x}) = {} & \mathfrak{X}(\varepsilon, \boldsymbol{x})(\boldsymbol{v}^{(0)}(\boldsymbol{x}) + \varepsilon \boldsymbol{v}^{(1)}(\boldsymbol{x})) \\ & + \sum_\pm \chi(r_\pm)\{\boldsymbol{v}^{(0)}(\pm a, 0) + \varepsilon \boldsymbol{V}(\pm a, 0) + \varepsilon^2 \boldsymbol{Z}^\pm(\boldsymbol{Y}^\pm(\varepsilon, \boldsymbol{x}))\} \\ & - \sum_\pm \mathfrak{X}(\varepsilon, r)\chi(r_\pm)\{\boldsymbol{v}^{(0)}(\pm a, 0) + \\ & + \varepsilon[\boldsymbol{V}(\pm a, 0) + r_\pm^{1/2} \boldsymbol{\xi}^\pm(\Theta^\pm(\varepsilon, r_\pm, \varphi_\pm))]\}, \quad \boldsymbol{x} \in \Omega_\varepsilon, \end{aligned} \tag{2.5.53}$$

$$\begin{aligned} \boldsymbol{U}^\circ(\varepsilon, \boldsymbol{x}) = {} & \mathfrak{X}(\varepsilon, \boldsymbol{x})(\boldsymbol{w}^{(0)}(x_1) + \varepsilon \boldsymbol{w}^{(1)}(x_1, \zeta)) \\ & + \sum_\pm \mathfrak{X}(\varepsilon, \boldsymbol{x}) \Big\{\boldsymbol{w}^{(0)}(\pm a) \\ & + \varepsilon \boldsymbol{V}(\pm a, 0) + \varepsilon^2 \boldsymbol{Z}^{\circ,\pm}\left(\frac{x_1}{\varepsilon^2}, \eta^\pm(\varepsilon, \boldsymbol{x})\right)\Big\} \\ & - \sum_\pm \mathfrak{X}(\varepsilon, \boldsymbol{x})\chi(r_\pm)\{\boldsymbol{w}^{(0)}(\pm a) + \varepsilon[\boldsymbol{w}^{(1)}(\pm a, \zeta(\pm a, x_2)) \\ & + (x_1 \mp a)\partial_1 \boldsymbol{w}^{(1)}(\pm a, \zeta(\pm a, x_2))]\}, \end{aligned} \tag{2.5.54}$$

which is consistent with the method of matched asymptotic expansions (see, for example, [32], [123], [25]).

Direct substitution of the vector fields (2.5.53), (2.5.54) into the original boundary value problem and the use of the Korn inequality show that

$$\epsilon(\boldsymbol{u} - \boldsymbol{U}) = O(\varepsilon^2 |\ln \varepsilon|^2),$$

where ϵ is the elastic energy functional.

2.5.7 Physical interpretation of the solution

The solution of the problem on a uniformly loaded elastic plane with an elliptic inclusion is well known (see, for example, [24]). One can see that for a thin elliptical inclusion the second-order terms of the asymptotic expansions of the stress tensor components have the square root singularity at the end points. The analysis presented in this section confirms this fact for arbitrary thin oval inclusions.

The leading order term of the stress field in the compound region $\Omega_\varepsilon \cup G_\varepsilon$ corresponds to the solution $\boldsymbol{v}^{(0)}$ of the boundary value problem in a homogeneous region Ω. The components of $\boldsymbol{v}^{(0)}$ and their derivatives are smooth functions in Ω. However, the derivatives of the second-order approximation $\boldsymbol{v}^{(1)}$ have the square root singularities at the end points $O_\pm$. It is important to mention that the singularity exponent is specified entirely by the geometry of the interface boundary, and these singularities are generated by the nonhomogeneous jump conditions, which can be written as

$$[\boldsymbol{v}^{(1)}](x_1) = H(x_1)\boldsymbol{\mathcal{V}}(x_1), [\boldsymbol{\sigma}^{(2)}(\boldsymbol{v}^{(1)}; x_1)] = \partial_1(H(x_1)\boldsymbol{\mathcal{T}}(x_1)), \quad (2.5.55)$$

where $(\boldsymbol{\mathcal{V}}, \boldsymbol{\mathcal{T}}) \in (\mathbb{C}^\infty([-a, a]))^2$; H is the normalized thickness of the inclusion.

Since ∂G_ε is smooth, the stress components are finite everywhere in the region $\Omega_\varepsilon \cup G_\varepsilon$. Singularities that occur in the second-order approximation are related to the "outer expansion" which should be matched with the boundary layer. The vectors $\varepsilon^2 \boldsymbol{Z}^\pm$ generate stress components bounded, as $\varepsilon \to 0$. Consequently, the components of the stress tensor $\boldsymbol{\sigma}(\boldsymbol{u}; \varepsilon, \boldsymbol{x})$ are bounded in $\Omega_\varepsilon \cup G_\varepsilon$ by a constant independent of ε. It is important to mention that the above results were proved with the assumption that the Lamé constants λ, μ and λ°, μ° have the same order of magnitude, as $\varepsilon \to 0$.

2.6 Exercises

1. Following formulae of Section 2.1.2, obtain the asymptotic representation of the stress intensity factors for the case of a "wavy crack" in two dimensions, where $h_-(x_1) = -h_+(x_1)$ (see (2.1.1)).

2. Consider a thin elliptical cavity in an elastic plane. Assume that the two–dimensional displacement vector satisfies the homogeneous Lamé system and homogeneous traction boundary conditions. At infinity the displacement field is linear (it corresponds to a load of constant intensity). Following the idea of Section 2.2, show that in the frame of the method of matched asymptotic expansions the boundary layer, that occurs near the ends of the thin cavity, corresponds to the plane strain problem on a semi–infinite parabolic cavity; the solution of this auxiliary problem can be found in the book [85]. The papers [128], [129] can be recommended for additional reading.

3

Domains with Conical and Cylindrical Boundaries

Boundary value problems, posed in domains with conical points on the boundary or with conical or cylindrical extensions at infinity, are considered in this Chapter. Section 3.1 deals with the structure of the stress singularity in the vicinity of conical inclusions and cavities and with the fracture phenomena in an elastic half–space with a conical notch. In Section 3.2 we look at the singularity exponents for the stress components at the vertex of an angular crack with the angle being close to 0, π or 2π. The first two sections are based on the results of [38], [59], [80]–[82]. Sections 3.3 and 3.4 are related to the asymptotic structure of solutions to boundary value problems of linear elasticity posed in thin layers. This asymptotic algorithm allows one to derive the equations of thin plates, and it provides the rigorous justification of the engineering approach. The papers [3] and [49] can be recommended for additional reading. On the basis of the results of Sections 3.3 and 3.4 we construct the REDUCE program (see Appendix) that can be effectively used for the asymptotic study of the elasticity problems in thin domains. Section 3.5 includes the analysis of the Lekhnitskii problem on an elastic half–space subjected to the gravity forces and containing a semi–infinite cylindrical shaft. The discussion is based on the classical work [46], [47] and on the papers [2], [4].

3.1 Stress-strain state in a neighbourhood of conical inclusions and cavities

In the present section we perform an asymptotic analysis of the stress and displacement fields in the vicinity of the apices of inclusions or cavities of the conical shape. We assume that the set of intersection of the cone and the unit sphere (with the centre at the vertex) has a small diameter ε, $0 < \varepsilon \ll 1$, and derive the leading term of the asymptotic approximation $\varepsilon^2\Lambda_2 + O(\varepsilon^3)$ of the stress singularity exponent near the vertex of the cone. As an example, we consider the axisymmetric problem for an elastic half-space with a surface-breaking thin conical cavity.

3.1.1 Spectral problem

Let k_ε denote a thin cone $\{\boldsymbol{x} \in \mathbb{R}^3 : x_3 > 0,\ \varepsilon^{-1}x_3^{-1}\boldsymbol{x}' \in g,\ \boldsymbol{x}' = (x_1, x_2)\}$, ε be a small positive parameter, and g be a plane domain bounded by a simple smooth contour ∂g. Assume that the cones k_ε and $K_\varepsilon = \mathbb{R}^3 \backslash \overline{k}_\varepsilon$ are occupied by elastic isotropic materials with Lamé constants λ°, μ° and λ, μ, respectively, and on the interface boundary the continuity conditions of displacement and tractions are prescribed. The displacement fields $\boldsymbol{u}$ (within the matrix K_ε) and $\boldsymbol{u}^\circ$ (within the conical inclusion k_ε) satisfy the problem

$$\boldsymbol{L}(\frac{\partial}{\partial \boldsymbol{x}})\boldsymbol{u}(\varepsilon, \boldsymbol{x}) = \boldsymbol{0},\ \boldsymbol{x} \in K_\varepsilon,\ \ \boldsymbol{L}^\circ(\frac{\partial}{\partial \boldsymbol{x}})\boldsymbol{u}^\circ(\varepsilon, \boldsymbol{x}) = \boldsymbol{0},\ \boldsymbol{x} \in k_\varepsilon, \tag{3.1.1}$$

$$\boldsymbol{u}(\varepsilon, \boldsymbol{x}) = \boldsymbol{u}^\circ(\varepsilon, \boldsymbol{x}),\ \ \boldsymbol{\sigma}^{(n)}(\boldsymbol{u}; \varepsilon, \boldsymbol{x}) = \boldsymbol{\sigma}^{\circ,(n)}(\boldsymbol{u}^\circ; \varepsilon, \boldsymbol{x}),\ \ \boldsymbol{x} \in \partial k_\varepsilon. \tag{3.1.2}$$

Here the quantities related to the inclusion k_ε are marked by "$\circ$".

Near the vertex of the cone k_ε the displacement fields $\boldsymbol{u}$ and $\boldsymbol{u}^\circ$ admit the representation

$$\boldsymbol{u}(\varepsilon, \boldsymbol{x}) = \rho^{\Lambda(\varepsilon)}\boldsymbol{v}(\varepsilon, \theta, \varphi),\ \ \boldsymbol{u}^\circ(\varepsilon, \boldsymbol{x}) = \rho^{\Lambda(\varepsilon)}\boldsymbol{v}^\circ(\varepsilon, \theta, \varphi),$$

where (ρ, θ, φ) are spherical coordinates, $\rho = ||\boldsymbol{x}||$, $\theta \in [0, \pi]$, $\varphi \in [0, 2\pi)$. The exponent Λ and the vectors $\boldsymbol{v}, \boldsymbol{v}^\circ$ are the solutions to the spectral problem on the unit sphere $\mathbb{S} = \{\boldsymbol{x} \in \mathbb{R}^3 : ||\boldsymbol{x}|| = 1\}$

$$\mathcal{P}(\Lambda(\varepsilon))\boldsymbol{v} = \boldsymbol{0} \ \text{ on } \ \mathbb{S}\backslash\overline{g}_\varepsilon, \tag{3.1.3}$$

$$\mathcal{P}^\circ(\Lambda(\varepsilon))\boldsymbol{v}^\circ = \boldsymbol{0} \ \text{ on } \ g_\varepsilon, \tag{3.1.4}$$

$$v = v^\circ, \; \mathcal{Q}(\Lambda(\varepsilon))v = \mathcal{Q}^\circ(\Lambda(\varepsilon))v^\circ \quad \text{on} \quad \partial g_\varepsilon. \tag{3.1.5}$$

The system (3.1.3)–(3.1.5) can be obtained from (3.1.1), (3.1.2) by the transformation to the spherical coordinates, if we denote the matrix differential operator of the Lamé system in spherical coordinates by $\rho^{-2}\mathcal{P}(\theta, \varphi, \rho\partial/\partial\rho, \; \partial/\partial\theta, \; \partial/\partial\varphi)$ and the traction vector on the surface ∂k_ε by $\rho^{-1}\mathcal{Q}(\theta, \varphi, \rho\partial/\partial\rho, \; \partial/\partial\theta, \; \partial/\partial\varphi)\boldsymbol{u}$. To simplify the notations we omit the arguments θ, φ as well as $\partial/\partial\theta, \; \partial/\partial\varphi$. In (3.1.3)–(3.1.5) g_ε is the intersection set of the cone k_ε and the unit sphere $\mathbb{S}$.

An axisymmetric problem on a conical inclusion with a circular cross-section has been investigated in papers [11], [36], [100], [101], [103], [119], [122] (they contain the transcendental equations with respect to the exponent Λ and the tables of numerical solutions).

Here with the use of the asymptotic algorithm presented in [55] we analyze several first positive eigenvalues of spectral problem (3.1.3)–(3.1.5), as $\varepsilon \to 0$.

As $\varepsilon \to 0$, the domain g_ε vanishes in the limit and the problem (3.1.3)–(3.1.5) is transformed into a system of equations on the entire sphere $\mathbb{S}$

$$\mathcal{P}(\Lambda_0)\Phi = 0. \tag{3.1.6}$$

The eigenvalues Λ_0 of the spectral problem (3.1.6) are integers, and the eigenvectors $\boldsymbol{\Phi}$ are traces on $\mathbb{S}$ of homogeneous vector polynomials $\boldsymbol{V}^{(m,j)}$ of degree m satisfying the Lamé system, $m = 0, 1, 2, ..., \; j = 1, 2, ..., 3(2m+1)$, or traces of the fields $\boldsymbol{V}^{(1,j)}(\partial/\partial\boldsymbol{x}) \; \mathbf{T}(\boldsymbol{x})$, where $\mathbf{T}$ is the Somigliana tensor.

We shall seek the solutions of the problem (3.1.1), (3.1.2) with a finite elastic energy, and therefore, only those fields in which $\Lambda(\varepsilon) > -1/2$ will be analyzed. Namely, we consider the perturbation of two first eigenvalues $\Lambda_0 = 0$ and $\Lambda_0 = 1$ of the system (3.1.6).

The vectors $\boldsymbol{V}^{(0,j)}$, $j = 1, 2, 3$, are the rigid body displacements. They correspond to the eigenvalue $\Lambda_0 = 0$ of the limit problem (3.1.6) and satisfy the problem (3.1.3)–(3.1.5), with $\Lambda(\varepsilon) = 0$. It implies that the eigenvalue $\Lambda_0 = 0$ is not perturbed. The vector polynomials $\boldsymbol{V}^{(1,j)}$ of the first order have the form

$$\boldsymbol{V}^{(1,j)}(\boldsymbol{x}) = x_j \boldsymbol{e}^{(j)}, \; j = 1, 2, 3, \; \boldsymbol{V}^{(1,4)}(\boldsymbol{x}) = 2^{-1/2}(x_2, x_1, 0)^T,$$

$$\boldsymbol{V}^{(1,5)}(\boldsymbol{x}) = 2^{-1/2}(0, x_3, x_2)^T, \; \boldsymbol{V}^{(1,6)}(\boldsymbol{x}) = 2^{-1/2}(x_3, 0, x_1)^T; \tag{3.1.7}$$

$$\boldsymbol{V}^{(1,7)}(\boldsymbol{x}) = 2^{-1/2}(x_2, -x_1, 0)^T, \; \boldsymbol{V}^{(1,8)}(\boldsymbol{x}) = 2^{-1/2}(0, x_3, -x_2)^T,$$

$$\boldsymbol{V}^{(1,9)}(\boldsymbol{x}) = 2^{-1/2}(-x_3, 0, x_1)^T. \tag{3.1.8}$$

The rotations $\boldsymbol{V}^{(1,7)}$, $\boldsymbol{V}^{(1,8)}$, $\boldsymbol{V}^{(1,9)}$ are the solutions to the problem (3.1.3)–(3.1.5), with $\Lambda(\varepsilon) = 1$. The fields (3.1.7) satisfy the equations (3.1.3), (3.1.4) and the continuity condition of displacements on ∂g_ε, but they leave a discrepancy in the traction condition (3.1.5). This discrepancy can be obtained with the use of the representations for the stresses $\sigma_{jk}^{(i)} := \sigma_{jk}(\boldsymbol{V}^{(1,i)}; \boldsymbol{x})$:

$$\begin{gathered} \sigma_{ii}^{(i)} = 2\mu + \lambda, \ \sigma_{ii}^{(j)} = \lambda, \ i \neq j, \ i, \ j = 1,2,3; \\ \sigma_{12}^{(4)} = \sigma_{23}^{(5)} = \sigma_{13}^{(6)} = \sqrt{2}\mu. \end{gathered} \tag{3.1.9}$$

Similar relations are valid for the stress components in an elastic inclusion. Note that in (3.1.9) we did not present the components which are equal to zero.

Let us take the value $\Lambda_0 = 1$ and linear combinations

$$\boldsymbol{\Phi}(\theta,\varphi) = \sum_{j=1}^{9} c_j \boldsymbol{\Phi}^{(1,j)}(\theta,\varphi), \quad \boldsymbol{\Phi}^\circ(\theta,\varphi) = \sum_{j=1}^{9} c_j^\circ \boldsymbol{\Phi}^{(1,j)}(\theta,\varphi), \tag{3.1.10}$$

with constant c_j, c_j°, as the leading order approximation to the solution of the problem (3.1.3)–(3.1.5). In order to compensate the discrepancy left by the fields (3.1.10) in the traction condition (3.1.5), in the next section we shall construct the boundary layer in the vicinity of the conical inclusion k_ε.

3.1.2 Boundary layer

In a neighbourhood of the point $\mathcal{N} = (0,0,1)$ on the unit sphere $\mathbb{S}$ we introduce new scaled variables

$$\boldsymbol{\xi} = \varepsilon^{-1}\boldsymbol{\eta},$$

with

$$\boldsymbol{\eta} = (\eta_1, \eta_2) = \frac{\boldsymbol{x}'}{x_3},$$

and represent the unit normal vector $\boldsymbol{n}$ on ∂k_ε by the relation

$$\boldsymbol{n} = \frac{1}{\sqrt{1 + \varepsilon^2(\boldsymbol{\xi}\cdot\boldsymbol{\nu})^2}} \, (\nu_1, \nu_2, -\varepsilon\boldsymbol{\xi}\cdot\boldsymbol{\nu})^T,$$

where $\boldsymbol{\nu} = (\nu_1, \nu_2)^T$ denotes the unit inward normal vector on ∂g.

In the stretched coordinates $\boldsymbol{\xi}$ the domain $g_\varepsilon \subset \mathbb{S}$ is transformed to $g \subset \mathbb{R}^2$, and the Lamé operator L and the operator $B = \boldsymbol{\sigma n}$ of traction condition at the interface are specified by

$$L\left(\frac{\partial}{\partial \boldsymbol{x}}\right) \left(\rho^{\{1+O(\varepsilon^2)\}} \boldsymbol{\Psi}(\boldsymbol{\xi})\right)\Big|_{\|\boldsymbol{x}\|=1} = \varepsilon^{-2} \mathcal{L}_0 \left(\frac{\partial}{\partial \boldsymbol{\xi}}\right) \boldsymbol{\Psi}(\boldsymbol{\xi}) \\ +\varepsilon^{-1} \mathcal{L}_1 \left(\boldsymbol{\xi}, \frac{\partial}{\partial \boldsymbol{\xi}}\right) \boldsymbol{\Psi}(\boldsymbol{\xi}) + O(1), \tag{3.1.11}$$

$$B\left(\frac{\partial}{\partial \boldsymbol{x}}, \boldsymbol{n}\right) \left(\rho^{\{1+O(\varepsilon^2)\}} \boldsymbol{\Psi}(\boldsymbol{\xi})\right)\Big|_{\|\boldsymbol{x}\|=1} = \varepsilon^{-1} \mathcal{B}_0 \left(\frac{\partial}{\partial \boldsymbol{\xi}}, \boldsymbol{\nu}\right) \boldsymbol{\Psi}(\boldsymbol{\xi}) \\ +\mathcal{B}_1 \left(\boldsymbol{\xi}, \frac{\partial}{\partial \boldsymbol{\xi}}, \boldsymbol{\nu}\right) \boldsymbol{\Psi}(\boldsymbol{\xi}) + O(\varepsilon), \tag{3.1.12}$$

where

$$\begin{gathered}
\mathcal{L}_0^{11}(\zeta_1, \zeta_2) = (\lambda + 2\mu)\zeta_1^2 + \mu\zeta_2^2, \\
\mathcal{L}_0^{12}(\zeta_1, \zeta_2) = \mathcal{L}_0^{21}(\zeta_1, \zeta_2) = (\lambda + \mu)\zeta_1\zeta_2, \\
\mathcal{L}_0^{22}(\zeta_1, \zeta_2) = (\lambda + 2\mu)\zeta_2^2 + \mu\zeta_1^2, \ \mathcal{L}_0^{33}(\zeta_1, \zeta_2) = \mu(\zeta_1^2 + \zeta_2^2), \\
\mathcal{L}_1^{13}(\boldsymbol{\xi}; \zeta_1, \zeta_2) = \mathcal{L}_1^{31}(\boldsymbol{\xi}; \zeta_1, \zeta_2) = -(\lambda + \mu)(\xi_1\zeta_1^2 + \xi_2\zeta_1\zeta_2), \\
\mathcal{L}_1^{23}(\boldsymbol{\xi}; \zeta_1, \zeta_2) = \mathcal{L}_1^{32}(\boldsymbol{\xi}; \zeta_1, \zeta_2) = -(\lambda + \mu)(\xi_1\zeta_1\zeta_2 + \xi_2\zeta_2^2); \\
\mathcal{B}_0^{11}(\boldsymbol{\nu}; \zeta_1, \zeta_2) = (\lambda + 2\mu)\nu_1\zeta_1 + \mu\nu_2\zeta_2, \\
\mathcal{B}_0^{22}(\boldsymbol{\nu}; \zeta_1, \zeta_2) = \mu\nu_1\zeta_1 + (\lambda + 2\mu)\nu_2\zeta_2, \\
\mathcal{B}_0^{12}(\boldsymbol{\nu}; \zeta_1, \zeta_2) = \lambda\nu_1\zeta_2 + \mu\nu_2\zeta_1, \\
\mathcal{B}_0^{21}(\boldsymbol{\nu}; \zeta_1, \zeta_2) = \mu\nu_1\zeta_2 + \lambda\nu_2\zeta_1, \\
\mathcal{B}_0^{33}(\boldsymbol{\nu}; \zeta_1, \zeta_2) = \mu(\nu_1\zeta_1 + \nu_2\zeta_2), \\
\mathcal{B}_1^{j3}(\boldsymbol{\xi}, \boldsymbol{\nu}; \zeta_1, \zeta_2) = -\lambda\nu_j(\xi_1\zeta_1 + \xi_2\zeta_2 - 1) - \mu\boldsymbol{\xi} \cdot \boldsymbol{\nu}\zeta_j, \\
\mathcal{B}_1^{3j}(\boldsymbol{\xi}, \boldsymbol{\nu}; \zeta_1, \zeta_2) = -\mu\nu_j(\xi_1\zeta_1 + \xi_2\zeta_2 - 1) - \lambda\boldsymbol{\xi} \cdot \boldsymbol{\nu}\zeta_j, \ j = 1, 2.
\end{gathered} \tag{3.1.13}$$

Denote the leading part of the boundary layer within the matrix K_ε by $\varepsilon \boldsymbol{w}^{(1)}(\boldsymbol{\xi})$ and within the inclusion k_ε by $\varepsilon \boldsymbol{w}^{\circ,(1)}(\boldsymbol{\xi})$. Relations (3.1.11), (3.1.13) and (3.1.9) yield the equations and interface conditions for the fields $\boldsymbol{w}^{(1)}$ and $\boldsymbol{w}^{\circ,(1)}$:

$$\mathcal{L}_0 \left(\frac{\partial}{\partial \boldsymbol{\xi}}\right) \boldsymbol{w}^{(1)}(\boldsymbol{\xi}) = \mathbf{0}, \ \boldsymbol{\xi} \in \mathbb{R}^2 \backslash \overline{g},$$

$$\mathcal{L}_0^\circ \left(\frac{\partial}{\partial \boldsymbol{\xi}}\right) \boldsymbol{w}^{\circ,(1)}(\boldsymbol{\xi}) = \mathbf{0}, \ \boldsymbol{\xi} \in g, \tag{3.1.14}$$

$$\mathcal{B}_0\left(\frac{\partial}{\partial \boldsymbol{\xi}}, \boldsymbol{\nu}\right) \boldsymbol{w}^{(1)}(\boldsymbol{\xi}) - \mathcal{B}_0^\circ\left(\frac{\partial}{\partial \boldsymbol{\xi}}, \boldsymbol{\nu}\right) \boldsymbol{w}^{\circ,(1)}(\boldsymbol{\xi})$$

$$= -\sum_{j=1}^{6} c_j \boldsymbol{\Psi}^{(j)}(\boldsymbol{\xi}, \boldsymbol{\nu}), \ \boldsymbol{w}^{(1)}(\boldsymbol{\xi}) = \boldsymbol{w}^{\circ(1)}(\boldsymbol{\xi}), \ \boldsymbol{\xi} \in \partial g, \tag{3.1.15}$$

where

$$\begin{gathered}
\boldsymbol{\Psi}^{(1)}(\boldsymbol{\xi}, \boldsymbol{\nu}) = ((\lambda + 2\mu - \lambda^\circ - 2\mu^\circ)\nu_1, \ (\lambda - \lambda^\circ)\nu_2, \ 0)^T, \\
\boldsymbol{\Psi}^{(2)}(\boldsymbol{\xi}, \boldsymbol{\nu}) = ((\lambda - \lambda^\circ)\nu_1, \ (\lambda + 2\mu - \lambda^\circ - 2\mu^\circ)\nu_2, \ 0)^T, \\
\boldsymbol{\Psi}^{(3)}(\boldsymbol{\xi}, \boldsymbol{\nu}) = (\lambda - \lambda^\circ)(\nu_1, \nu_2, 0)^T, \\
\boldsymbol{\Psi}^{(4)}(\boldsymbol{\xi}, \boldsymbol{\nu}) = 2^{1/2}(\mu - \mu^\circ)(\nu_2, \nu_1, 0)^T, \\
\boldsymbol{\Psi}^{(5)}(\boldsymbol{\xi}, \boldsymbol{\nu}) = 2^{1/2}(\mu - \mu^\circ)(0, 0, \nu_2)^T, \\
\boldsymbol{\Psi}^{(6)}(\boldsymbol{\xi}, \boldsymbol{\nu}) = 2^{1/2}(\mu - \mu^\circ)(0, 0, \nu_1)^T.
\end{gathered} \tag{3.1.16}$$

Due to (3.1.13), the boundary value problem (3.1.14), (3.1.15) can be split up into the plane strain and anti-plane shear problems. Since the average values of (3.1.16) over ∂g equal zero, a static solution $\boldsymbol{w}^{(1)}$, $\boldsymbol{w}^{\circ,(1)}$ of (3.1.14), (3.1.15) exists, and the vector-valued function $\boldsymbol{w}^{(1)}$ vanishes at infinity.

In accordance with the results of [38], the field $\boldsymbol{w}^{(1)}$ is specified by

$$\begin{gathered}
\boldsymbol{w}^{(1)}(\boldsymbol{\xi}) = \boldsymbol{\Upsilon}^{(1)}(\boldsymbol{\xi}) + O(\|\boldsymbol{\xi}\|^{-2}) \\
= \sum_{j=1}^{6} c_j \sum_{i=1}^{5} \alpha_i^{(j)} \boldsymbol{W}^{(i)}(\partial/\partial\boldsymbol{\xi})\boldsymbol{\Gamma}(\boldsymbol{\xi}) + O(\|\boldsymbol{\xi}\|^{-2}), \text{ as } \|\boldsymbol{\xi}\| \to \infty; \\
\boldsymbol{W}^{(1)}(\boldsymbol{\xi}) = (\xi_1, 0, 0)^T, \ \boldsymbol{W}^{(2)}(\boldsymbol{\xi}) = (0, \xi_2, 0)^T, \\
\boldsymbol{W}^{(3)}(\boldsymbol{\xi}) = 2^{-1/2}(\xi_2, \xi_1, 0)^T, \\
\boldsymbol{W}^{(4)}(\boldsymbol{\xi}) = (0, 0, \xi_1)^T, \ \boldsymbol{W}^{(5)}(\boldsymbol{\xi}) = (0, 0, \xi_2)^T; \\
\boldsymbol{\Gamma}(\boldsymbol{\xi}) = (\gamma_{ij}(\boldsymbol{\xi}))_{i,j=1}^3, \\
\gamma_{ij}(\boldsymbol{\xi}) = [4\pi\mu(\lambda + 2\mu)]^{-1}(-\delta_{ij}(\lambda + 3\mu) \ \ln \ \|\boldsymbol{\xi}\| + (\lambda + \mu) \ \xi_i\xi_j\|\boldsymbol{\xi}\|^{-2}), \\
\gamma_{3j}(\boldsymbol{\xi}) = \gamma_{j3}(\boldsymbol{\xi}) = 0, \ i, \ j = 1, 2; \ \gamma_{33}(\boldsymbol{\xi}) = -(2\pi\mu)^{-1} \ \ln \ \|\boldsymbol{\xi}\|.
\end{gathered} \tag{3.1.17}$$

The coefficients $\alpha_i^{(j)}$ can be written in terms of components of the Pólya–Szegö matrix $\boldsymbol{m} = (m_{ik})_{i,k=1}^5$ defined in Chapter 5 (see also Chapter 1). The components m_{ik} are the coefficients near $\boldsymbol{W}^{(i)}\left(\frac{\partial}{\partial \boldsymbol{\xi}}\right) \boldsymbol{\Gamma}(\boldsymbol{\xi})$ in the asymptotic representation of the form (3.1.17) for the special solutions $\boldsymbol{Z}^{(i)}$ of the problem (3.1.14) and (3.1.15) with the following vectors in the right–hand sides of (3.1.15)

$$\begin{gathered}((\lambda^\circ + 2\mu^\circ - \lambda - 2\mu)\nu_1, (\lambda^\circ - \lambda)\nu_2, 0)^T, \\ ((\lambda - \lambda^\circ)\nu_1, (\lambda^\circ + 2\mu^\circ - \lambda - 2\mu)\nu_2, 0)^T, \\ 2^{1/2}(\mu^\circ - \mu)(\nu_2, \nu_1, 0)^T,\ (\mu^\circ - \mu)(0,0,\nu_1)^T,\ (\mu^\circ - \mu)(0,0,\nu_2)^T.\end{gathered} \tag{3.1.18}$$

Note that the matrix $\boldsymbol{m}$ is negative (positive) definite for sufficiently soft (rigid) inclusions of non-zero volume. The above-mentioned connection between $\alpha_k^{(i)}$ and m_{jk} is given by the formulae

$$\begin{gathered}\alpha_k^{(j)} = m_{jk},\ j = 1,2;\ \alpha_k^{(3)} = (\lambda - \lambda^\circ)[2(\mu - \mu^\circ + \lambda - \lambda^\circ)]^{-1} \\ \times(m_{1k} + m_{2k}),\ \alpha_k^{(4)} = m_{3k},\ \alpha_k^{(j)} = 0,\ j = 5,6,\ k = 1,2,3; \\ \alpha_4^{(j)} = \alpha_5^{(j)} = 0,\ j = 1,2,3,4;\ \alpha_k^{(k+1)} = 2^{1/2} m_{45},\ k = 4,5; \\ \alpha_4^{(6)} = 2^{1/2} m_{44},\ \alpha_5^{(5)} = 2^{1/2} m_{55}.\end{gathered}$$

The components of $\boldsymbol{w}^{(1)}$ decay at infinity like $O(\|\boldsymbol{\xi}\|^{-1})$, and therefore, the boundary layer $\varepsilon\chi(\theta)\boldsymbol{w}^{(1)}(\varepsilon^{-1}\boldsymbol{\eta})$ leaves a discrepancy of order $O(\varepsilon^2)$ in (3.1.3) for $\Lambda = 1$. Here χ is an infinitely smooth cut-off function such that

$$\chi(\theta) = 1 \text{ for } \theta \in [0, \pi/6], \quad \text{and} \quad \chi(\theta) = 0 \text{ for } \theta \in [\pi/3, \pi].$$

This function is introduced because the boundary layer is defined on the upper semi-sphere only.

We seek the asymptotic approximation to the solution of the problem (3.1.3)–(3.1.5) in the form

$$\Lambda(\varepsilon) \sim 1 + \varepsilon^2 \Lambda_2, \tag{3.1.19}$$

$$\begin{gathered}\boldsymbol{v}(\varepsilon, \theta, \varphi) \sim \boldsymbol{\Phi}(\theta, \varphi) + \varepsilon\chi(\theta)\boldsymbol{w}^{(1)}(\varepsilon^{-1}\boldsymbol{\eta}) \\ +\varepsilon^2\boldsymbol{\Phi}^{(2)}(\theta, \varphi) + \varepsilon^2\chi(\theta)\boldsymbol{w}^{(2)}(\varepsilon^{-1}\boldsymbol{\eta}).\end{gathered}$$

Due to (3.1.11), (3.1.12) and (3.1.9), the vector-valued function $\boldsymbol{w}^{(2)}$ satisfies the boundary value problem

$$\mathcal{L}_0\boldsymbol{w}^{(2)} + \mathcal{L}_1\boldsymbol{w}^{(1)} = \boldsymbol{0} \text{ in } \mathbb{R}^2\backslash\overline{g}, \tag{3.1.20}$$

$$\mathcal{L}_0^\circ\boldsymbol{w}^{\circ,(2)} + \mathcal{L}_1^\circ\boldsymbol{w}^{\circ,(1)} = \boldsymbol{0} \text{ in } g, \tag{3.1.21}$$

$$\boldsymbol{w}^{(2)} = \boldsymbol{w}^{\circ,(2)},$$

$$\mathcal{B}_0 \boldsymbol{w}^{(2)} - \mathcal{B}_0^\circ \boldsymbol{w}^{\circ,(2)} = \mathcal{B}_1^\circ \boldsymbol{w}^{\circ,(1)} - \mathcal{B}_1 \boldsymbol{w}^{(1)} + \sum_{j=1}^{6} c_j \boldsymbol{\Psi}^{(1,j)} \text{ on } \partial g, \quad (3.1.22)$$

with

$$\begin{gathered} \boldsymbol{\Psi}^{(1,1)}(\boldsymbol{\xi}) = \boldsymbol{\Psi}^{(1,2)}(\boldsymbol{\xi}) = (0,0,(\lambda-\lambda^\circ)\boldsymbol{\xi}\cdot\boldsymbol{\nu})^T, \\ \boldsymbol{\Psi}^{(1,3)}(\boldsymbol{\xi}) = (0,0,(\lambda+2\mu-\lambda^\circ-2\mu^\circ)\boldsymbol{\xi}\cdot\boldsymbol{\nu})^T, \\ \boldsymbol{\Psi}^{(1,4)}(\boldsymbol{\xi}) = \mathbf{0}, \ \boldsymbol{\Psi}^{(1,5)}(\boldsymbol{\xi}) = 2^{1/2}(0,(\mu-\mu^\circ)\boldsymbol{\xi}\cdot\boldsymbol{\nu},0)^T, \\ \boldsymbol{\Psi}^{(1,6)}(\boldsymbol{\xi}) = 2^{1/2}((\mu-\mu^\circ)\boldsymbol{\xi}\cdot\boldsymbol{\nu},0,0)^T. \end{gathered} \quad (3.1.23)$$

At infinity the asymptotic behaviour of the field $\boldsymbol{w}^{(2)}$ is specified by the following statement.

Proposition 1. *Every solution $\boldsymbol{w}^{(2)}$ of (3.1.20), that allows the estimate $O(\|\boldsymbol{\xi}\|^\delta)$ for $\delta \in (0,1)$, has the asymptotic form*

$$\begin{aligned} \boldsymbol{w}^{(2)}(\boldsymbol{\xi}) &= \boldsymbol{\Upsilon}^{(2)}(\boldsymbol{\xi}) + O(\|\boldsymbol{\xi}\|^{-1}) \\ &= \boldsymbol{a}\boldsymbol{\Gamma}(\boldsymbol{\xi}) + \boldsymbol{b} + \boldsymbol{\Xi}(\varphi) + O(\|\boldsymbol{\xi}\|^{-1}), \ \|\boldsymbol{\xi}\| \to \infty; \end{aligned} \quad (3.1.24)$$

$$\boldsymbol{\Xi}(\boldsymbol{\xi}) = (\pi\mu)^{-1}(\tilde{\Xi}_1, \tilde{\Xi}_2, \tilde{\Xi}_3), \quad (3.1.25)$$

$$\tilde{\Xi}_j(\boldsymbol{\xi}) = \varkappa^{-1}\sum_{k=5}^{6} c_k \sum_{i=1}^{2} \alpha_{3+i}^{(k)} \xi_i \xi_j \|\boldsymbol{\xi}\|^{-2}, \ j=1,2;$$

$$\tilde{\Xi}_3(\boldsymbol{\xi}) = (\varkappa+1)^{-1}\sum_{j=1}^{4} c_j (\alpha_1^{(j)}\xi_1^2\|\boldsymbol{\xi}\|^{-2}$$

$$+\alpha_2^{(j)}\xi_2^2\|\boldsymbol{\xi}\|^{-2} + 2^{1/2}\alpha_3^{(j)}\xi_1\xi_2\|\boldsymbol{\xi}\|^{-2}),$$

$$\varkappa = (\lambda+3\mu)(\lambda+\mu)^{-1}.$$

Proof. By virtue of (3.1.17) $\boldsymbol{\Upsilon}^{(1)}$ is a homogeneous vector function of degree -1. Since

$$\boldsymbol{\xi}\cdot\nabla_{\xi}\partial/\partial\xi_j = \partial/\partial\xi_j \|\boldsymbol{\xi}\| \partial/\partial\|\boldsymbol{\xi}\| - \partial/\partial\xi_j,$$

then according to (3.1.13)

$$\mathcal{L}_1(\boldsymbol{\xi}, \partial/\partial\boldsymbol{\xi})\boldsymbol{\Upsilon}^{(1)} = 2(\lambda+\mu)(\Upsilon_{3,1}^{(1)}, \Upsilon_{3,2}^{(1)}, \Upsilon_{1,1}^{(1)} + \Upsilon_{2,2}^{(1)})^T$$

$$+O(\|\boldsymbol{\xi}\|^{-3}) = \|\boldsymbol{\xi}\|^{-2}\boldsymbol{\Theta}(\varphi) + O(\|\boldsymbol{\xi}\|^{-3}).$$

Here and later the subscript k after a comma denotes differentiation with respect to ξ_k. Looking for a particular solution of the equation

$$\mathcal{L}_0\, \boldsymbol{\Xi} = -\frac{1}{\|\boldsymbol{\xi}\|^2}\, \boldsymbol{\Theta},$$

we obtain equalities (3.1.25). Note that

$$\mathcal{L}_0(\boldsymbol{a}\boldsymbol{\Gamma} + \boldsymbol{b}) = \mathbf{0} \quad \text{for} \quad \boldsymbol{\xi} \neq \mathbf{0},$$

and the representation (3.1.24) follows from the results [38], [62]. □

The solution of (3.1.20) – (3.1.22) is defined up to an arbitrary additive constant vector (one can choose an arbitrary $\boldsymbol{b}$ in (3.1.24)). Further, it will be convenient to take

$$\boldsymbol{b} = -(\ln\ \varepsilon)(2\pi\mu)^{-1}(\varkappa(\varkappa+1)^{-1}\alpha_1,\ \varkappa(\varkappa+1)^{-1}\alpha_2, \alpha_3)^T.$$

This choice yields that the quantity $\boldsymbol{\Upsilon}^{(2)}$, represented in coordinates $\boldsymbol{\eta} = \varepsilon\boldsymbol{\xi}$, is independent of ε.

The components of the vector $\boldsymbol{a}$ are specified by

Proposition 2. *The equalities*

$$a_k = \sum_{j=1}^{6} c_j \beta_k^{(j)},\ k = 1, 2, 3;\ \beta_1^{(p)} = 4\mu(\lambda + 3\mu)^{-1}\alpha_4^{(p)}, \tag{3.1.26}$$

$$\beta_2^{(p)} = 4\mu(\lambda + 3\mu)^{-1}\alpha_5^{(p)},\ p = 5, 6;\ \beta_3^{(j)} = (A + B)(\alpha_1^{(j)}$$

$$+\alpha_2^{(j)} + 2\delta_{j3}[(\lambda - \lambda^\circ)A - (\lambda + 2\mu - \lambda^\circ - 2\mu^\circ)]\text{mes}g,\ j = 1, 2, 3, 4;$$

$$A = (\lambda - \lambda^\circ)(\lambda + \mu - \lambda^\circ - \mu^\circ)^{-1},\ B = -(\lambda + \mu)(\lambda + 2\mu)^{-1},$$

are valid.

Proof. Consider the dot product of a unit vector $\boldsymbol{e}^{(i)}$ and the system (3.1.20), (3.1.21), integrate by parts in a circular domain D_R of a radius

R and take the limit, as $R \to \infty$. We derive

$$\int_{D_R \setminus g} \mathbf{e}^{(i)} \cdot (\mathcal{L}_0 \boldsymbol{w}^{(2)} + \mathcal{L}_1 \boldsymbol{w}^{(1)}) d\boldsymbol{\xi} + \int_g \mathbf{e}^{(i)} \cdot (\mathcal{L}_0^\circ \boldsymbol{w}^{\circ,(2)} + \mathcal{L}_1^\circ \boldsymbol{w}^{\circ,(1)}) d\boldsymbol{\xi}$$
$$= \int_{\partial D_R} \mathbf{e}^{(i)} \cdot (\mathcal{B}_0 \boldsymbol{w}^{(2)} + \mathcal{B}_1 \boldsymbol{w}^{(1)}) dl + \int_{\partial g} \mathbf{e}^{(i)} \cdot \sum_{j=1}^{6} c_j \boldsymbol{\Psi}^{(1,j)} dl + 2 I_i;$$
$$I_j = \mu R^{-1} \int_{\partial D_R} \xi_j w_3^{(1)} dl + (\mu - \mu^\circ) \int_{\partial g} w_3^{(1)} \nu_j dl, \; j = 1, 2, \tag{3.1.27}$$
$$I_3 = \sum_{j=1}^{2} \left\{ \lambda R^{-1} \int_{\partial D_R} \xi_j w_j^{(1)} dl + (\lambda - \lambda^\circ) \int_{\partial g} w_j^{(1)} \nu_j dl \right\} .$$

Here $\mathcal{B}_0$ and $\mathcal{B}_1$ are operators given by (3.1.13) with the normal vector $\boldsymbol{\nu}$ replaced by the vector $(\cos\varphi, \; \sin\varphi)^T$. In order to evaluate the first two integrals on the right-hand side of (3.1.27), we note that the first one equals

$$\int_{\partial D_R} \mathbf{e}^{(i)} \cdot \mathcal{B}_0 \boldsymbol{a} \Gamma dl + \int_{\partial D_R} \mathbf{e}^{(i)} \cdot (\mathcal{B}_0 \Xi + \mathcal{B}_1 \Upsilon^{(1)}) dl + o(1), \tag{3.1.28}$$

where

$$\int_{\partial D_R} \mathbf{e}^{(i)} \cdot \mathcal{B}_0 \boldsymbol{a} \Gamma dl = - \int_{D_R} \mathbf{e}^{(i)} \cdot \boldsymbol{a} \delta(\boldsymbol{\xi}) d\boldsymbol{\xi} = -a_i,$$

$$\int_{\partial g} \xi_j \nu_k dl = -\delta_{jk} \mathrm{mes} g. \tag{3.1.29}$$

Formulae (3.1.13) and (3.1.17), (3.1.25) yield the second integral in (3.1.28). For evaluation of the integrals I_j the components of the normal vector can be represented in terms of

$$(\mathcal{B}_0 - \mathcal{B}_0^\circ) \xi_j \mathbf{e}^{(3)}, \; (\mathcal{B}_0 - \mathcal{B}_0^\circ) \xi_j \mathbf{e}^{(j)}, \; j = 1, 2,$$

and the Betti formula can be used together with the expansion of the vector $\boldsymbol{w}^{(1)}$ at infinity. Using (3.1.29) we consequently obtain (3.1.26). □

3.1.3 Second-order term Λ_2

Consider the quantities $\boldsymbol{\Phi}^{(2)}$ and Λ_2 from the asymptotic approximation (3.1.19). The following relation is used:

$$\mathcal{P}(1 + \varepsilon^2 \Lambda_2) \sim \mathcal{P}(1) + \varepsilon^2 \Lambda_2 \mathcal{P}'(1),$$

where $\mathcal{P}(1)$ admits the representation

$$\mathcal{P}(1) = \mathcal{L}_0(\partial/\partial\boldsymbol{\eta}) + \mathcal{L}_1(\boldsymbol{\eta}, \partial/\partial\boldsymbol{\eta}) + \mathcal{L}_2(\boldsymbol{\eta}, \partial/\partial\boldsymbol{\eta}) \qquad (3.1.30)$$

in the vicinity of $\mathcal{N}$. Note that the matrix differential operator $\mathcal{L}_2$ has the coefficients of order $O(||\boldsymbol{\eta}||^k)$ near the k–order derivatives. Taking into account the terms $O(\varepsilon^2)$ in (3.1.3) (related to the boundary layer), we derive the following system with respect to the vector $\boldsymbol{\Phi}^{(2)}$ and the quantity Λ_2

$$\mathcal{P}(1)\boldsymbol{\Phi}^{(2)} = -\Lambda_2\mathcal{P}'(1)\boldsymbol{\Phi} - \mathbf{F} \quad \text{on } \mathbb{S}, \qquad (3.1.31)$$

where

$$\mathbf{F} = \mathcal{L}_2\chi\boldsymbol{\Upsilon}^{(1)} + (\mathcal{P}(1) - \mathcal{L}_0)\chi\boldsymbol{\Upsilon}^{(2)} + [\mathcal{L}_0 + \mathcal{L}_1, \chi]\boldsymbol{\Upsilon}^{(1)} + [\mathcal{L}_0, \chi]\boldsymbol{\Upsilon}^{(2)};$$

here $[\mathcal{A}, \mathcal{B}] = \mathcal{A}\mathcal{B} - \mathcal{B}\mathcal{A}$ is the commutator of operators $\mathcal{A}$ and $\mathcal{B}$.

Proposition 3.

1. *The system*

$$\mathcal{P}(1)\boldsymbol{V} = \boldsymbol{F}_* \quad on\ \mathbb{S}$$

is solvable if and only if the following orthogonality conditions hold

$$\int_{\mathbb{S}} \boldsymbol{F}_* \cdot \mathbf{Y}^{(1,j)} ds = 0, \quad j = 1, 2, ..., 9, \qquad (3.1.32)$$

where $\mathbf{Y}^{(1,j)}$ *are traces of the fields* $\mathbf{V}^{(1,j)}\left(\frac{\partial}{\partial \boldsymbol{x}}\right)\boldsymbol{T}(\boldsymbol{x})$ *on the sphere* $\mathbb{S}$. *The solution* $\boldsymbol{V}$ *is defined up to an arbitrary constant vector term* $\mathbf{c}$.

2. *The following equalities hold:*

$$\int_{\mathbb{S}} \mathbf{Y}^{(1,k)} \cdot \mathcal{P}'(1)\boldsymbol{\Phi}^{(1,j)} ds = -\delta_{jk}, \; j, k = 1, 2, ..., 9, \qquad (3.1.33)$$

$$\int_{\mathbb{S}} \mathbf{Y}^{1,j} \cdot \boldsymbol{F} ds = \sum_{k=1}^{9} M_{jk} c_k, \; j = 1, 2, ..., 9, \qquad (3.1.34)$$

with

$$\begin{gathered}
M_{1j} = q[-(2-\varkappa)\alpha_1^{(j)} - \alpha_2^{(j)} + \beta_3^{(j)} \\
+2(\varkappa+1)^{-1}(\alpha_1^{(j)} + \alpha_2^{(j)})],\ M_{2j} = q[-\alpha_1^{(j)} - (2-\varkappa)\alpha_2^{(j)} \\
+\beta_3^{(j)} + 2(\varkappa+1)^{-1}(\alpha_1^{(j)} + \alpha_2^{(j)})],\ M_{3j} = -q(\varkappa+1)\beta_3^{(j)}, \\
M_{4j} = -q(1-\varkappa)\alpha_3^{(j)},\ j = 1,2,3,4; \\
M_{5p} = 2^{1/2}q[(3+\varkappa)\alpha_5^{(p)} + (1-\varkappa)\beta_2^{(p)}], \qquad (3.1.35) \\
M_{6p} = 2^{-1/2}q[(3+\varkappa)\alpha_4^{(p)} + (1-\varkappa)\beta_1^{(p)}], \\
M_{8p} = -2^{1/2}q(\varkappa+1)[(2+\lambda\mu^{-1})\alpha_5^{(p)} + \beta_2^{(p)}], \\
M_{9p} = 2^{-1/2}q(\varkappa+1)[(2+\lambda\mu^{-1})\alpha_4^{(p)} + \beta_1^{(p)}],\ p = 5,6; \\
q = (\lambda+\mu)[8\pi\mu(\lambda+2\mu)]^{-1}.
\end{gathered}$$

Proof. The differential operator of the Lamé system is formally self–adjoint which yields

$$\mathcal{P}^*(\Lambda) = \mathcal{P}(-1-\overline{\Lambda}).$$

Consequently, the first assertion follows from the statements concerning solutions of the homogeneous Lamé system and presented in the first section.

To verify (3.1.33) we introduce a function

$$\zeta(\|\boldsymbol{x}\|) \in C_0^\infty(\mathcal{D}_1),$$

which equals 1 in the vicinity of the origin. Here

$$\mathcal{D}_d = \{\boldsymbol{x} \in \mathbb{R}^3 :\ \|\boldsymbol{x}\| < d\}$$

is a ball of a radius d with the centre at the origin. The definition of the Somigliana tensor yields

$$\begin{aligned}
&\int_{\mathcal{D}_1} \boldsymbol{V}^{(1,k)}\left(\frac{\partial}{\partial \boldsymbol{x}}\right)\mathbf{T}(\boldsymbol{x}) \cdot L\left(\frac{\partial}{\partial \boldsymbol{x}}\right)(\zeta(\rho)\boldsymbol{V}^{(1,j)}(\boldsymbol{x}))d\boldsymbol{x} \\
&= \int_{\mathcal{D}_1} \zeta(\rho)\boldsymbol{V}^{(1,j)}(\boldsymbol{x}) \cdot L\left(\frac{\partial}{\partial \boldsymbol{x}}\right)\boldsymbol{V}^{(1,k)}\left(\frac{\partial}{\partial \boldsymbol{x}}\right)\mathbf{T}(\boldsymbol{x})d\boldsymbol{x} \qquad (3.1.36) \\
&= \boldsymbol{V}^{(1,k)}(\partial/\partial \boldsymbol{x})\boldsymbol{V}^{(1,j)}(\mathbf{0}) = \delta_{jk}.
\end{aligned}$$

Relations (3.1.36), together with

$$\mathcal{P}\left(\rho\,\frac{\partial}{\partial\rho}\right)(\rho\boldsymbol{\Phi}^{(1,j)}) = \rho\mathcal{P}\left(1+\rho\,\frac{\partial}{\partial\rho}\right)\boldsymbol{\Phi}^{(1,j)},$$

$$\mathcal{P}(\Lambda+1) = \mathcal{P}(1) + \Lambda\mathcal{P}'(1) + 1/2\Lambda^2\mathcal{P}''(1),$$

yield (3.1.33) and

$$\int_{\mathcal{D}_1} \boldsymbol{V}^{(1,k)}\left(\frac{\partial}{\partial \boldsymbol{x}}\right)\mathbf{T}(\boldsymbol{x})\cdot L\left(\frac{\partial}{\partial \boldsymbol{x}}\right)(\zeta(\rho)\boldsymbol{V}^{(1,j)}(\boldsymbol{x}))d\boldsymbol{x}$$
$$= \lim_{\delta\to 0}\int_\delta^1\int_{\mathbb{S}} \rho^{-2}\mathbf{Y}^{(1,k)}(\theta,\varphi)\mathcal{P}\left(\rho\frac{\partial}{\partial\rho}\right)(\zeta(\rho)\rho\boldsymbol{\Phi}^{(1,j)}(\theta,\varphi))d\rho ds$$
$$= \lim_{\delta\to 0}\int_\delta^1\int_{\mathbb{S}}\left(\frac{\partial\zeta}{\partial\rho}\mathbf{Y}^{(1,k)}(\theta,\varphi)\mathcal{P}'(1)\boldsymbol{\Phi}^{(1,j)}(\theta,\varphi)\right.$$
$$-\frac{1}{2}\frac{\partial\zeta}{\partial\rho}\mathbf{Y}^{(1,k)}(\theta,\varphi)\mathcal{P}''(1)\boldsymbol{\Phi}^{(1,j)}(\theta,\varphi)$$
$$\left.+\frac{1}{2}\frac{\partial\zeta}{\partial\rho}\mathbf{Y}^{(1,k)}(\theta,\varphi)\mathcal{P}''(1)\boldsymbol{\Phi}^{(1,j)}(\theta,\varphi)\right)d\rho ds$$
$$= -\lim_{\delta\to 0}\zeta(\delta)\int_{\mathbb{S}}\mathbf{Y}^{(1,k)}(\theta,\varphi)\mathcal{P}'(1)\boldsymbol{\Phi}^{(1,j)}(\theta,\varphi)ds$$
$$= -\int_{\mathbb{S}}\mathbf{Y}^{(1,k)}(\theta,\varphi)\cdot\mathcal{P}'(1)\boldsymbol{\Phi}^{(1,j)}(\theta,\varphi)ds.$$

Equalities (3.1.34) follow from the following equations based on (3.1.30) and the first part of the Proposition 3:

$$\mathcal{P}(1)(\chi(\theta)(\boldsymbol{\Upsilon}^{(1)}(\boldsymbol{\eta})+\boldsymbol{\Upsilon}^{(2)}(\boldsymbol{\eta}))) = \boldsymbol{F}(\boldsymbol{\eta}) - \chi(\theta)\sum_{j=1}^{6}c_j$$
$$\times\left(\sum_{i=1}^{5}\alpha_i^{(j)}\boldsymbol{W}^{(i)}\left(\frac{\partial}{\partial\boldsymbol{\eta}}\right)+\sum_{i=1}^{3}(\beta_i^{(j)}+d_i^{(j)})\boldsymbol{e}^{(i)}\right)\delta(\boldsymbol{\eta}),$$
$$d_i^{(p)} = -(\lambda+\mu)\mu^{-1}\alpha_{3+i}^{(p)},\ l=1,2,\ p=5,6;$$
$$d_3^{(j)} = -(\lambda+\mu)(\lambda+2\mu)^{-1}(\alpha_1^{(j)}+\alpha_2^{(j)}),$$
$$\int_{\mathbb{S}}\boldsymbol{\Upsilon}^{(1,k)}\cdot\mathcal{P}(1)(\chi\boldsymbol{\Upsilon}^{(1)}+\chi\boldsymbol{\Upsilon}^{(2)})ds = 0.$$

□

It follows from (3.1.33), (3.1.34) that the orthogonality conditions (3.1.32) for

$$\boldsymbol{F}_* = -\boldsymbol{F} - \Lambda_2\mathcal{P}'(1)\boldsymbol{\Phi}$$

can be represented by a system of linear algebraic equations with a spectral parameter Λ_2

$$M\boldsymbol{c} = \Lambda_2\boldsymbol{c}. \tag{3.1.37}$$

Thus, Λ_2 is an eigenvalue of the matrix M with elements (3.1.35), and the vector $\boldsymbol{c}$ of coefficients of the linear combination (3.1.10) is the corresponding eigenvector.

3.1.4 Examples

A "thin" angular crack.

Let $k_\varepsilon = \mathbb{R}^3 \backslash \{\boldsymbol{x} : x_2 = 0,\ x_3 \geq 0,\ |x_1| \leq \varepsilon x_3\}$. The corresponding set $\overline{g}_\varepsilon$ on the unit sphere $\mathbb{S}$ is the arc of length $2\tan^{-1}\varepsilon$. The Pólya–Szegö matrix for a crack consists of two blocks

$$-\frac{\pi(\lambda+2\mu)}{2\mu(\lambda+\mu)}\begin{pmatrix} \lambda^2 & (\lambda+2\mu)\lambda \\ (\lambda+2\mu)\lambda & (\lambda+2\mu)^2 \end{pmatrix},$$

$$\operatorname{diag}\left(-\pi\mu\,\frac{\lambda+2\mu}{\lambda+\mu};0;-\pi\mu\right). \tag{3.1.38}$$

The 9×9 matrix M has the block diagonal structure, and $\Lambda_2^{(7)} = \Lambda_2^{(8)} = \Lambda_2^{(9)} = 0$. Also, using (3.1.38) and (3.1.35) we find that

$$\Lambda_2^{(5)} = -\frac{2\lambda^2+9\mu\lambda+5\mu^2}{4(\lambda+2\mu)(\lambda+3\mu)},\ \Lambda_2^{(6)} = 0.$$

The remaining 4×4 block is

$$\frac{\mu^{-2}}{16}\begin{pmatrix} \lambda t_1 & \lambda t_2 & -\lambda t_3 & 0 \\ (\lambda+2\mu)t_1 & (\lambda+2\mu)t_2 & -(\lambda+2\mu)t_3 & 0 \\ \lambda t_1 & \lambda t_2 & -\lambda t_3 & 0 \\ 0 & 0 & 0 & -4\mu^3(\lambda+\mu)^{-1} \end{pmatrix},$$

with

$$t_1 = 2\mu+\lambda(1-\varkappa),\ t_2 = (2-\varkappa)(\lambda+2\mu)-\lambda,\ t_3 = 4(\lambda+\mu)-2\lambda(\varkappa+1).$$

Therefore,

$$\Lambda_2^{(1)} = \Lambda_2^{(2)} = 0,\ \Lambda_2^{(3)} = -1/4,\ \Lambda_2^{(4)} = -\mu[4(\lambda+\mu)]^{-1}.$$

We emphasize that the stresses in problems concerning the tension at infinity of a space with a narrow crack by the forces σ_{33}^∞, σ_{11}^∞ or σ_{13}^∞

are constant and therefore have no singularities. Finally, $\Lambda_2^{(3)} < 0$, and $\Lambda_2^{(4)} \in (-1/4, 0)$, $\Lambda_2^{(5)} \in (-1/2, -5/24)$.

Circular cone.

Let $k_\varepsilon = \{x : x_3 > 0, \ \|x'\| < \varepsilon x_3\}$. Then g is a unit circle and the corresponding Pólya–Szegö matrix is

$$-\frac{\pi(\lambda+2\mu)}{\mu}\begin{pmatrix} \lambda+\mu\varkappa & \lambda+\mu(2-\varkappa) & 0 \\ \lambda+\mu(2-\varkappa) & \lambda+\mu\varkappa & 0 \\ 0 & 0 & 2\mu(\varkappa-1) \end{pmatrix}, \ -2\pi\mu\begin{pmatrix} 1 & 0 \\ 0 & 1 \end{pmatrix}.$$

Four eigenvalues of the matrix M are specified by the formulae

$$\Lambda_2^{(3)} = \Lambda_2^{(4)} = -\mu(\lambda+\mu)^{-1} \in (-1, 0),$$
$$\Lambda_2^{(5)} = \Lambda_2^{(6)} = -(2\lambda^2 + 9\mu\lambda + 5\mu^2)[2(\lambda+2\mu)(\lambda+3\mu)]^{-1} \in (-1; -5/12).$$

These correspond to non-axisymmetric solutions. The axisymmetric components yield the singularity exponents $\varepsilon^2\Lambda_2^{(i)} + O(\varepsilon^3)$, $i = 1, 2$, where

$$\Lambda_2^{(1)} = 0, \ \Lambda_2^{(2)} = -\frac{5\lambda^2 + 9\mu\lambda + 2\mu^2}{4(\lambda+\mu)(\lambda+2\mu)} \in \left(-\frac{5}{4}, -\frac{1}{4}\right). \tag{3.1.39}$$

Formulae (3.1.19), (3.1.39) agree with numerical solutions presented in [11, p. 962] and [103, p. 322] (see Fig. 3.4, zone $\alpha \sim \pi$).

It turns out that the exponent of the stress singularity for a non-axisymmetric loading can have a higher order than for the axisymmetric case.

3.1.5 Conical cavity in an elastic half-space

Let k_ε be a circular cone $\{x : \theta < \sin^{-1}\varepsilon\}$, and

$$\mathbb{R}_1^3 = \{x : x_3 < 1\}, \ \Omega_\varepsilon = \mathbb{R}_1^3 \setminus \overline{k}_\varepsilon.$$

Consider the deformation of a half-space Ω_ε with a conical recess subjected to axisymmetric normal load with densities p and q applied to the surface $\partial\Omega_\varepsilon$ in a neighbourhood of the cavity edge (see Fig. 3.1).

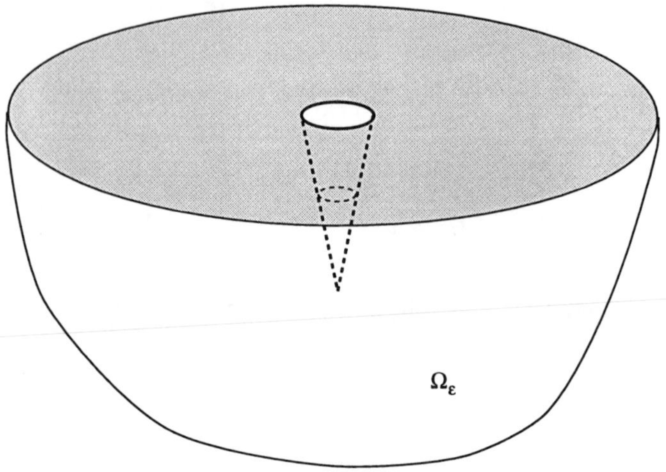

Fig. 3.1: A thin conical cavity in an elastic half-space.

In a neighbourhood of the point $\mathcal{N}$ we introduce stretched coordinates

$$\boldsymbol{y} = (y_1, y_2, y_3),$$

with

$$y_j = \frac{x_j}{\varepsilon x_3}, \; j = 1, 2; \; y_3 = \frac{x_3 - 1}{\varepsilon}.$$

Assume that the traction

$$q(\varepsilon, \boldsymbol{x}) = \varepsilon^{-2} q_0(r_y), \; r_y = \sqrt{y_1^2 + y_2^2},$$

is applied to the surface $\{\boldsymbol{x} : x_3 = 1\}$, while the load

$$p(\varepsilon, \boldsymbol{x}) = \varepsilon^{-3} p_0(y_3)$$

is applied to $\partial k_\varepsilon \cap \mathbb{R}^3_1$. Here q_0, p_0 are bounded functions. In the limit $\varepsilon \to 0$ a neighbourhood of a surface edge of the cavity will be described in $\boldsymbol{y}$ coordinates as a half-space with a semi-infinite cylinder $\mathbf{C} = \{\boldsymbol{y} : r_y < 1,\ y_3 < 0\}$.

Assume that the displacement vector $\boldsymbol{u}$ satisfies the homogeneous Lamé system (body force densities are equal to zero). Due to the axial symmetry,

$$u_\varphi = 0,\ \sigma_{\varphi r}(\boldsymbol{u}; \boldsymbol{x}) = \sigma_{\varphi z}(\boldsymbol{u}; \boldsymbol{x}) = 0,$$

where (r, φ, z) are cylindrical coordinates. The boundary conditions on the surface ∂k_ε and on the boundary of the half-space have the form

$$\sigma_{\theta\theta}(\boldsymbol{u}; \boldsymbol{x}) = -p(\varepsilon, \boldsymbol{x}),\ \sigma_{\rho\theta}(\boldsymbol{u}; \boldsymbol{x}) = \sigma_{\theta\varphi}(\boldsymbol{u}; \boldsymbol{x}) = 0,\ \boldsymbol{x} \in \partial k_\varepsilon \cap \mathbb{R}^3_1, \tag{3.1.40}$$

$$\sigma_{zz}(\boldsymbol{u}; \boldsymbol{x}) = -q(\varepsilon, \boldsymbol{x}),\ \sigma_{zr}(\boldsymbol{u}; \boldsymbol{x}) = -\sigma_{z\varphi}(\boldsymbol{u}; \boldsymbol{x}) = 0,\ \boldsymbol{x} \in \partial\mathbb{R}^3_1 \cap \Omega_\varepsilon. \tag{3.1.41}$$

The formulation for an elastic half-space yields an auxiliary limit problem which describes the stress–strain state outside a neighbourhood of the conical notch. Also, the zone of application of the load p and q is assumed to be small. Further, we perform the analysis of a boundary layer near the edges of k_ε. A two-dimensional boundary layer will be needed to obtain a solution of an auxiliary plane-strain problem that occurs near the conical surface (see [53, Section 2.2] and [94, Chapter 4]).

3.1.6 Half-space with a cylindrical cavity

In new coordinates $\boldsymbol{y}$ the domain Ω_ε will be transformed to $\mathbb{R}^3_0 \backslash \overline{\mathbf{C}}$, as $\varepsilon \to 0$. Let L be the operator of the Lamé system, and B and Γ the operators of the boundary conditions (3.1.40) and (3.1.41). In the coordinates $\boldsymbol{y}$ these operators are split into formal series in powers of ε :

$$L(\partial/\partial\boldsymbol{x})\boldsymbol{\Psi}(\boldsymbol{y}) = \varepsilon^{-2} L(\partial/\partial\boldsymbol{y})\boldsymbol{\Psi}(\boldsymbol{y}) + \varepsilon^{-1} L_1(\boldsymbol{y}, \partial/\partial\boldsymbol{y})\boldsymbol{\Psi}(\boldsymbol{y}) + \ldots, \tag{3.1.42}$$

$$B(\partial/\partial\boldsymbol{x})\boldsymbol{\Psi}(\boldsymbol{y}) = \varepsilon^{-1} B_0(\boldsymbol{y}, \partial/\partial\boldsymbol{y})\boldsymbol{\Psi}(\boldsymbol{y}) + B_1(\boldsymbol{y}, \partial/\partial\boldsymbol{y})\boldsymbol{\Psi}(\boldsymbol{y}) + \ldots,$$

$$\Gamma(\partial/\partial\boldsymbol{x})\boldsymbol{\Psi}(\boldsymbol{y}) = \varepsilon^{-1}\Gamma(\partial/\partial\boldsymbol{y})\boldsymbol{\Psi}(\boldsymbol{y}) + \Gamma_1(\boldsymbol{y}, \partial/\partial\boldsymbol{y})\boldsymbol{\Psi}(\boldsymbol{y}) + \ldots,$$

$$L_1\left(\boldsymbol{y}, \frac{\partial}{\partial\boldsymbol{y}}\right) = -2y_3 L\left(\frac{\partial}{\partial y_1}, \frac{\partial}{\partial y_2}, 0\right)$$

$$-\left(y_1\frac{\partial}{\partial y_1}+y_2\frac{\partial}{\partial y_2}\right)L'\left(\frac{\partial}{\partial \boldsymbol{y}}\right)-\left(y_3\frac{\partial}{\partial y_3}+1\right)L'\left(\frac{\partial}{\partial y_1},\frac{\partial}{\partial y_2},0\right);$$

$$B_0\left(\boldsymbol{y},\frac{\partial}{\partial \boldsymbol{y}}\right)=\cos\varphi\boldsymbol{\sigma}^{(1)}+\sin\varphi\boldsymbol{\sigma}^{(2)},$$

$$B_1\left(\boldsymbol{y},\frac{\partial}{\partial \boldsymbol{y}}\right)=-y_3B_0\left(\boldsymbol{y},\frac{\partial}{\partial y_1},\frac{\partial}{\partial y_2},0\right)$$

$$-\left(y_1\frac{\partial}{\partial y_1}+y_2\frac{\partial}{\partial y_2}\right)B_0'\left(\boldsymbol{y},\frac{\partial}{\partial \boldsymbol{y}}\right)-\boldsymbol{\sigma}^{(3)};$$

$$\Gamma\left(\frac{\partial}{\partial \boldsymbol{y}}\right)=\boldsymbol{\sigma}^{(3)},\ \Gamma_1\left(\boldsymbol{y},\frac{\partial}{\partial \boldsymbol{y}}\right)=-y_3\Gamma\left(\frac{\partial}{\partial y_1},\frac{\partial}{\partial y_2},0\right)$$

$$-\Gamma\left(0,0,y_1\frac{\partial}{\partial y_1}+y_2\frac{\partial}{\partial y_2}\right);\ \boldsymbol{\sigma}^{(j)}(\boldsymbol{\Psi};\boldsymbol{y})=(\sigma_{jk}(\boldsymbol{\Psi};\boldsymbol{y}))_{k=1}^3.$$

The boundary layer in the vicinity of $\mathcal{N}$ is represented by

$$\varepsilon^{-2}\boldsymbol{w}^{(0)}(\boldsymbol{y})+\varepsilon^{-1}\boldsymbol{w}^{(1)}(\boldsymbol{y}).$$

Following (3.1.42) and (3.1.40), (3.1.41), the vector-valued function $\boldsymbol{w}^{(0)}$ satisfies the equations

$$L(\partial/\partial\boldsymbol{y})\boldsymbol{w}^{(0)}(\boldsymbol{y})=\boldsymbol{0},\ \boldsymbol{y}\in\mathbb{R}_0^3\backslash\overline{\mathbf{C}},\ \Gamma(\partial/\partial\boldsymbol{y})\boldsymbol{w}^{(0)}(\boldsymbol{y})=\boldsymbol{0},\ \boldsymbol{y}\in\partial\mathbb{R}_0^3\backslash\mathbf{C},$$

$$B_0(\boldsymbol{y},\partial/\partial\boldsymbol{y})\boldsymbol{w}^{(0)}(\boldsymbol{y})=-p_0(y_3)(\cos\varphi,\sin\varphi,0)^T,\ \boldsymbol{y}\in\partial\mathbf{C}\cap\mathbb{R}_0^3. \quad (3.1.43)$$

Similar problems were studied in [2], [47] (see also Section 3.5). Here we need only axisymmetric solutions. The asymptotic approximation of the vector field $\boldsymbol{w}^{(0)}$ has the form

$$\boldsymbol{w}^{(0)}(\boldsymbol{y})=c_3\boldsymbol{T}^{(3)}(\boldsymbol{y})+c_1\left(\frac{\partial}{\partial y_1}\boldsymbol{T}^{(1)}(\boldsymbol{y})+\frac{\partial}{\partial y_2}\boldsymbol{T}^{(2)}(\boldsymbol{y})\right) \quad (3.1.44)$$

$$+O(\|\boldsymbol{y}\|^{-3}ln\|\boldsymbol{y}\|),\ \|\boldsymbol{y}\|\to\infty,$$

where $\mathbf{T}^{(j)}$, $j=1,2,3$, are displacements in an elastic half-space subjected to a unit concentrated force in $\boldsymbol{e}^{(j)}$ direction (see, for example, [99, p. 237]).

The external load from (3.1.43) is self-balanced, which yields $c_3 = 0$. We use the method [62] to evaluate the constant c_1. Consider the polynomial axisymmetric solution of the homogeneous boundary value problem for the Lamé operator in a half-space

$$\boldsymbol{V}(\boldsymbol{y}) = \frac{1}{\sqrt{2}}\left(y_1, y_2, -\frac{2\lambda}{\lambda+2\mu}\,y_3\right)^T.$$

The discrepancy in the boundary conditions on $\partial\mathbf{C}\cap\mathbb{R}^3_0$ is

$$\frac{\sqrt{2}\mu(3\lambda+2\mu)}{\lambda+2\mu}\,(\cos\varphi, \sin\varphi, 0)^T.$$

This error is compensated by the axisymmetric field $\alpha\boldsymbol{Y}$, with

$$Y_k(\boldsymbol{y}) = (2\mu)^{-1} r_y^{-2} y_k,\ k=1,2;\ Y_3(\boldsymbol{y}) = 0, \tag{3.1.45}$$

$$\alpha = \sqrt{2}\mu(3\lambda+2\mu)(\lambda+2\mu)^{-1},$$
$$\sigma_{\varphi\varphi}(\boldsymbol{Y};\boldsymbol{y}) = -\sigma_{rr}(\boldsymbol{Y};\boldsymbol{y}) = r_y^{-2}, \tag{3.1.46}$$
$$\sigma_{zz}(\boldsymbol{Y}) = \sigma_{r\varphi}(\boldsymbol{Y}) = \sigma_{rz}(\boldsymbol{Y}) = \sigma_{\varphi z}(\boldsymbol{Y}) = 0.$$

Since $\Gamma\left(\frac{\partial}{\partial \mathbf{y}}\right)\boldsymbol{Y} = \mathbf{0}$ on $\partial\mathbb{R}^3_0\backslash\mathbf{C}$, and due to (3.1.46), the vector

$$\boldsymbol{\zeta} = \boldsymbol{V} + \alpha\boldsymbol{Y}$$

satisfies the homogeneous boundary value problem (3.1.43).

The following statement holds.

Proposition 4. *The non-zero coefficient of the asymptotic approximation* (3.1.44) *is specified by*

$$c_1 = -4\pi\frac{\lambda+\mu}{\lambda+2\mu}\,P,\ P = \int_{-\infty}^{0} p_0(t)dt. \tag{3.1.47}$$

Proof. Let $\mathcal{D}_R = \{\boldsymbol{y} : \|\boldsymbol{y}\| < R\}$. Substitute the fields $\boldsymbol{w}^{(0)}$ and $\boldsymbol{\zeta}$ into the Betti formula for the domain $(\mathcal{D}_R\cap\mathbb{R}^3_0)\backslash\mathbf{C}$. Taking into account the boundary conditions on $\partial\mathbb{R}^3_0$, we have

$$\int_{\mathcal{S}_1}\boldsymbol{\zeta}\cdot\boldsymbol{\sigma}^{(n)}(\boldsymbol{w}^{(0)}) - \boldsymbol{w}^{(0)}\cdot\boldsymbol{\sigma}^{(n)}(\boldsymbol{\zeta})ds \tag{3.1.48}$$

$$= \int_{\mathcal{S}_2} \zeta \cdot \sigma^{(n)}(w^{(0)}) - w^{(0)} \cdot \sigma^{(n)}(\zeta) ds,$$

$$\sigma^{(n)} = \sigma n, \ \mathcal{S}_1 = (\partial \mathrm{C} \cap \mathbb{R}^3_0) \cap \mathcal{D}_R, \ \mathcal{S}_2 = (\partial \mathcal{D}_R \cap \mathbb{R}^3_0) \backslash \mathrm{C},$$

where n is the unit outward normal vector. The integrals in the left-hand side of (3.1.48) can be extended to $\partial \mathrm{C} \cap \mathbb{R}^3_0$. Due to (3.1.44), the right-hand side of (3.1.48) is evaluated by

$$c_1 \int_{\partial \mathcal{D}_R \cap \mathbb{R}^3_0} \{V(y) \cdot \sigma^{(n)}(\mathbf{T}^{(1)}_{,1} + \mathbf{T}^{(2)}_{,2}; y)$$

$$-(\mathbf{T}^{(1)}_{,1}(y) + \mathbf{T}^{(2)}_{,2}(y)) \cdot \sigma^{(n)}(V; y)\} ds_y$$

$$= -c_1 \int_{\mathcal{D}_R \cap \mathbb{R}^3_0} 2^{1/2} V(y) \cdot V(\partial/\partial y_1, \ \partial/\partial y_2, 0)$$

$$\times \delta(y_1, y_2, 0) dy_1 dy_2 = 2^{1/2} c_1,$$

with error $o(1)$, as $R \to \infty$.

Taking the limit $R \to \infty$ and evaluating the integral over $\partial \mathrm{C} \cap \mathbb{R}^3_0$, we obtain (3.1.47). □

Consider the next term of the asymptotic approximation of a solution of the boundary layer type. It follows from (3.1.42) that the vector $w^{(1)}$ satisfies the boundary value problem

$$L(\partial/\partial y) w^{(1)}(y) = -L_1(y, \partial/\partial y) w^{(0)}(y), y \in \mathbb{R}^3_0 \backslash \overline{\mathrm{C}}, \tag{3.1.49}$$

$$B_0(y, \partial/\partial y) w^{(1)}(y) = -B_1(y, \partial/\partial y) w^{(0)}(y), y \in \partial \mathrm{C} \cap \mathbb{R}^3_0, \tag{3.1.50}$$

$$\Gamma(\partial/\partial y) w^{(1)}(y) = -q_0(r_y) e^{(3)} - \Gamma_1(y, \partial/\partial y) w^{(0)}(y), \tag{3.1.51}$$

$$y \in \partial \mathbb{R}^3_0 \backslash \mathrm{C}.$$

Asymptotic representation (3.1.44) yields that the right-hand sides of (3.1.49) and (3.1.50) are quantities of order $O(||y||^{-3})$ and $O(||y||^{-2})$, as $||y|| \to \infty$. Consequently, according to the results of [2], [38], [62],

$$w^{(1)}(y) = c_3 T^{(3)}(y) + \Upsilon(y) + O(||y||^{-2} \ln ||y||), \ ||y|| \to \infty, \tag{3.1.52}$$

where c_3 is constant, Υ is a particular solution of the problem

$$L \Upsilon = -c_1 L_1 \left(\frac{\partial}{\partial y_1} \mathbf{T}^{(1)} + \frac{\partial}{\partial y_2} \mathbf{T}^{(2)} \right) \text{ in } \mathbb{R}^3_0,$$

$$\Gamma\Upsilon = -c_1\Gamma_1\left(\frac{\partial}{\partial y_1}\mathbf{T}^{(1)} + \frac{\partial}{\partial y_2}\mathbf{T}^{(2)}\right) \text{ on } \partial\mathbb{R}^3_0\backslash O.$$

The factor c_3 in the asymptotic form (3.1.52) is evaluated by the formula

$$c_3 = -2\pi Q - 2\mu(\lambda+\mu)^{-1}c_1, \; Q = \int_1^{+\infty} q_0(t)dt. \qquad (3.1.53)$$

The derivation is similar to one presented in Proposition 4. The vector-valued function Υ is determined by the equality

$$\Upsilon(y) = y_3\{y_1\frac{\partial}{\partial y_1} + y_2\frac{\partial}{\partial y_2}\}\Xi(y),$$

and its components are homogeneous functions of degree -1.

3.1.7 Asymptotic approximation of the stress-strain state in Ω_ε

Return back to coordinates x. Formulae (3.1.44), (3.1.52) and (3.1.53) give that for x such that $\|x - \mathcal{N}\| = O(\sqrt{\varepsilon})$ the following relation holds:

$$\varepsilon^{-2}w^{(0)}(y) + \varepsilon^{-1}w^{(1)}(y) \sim c_3 T^{(3)}(x-\mathcal{N})$$

$$+ c_1\left(\frac{\partial}{\partial x_1}\mathbf{T}^{(1)}(x-\mathcal{N}) + \frac{\partial}{\partial x_2}\mathbf{T}^{(2)}(x-\mathcal{N})\right). \qquad (3.1.54)$$

Matching the three-dimensional boundary layer with the displacement field v, which approximates the solution u outside neighbourhood of k_ε, we deduce that v satisfies the boundary value problem

$$Lv = 0 \text{ in } \mathbb{R}^3_1, \; \Gamma v = c_3 e^{(3)}\delta + c_1(e^{(1)}\delta_{,1} + e^{(2)}\delta_{,2}) \text{ on } \partial\mathbb{R}^3_1,$$

where $e^{(j)}$, $j = 1,2,3$, are the unit vectors in $\mathbb{R}^3$, while the δ–function is concentrated at the point $x = \mathcal{N}$. Therefore, $v(x)$ agrees with the right-hand side of the relationship (3.1.54).

Thus, we found the asymptotic representation of the solution in two zones: in the vicinity of the boundary, where the external load is applied, and outside a neighbourhood of the cone k_ε.

Now we construct additional terms that take into account the stress singularity at the vertex of the cone. The vector v leaves a discrepancy in the homogeneous boundary condition (3.1.40):

$$\sigma_{\theta\theta}(\boldsymbol{v};\rho) = X(\rho) + O(\varepsilon), \quad \sigma_{\rho\theta}(\boldsymbol{v};\rho) = \sigma_{\theta\varphi}(\boldsymbol{v};\rho) = 0$$

$$\text{on } \partial k_\varepsilon \cap \mathbb{R}^3_1, \tag{3.1.55}$$

where

$$X(\rho) = \frac{\mu}{\lambda+\mu}\left(\frac{Q}{2(1-\rho)^2} - \frac{2}{\lambda+2\mu}\,P\left(\frac{\lambda+\mu}{(1-\rho)^3} + \frac{\mu}{(1-\rho)^2}\right)\right).$$

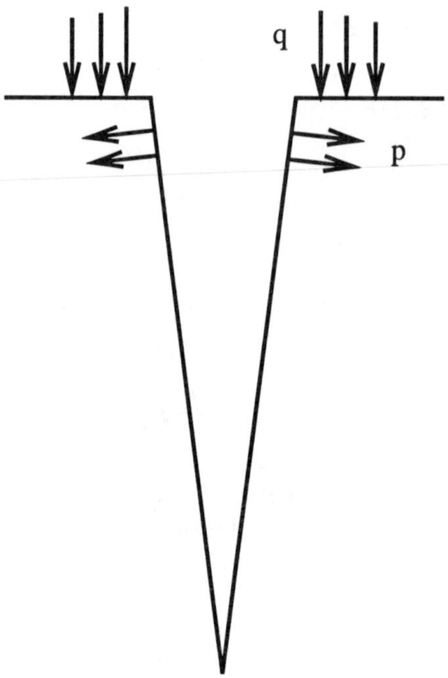

Fig. 3.2: An axisymmetric load applied in a neighbourhood of the conical cavity.

The right-hand side of (3.1.55) is "slowly varying" in z outside a neighbourhood of the point $\mathcal{N}$. To compensate the above discrepancy we construct a boundary layer $\varepsilon z \boldsymbol{w}(y_1, y_2, z)$. The components of $\boldsymbol{w}$ are solutions of the plane-strain and anti-plane shear problems in the domain $\mathbb{R}^2 \backslash D_1$.

In coordinates (y_1, y_2, z) relation (3.1.55) has the form

$$\sigma_{\theta\theta}(\boldsymbol{v}; \boldsymbol{y}, z) = X(z) + O(\varepsilon).$$

Thus, the traction boundary conditions for the vector $\mathcal{W} = (w_1, w_2)$ are specified by

$$\sigma_{\theta\theta}(\mathcal{W}; \boldsymbol{y}, z) = -X(z), \quad \sigma_{r\theta}(\mathcal{W}; \boldsymbol{y}, z) = 0 \quad \text{on} \quad \partial D_1. \tag{3.1.56}$$

The vector $\boldsymbol{w}$ admits the representation

$$\boldsymbol{w}(\boldsymbol{y}, z) = X(z)\boldsymbol{Y}(\boldsymbol{y}),$$

where $\boldsymbol{Y}$ is defined by (3.1.45).

For the axisymmetric field $\boldsymbol{u}$ the following asymptotic expansion in a neighbourhood of the vertex of the cone k_ε holds:

$$\boldsymbol{u}(\varepsilon, \boldsymbol{x}) = c^{(0)}(\varepsilon)\boldsymbol{e}^{(3)} + c^{(1)}(\varepsilon)\rho^{\Lambda^{(1)}(\varepsilon)}\boldsymbol{\Phi}^{(1)}(\varepsilon, \theta, \varphi) \tag{3.1.57}$$

$$+ c^{(2)}(\varepsilon)\rho^{\Lambda^{(2)}(\varepsilon)}\boldsymbol{\Phi}^{(2)}(\varepsilon, \theta, \varphi) + \ldots,$$

where $c^{(i)}(\varepsilon)$ are constants, $\Lambda^{(i)}(\varepsilon)$ are specified by (3.1.19), (3.1.39), as $\varepsilon \to 0$, the vector-functions $\boldsymbol{\Phi}^{(i)}$ of the angular variables are defined by

$$\begin{gathered}\rho\boldsymbol{\Phi}^{(i)}(0, \theta, \varphi) = b_1^{(i)}(x_1\boldsymbol{e}^{(1)} + x_2\boldsymbol{e}^{(2)}) + b_2^{(i)}x_3\boldsymbol{e}^{(3)},\\ b_1^{(1)} = b_2^{(2)} = 1, \; b_1^{(2)} = 0,\\ b_2^{(1)} = -(5\lambda^2 + 9\mu\lambda + 2\mu^2)[4(\lambda + \mu)(\lambda + 2\mu)]^{-1}.\end{gathered} \tag{3.1.58}$$

From (3.1.19), (3.1.39) and the asymptotic representation

$$\boldsymbol{u}(\varepsilon, \boldsymbol{x}) \sim \boldsymbol{v}(\boldsymbol{x}) + \varepsilon z \boldsymbol{w}(y_1, y_2, z),$$

we derive that the coefficients of the expansion (3.1.57) have the form

$$c^{(0)}(\varepsilon) = (2\lambda + 3\mu)(4\pi\mu(\lambda + \mu))^{-1}(c_3 - c_1) + O(\varepsilon), \tag{3.1.59}$$

$$c^{(i)}(\varepsilon) = \pm(v_{1,1}(0)b_2^{(j)} - v_{3,3}(0)b_1^{(j)}) + O(\varepsilon), \; i \neq j, \; i, \; j = 1, 2,$$

$$v_{1,1}(0) = (2\lambda + \mu)(8\pi\mu(\lambda + \mu))^{-1}(2c_1 - c_3),$$

$$v_{3,3}(0) = (2\pi\mu)^{-1}(c_1 - c_3).$$

On the basis of the above analysis we conclude the following.

1°. The exponent of the stress singularity at the vertex of a conical cavity has the order $O(\varepsilon^2)$. According to the criterion of Novozhilov [98], the fracture occurs when

$$\frac{1}{\text{mes}k}\int_k \sigma_{ss}(\boldsymbol{x})ds > \sigma_c,$$

where $k = \{\boldsymbol{x} : \rho < d,\ \theta = \theta_0\}$ is a conical surface, and σ_c denotes the critical stress. It yields that

$$\sigma_c < 2\Sigma(\theta_0)(\sin\theta_0)^{-1}(2 + \varepsilon^2\Lambda_2^{(2)} + O(\varepsilon^3))^{-1}$$

$$\times \exp(\varepsilon^2|\ln\ d|(\Lambda_2^{(2)} + O(\varepsilon))) = \Sigma(\theta_0)(\sin\theta_0)^{-1} + O(\varepsilon^2|\ln\ d|). \quad (3.1.60)$$

Here $\Sigma(\theta_0)$ is a quantity evaluated by formulae (3.1.57), (3.1.58) and (3.1.59), d is a so-called "parameter of structure" related to the distance from the vertex to the area of applied traction. Note that if the remainder term in (3.1.60) is small, then the order $O(\varepsilon^2)$ singularity exerts no influence; however for small d (i.e., $\varepsilon^2|\ln\ d| = O(1)$) the second term in (3.1.60) will be important.

2°. For the part of the conical surface ∂k_ε between the vertex O and the area of applied traction (see Fig. 3.2) the only non-zero stress components are $\sigma_{\rho\rho}$ and $\sigma_{\varphi\varphi}$ (it follows from the axial symmetry and formulae (3.1.40), (3.1.41)). Due to (3.1.46), (3.1.55), (3.1.56), the leading terms are

$$\begin{aligned} \sigma_{\rho\rho} = -6s(\rho);\ \sigma_{\varphi\varphi} = 2(1-2\nu)s(\rho),\ \nu = \lambda[2(\lambda+\mu)]^{-1}, \\ s(\rho) = Q(2(1-\rho)^2)^{-1} - P(1-\nu)^{-1}((1-\rho)^{-3} \\ + (1-2\nu)(1-\rho)^{-2}). \end{aligned} \quad (3.1.61)$$

Let the quantities P and Q be positive. If $QP^{-1} < 4$ then the stress $\sigma_{\rho\rho}$ is tensile and increases monotonically for $\rho \in (0,1)$; the stress $\sigma_{\varphi\varphi}$ is compressive. If $QP^{-1} \geq 4$, the stress $\sigma_{\rho\rho}$ is compressive in a neighbourhood of the apex, while $\sigma_{\varphi\varphi}$ is tensile. For $QP^{-1} > (5-4\nu)(1-\nu)^{-1} = \gamma_0$ there is a local maximum of $\sigma_{\varphi\varphi}$ at the point $\rho_0 = 1 - 3[(1-\nu)QP^{-1} - 2(1-2\nu)]^{-1}$ (see Fig. 3.3, where a graph of the function $P^{-1}s$ is shown for $\nu = 1/3$ and the parameter QP^{-1} equal to 6,10,13 (curves 1,2,3, respectively); the stresses $\sigma_{\varphi\varphi}$ and $\sigma_{\rho\rho}$ are evaluated from (3.1.61)).

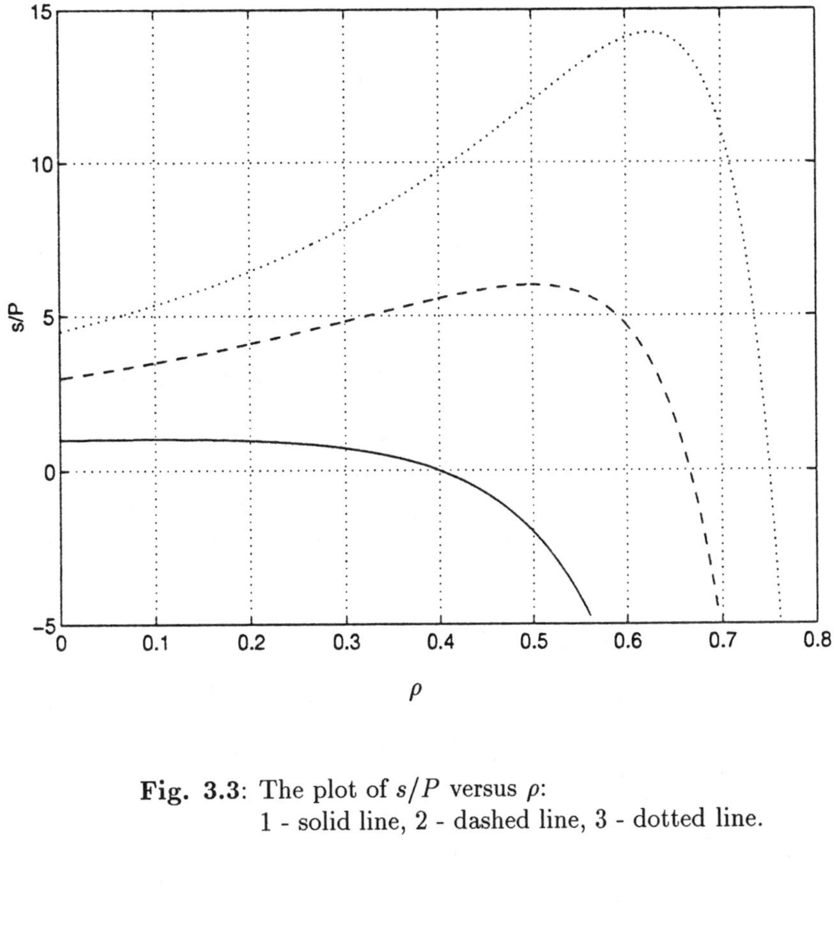

Fig. 3.3: The plot of s/P versus ρ:
1 - solid line, 2 - dashed line, 3 - dotted line.

We conclude that the fracture is possible at some distance from the vertex O when $\varepsilon^2|\ln\, d| \ll 1$; it is characterized by small surface-breaking cracks perpendicular to the surface $\{\rho = \rho_0,\ \theta = \sin^{-1}\varepsilon\}$. As the ratio Q/P increases, the point ρ_0 moves from the vertex O to the boundary of the half-space. This effect of a shift of a fracture zone was also observed in experiments (see [116], [124]).

3.2 Asymptotics of singularity exponents at the tip of an angular crack

The asymptotic expansion technique described above can be also applied to problems of mechanics of cracks, for example, when the cone k coincides with a plane crack $M_\alpha = \{\boldsymbol{x} \in \mathbb{R}^3 : x_3 = 0,\ |\varphi| \le \alpha/2\}$. In this case the spectral problem (3.1.3)–(3.1.5) reduces to

$$\mathcal{P}(\Lambda)\boldsymbol{v} = \mathbf{0} \text{ in } G_\alpha,\ \ \mathcal{Q}(\Lambda)\boldsymbol{v} = \mathbf{0} \text{ on } \partial G_\alpha, \tag{3.2.1}$$

with

$$G_\alpha = \mathbb{S} \setminus M_\alpha,\ \ \partial G_\alpha = \mathbb{S} \cap M_\alpha,$$

which is equivalent to the problem $\mathbf{L}(\rho^\Lambda \boldsymbol{v}) = \mathbf{0}$ in $\mathbb{R}^3 \setminus M_\alpha$, $\mathbf{B}(\rho^\Lambda \boldsymbol{v}) = 0$ on M_α.

In this section we shall construct the asymptotic representations of the eigenvalues $\Lambda(\alpha)$ of the spectral problem (3.2.1) when $\alpha \to 0$ (the complete cone $\mathbb{R}^3 \setminus O$ is disturbed), $\alpha \to \pi$ (an elastic space with a cut in the form of a half-plane) and $\alpha \to 2\pi$ (two semi-spheres). Because the asymptotic scheme is similar to the algorithm presented in Section 3.1, here we will give only the brief description of the results. For additional reading the papers [55], [62], [88], [113] and [10] are recommended.

3.2.1 Limit cases $\alpha \to 0$ and $\alpha \to 2\pi$

At $\alpha \to 2\pi$ the region G_α consists of two semi-spheres $\mathbb{S}_+$ and $\mathbb{S}_-$ connected by a thin bridge. In this case, in contrast to the problem considered in Section 3.1, the stress singularity exponent near the tip of the crack M_α is characterized by the perturbation of the eigenvalue $\Lambda = 0$ of the limit problem (3.2.1) on $\mathbb{S}_+$, $\mathbb{S}_-$. There are six eigenvectors $\boldsymbol{v}^{(j)} = \boldsymbol{e}^{(j)}$ and $\boldsymbol{v}^{(j+3)} = \text{sign } x_3\ \boldsymbol{e}^{(j)}$, $j = 1, 2, 3$, corresponding to the eigenvalue $\Lambda = 0$. The fields $\boldsymbol{e}^{(j)}$, $j = 1, 2, 3$, satisfy the problem (3.2.1) in the region $\mathbb{S} \setminus M_\alpha$ at $\alpha < 2\pi$. The other three eigenvectors are not continuous in $\mathbb{S} \setminus M_\alpha$. In the vicinity of the point $\mathcal{M} = (-1, 0, 0)$ on the unit sphere $\mathbb{S}$, we introduce new scaled variables $\boldsymbol{\xi} = (\xi_1, \xi_2) = \varepsilon^{-1}(\eta_1, \eta_2)$, with $\eta_1 = -x_2$, $\eta_2 = -x_3$, and construct the solution of the boundary layer type. In the stretched coordinates $\boldsymbol{\xi}$ the domain G_α is transformed to a plane with two collinear semi-infinite cracks $\Xi = \{\boldsymbol{\xi} \in \mathbb{R}^2 : \xi_2 = 0, |\xi_1| \ge 1\}$. The behaviour of the stress and displacement fields near the bridge is described in terms of the solutions to the plane-strain and anti-plane shear elasticity problems in the domain $\mathbb{R}^2 \setminus \Xi$. Here we do not discuss the details of the asymptotic analysis of these problems and refer the reader to the paper [88]. With the use of the results of [88] and the asymptotic algorithm described above, we obtain that the eigenvalues $\Lambda^{(k)}(\alpha)$ of the problem (3.2.1) corresponding to the normal opening and to the shear along the Ox_1 and Ox_2 axes (k = 1,2,3, respectively) are specified by

$$\Lambda^{(k)}(\alpha) = [2(2\ln 2 - \ln(2 - \alpha\pi^{-1}))]^{-1} + O(|\ln(2 - \alpha\pi^{-1})|^{-2}), \quad (3.2.2)$$

$$k = 1, 2, 3.$$

It should be observed that at a sufficiently small $2\pi - \alpha$ there are no other eigenvalues in the interval $(-1 + \delta,\ 1 - \delta)$.

The asymptotic formulae for the singularity exponents, as $\alpha \to 0$, have been obtained in Section 3.1. They are

$$\begin{gathered} \Lambda^{(k)}(\alpha) = 1 + \alpha^2\pi^{-2}\Lambda_2^{(k)} + O(\alpha^3), \ k = 1, 2, 3, \\ \Lambda_2^{(1)} = -1/4, \ \Lambda_2^{(2)} = -\mu[4(\lambda + \mu)]^{-1}, \\ \Lambda_2^{(3)} = -(2\lambda^2 + 9\mu\lambda + 5\mu^2)[4(\lambda + 2\mu)(\lambda + 3\mu)]^{-1}. \end{gathered} \quad (3.2.3)$$

Note that for a sufficiently small α there are no other eigenvalues in the interval $(-1, 2 - \delta)$ besides those listed above.

3.2.2 Variation of the eigenvalues for $\alpha \in (0, 2\pi)$

Suppose now that the angle α belongs to the interval $(0, 2\pi)$. An increase of α will make the cut $m_\alpha = \mathbb{S} \cap M_\alpha$ on the sphere $\mathbb{S}$ longer. Such perturbation of the region is regular. However, it is convenient to regard the domain $G_{\alpha(1+\varepsilon)}$ as a singular perturbation of the region G_α. Let $\mathbf{v}^{(j)}$, $j = 1, .., n$, be the eigenvectors of the problem (3.2.1) corresponding to the eigenvalue $\Lambda(\alpha)$. Assume that there are no generalized eigenvectors and denote the eigenvectors of the problem, adjoint to (3.2.1), by $\mathbf{V}^{(1)}, ..., \mathbf{V}^{(n)}$. (Note that the adjoint problem is the problem (3.2.1), with the complex parameter $\Lambda(\alpha)$ replaced by $-1 - \overline{\Lambda(\alpha)}$.) Let the fields $\mathbf{V}^{(i)}$ be normalized by the condition

$$\int_{G_\alpha} \overline{\mathbf{V}^{(j)}} \cdot \frac{d\mathcal{P}}{d\Lambda}(\Lambda)\mathbf{v}^{(k)} ds + \int_{\partial G_\alpha} \overline{\mathbf{V}^{(j)}} \cdot \frac{d\mathcal{Q}}{d\Lambda}(\Lambda)\mathbf{v}^{(k)} dl = \delta_{j,k},$$

$$j,\ k = 1, ..., n. \quad (3.2.4)$$

In a neighbourhood of the ends $N_\alpha^\pm$ of the cut m_α, the vectors $\boldsymbol{v}^{(j)}$, $\mathbf{V}^{(j)}$ admit the representations

$$\boldsymbol{v}^{(j)} = \mathbf{a}^\pm + \sum_{i=1}^{3} K_{i,\pm}^{(j)} r_\pm^{1/2} \boldsymbol{\Phi}^{(i)}(\varphi_\pm) + O(r_\pm),$$

$$\mathbf{V}^{(j)} = \mathbf{b}^\pm + \sum_{j=1}^{3} k_{i,\pm}^{(j)} r_\pm^{1/2} \boldsymbol{\Phi}^{(i)}(\varphi_\pm) + O(r_\pm),$$

where $(r_\pm, \varphi_\pm)$ are polar coordinates in the planes tangent to $\mathbb{S}$ at the points $\mathcal{N}_\alpha^\pm$, $K_{i,\pm}^{(j)}, k_{i,\pm}^{(j)}$ are constant, and the fields $\boldsymbol{\Phi}^{(i)}$ are specified by

$$(\Phi_r^{(1)}(\varphi), \Phi_\varphi^{(1)}(\varphi)) = (4\mu)^{-1}(2\pi)^{-1/2}((2\varkappa - 1)\cos(\varphi/2) - \cos(3\varphi/2),$$

$$-(2\varkappa + 1)\sin(\varphi/2) + \sin(3\varphi/2)), \quad \Phi_z^{(1)}(\varphi) = 0,$$

$$(\Phi_r^{(2)}(\varphi), \Phi_\varphi^{(2)}(\varphi)) = (4\mu)^{-1}(2\pi)^{-1/2}(3\sin(3\varphi/2) - (2\varkappa - 1)\sin(\varphi/2),$$

$$3\cos(3\varphi/2) - (2\varkappa + 1)\cos(\varphi/2)), \quad \Phi_z^{(2)}(\varphi) = 0,$$

$$\Phi_r^{(3)}(\varphi) = \Phi_\varphi^{(3)}(\varphi) = 0, \quad \Phi_z^{(3)}(\varphi) = \mu^{-1}(\pi/2)^{-1/2}\sin(\varphi/2).$$

Here $\varkappa = (\lambda + 3\mu)(\lambda + \mu)^{-1}$, and the axis Oz is directed along the ray passing the point $\mathcal{N}_\alpha^\pm$.

Far from the ends of the cut $m_{\alpha(1+\varepsilon)}$ the eigenvector of problem (3.2.1) in $G_{\alpha(1+\varepsilon)}$ can be found as a sum

$$\boldsymbol{v}(\varepsilon, \theta, \varphi) \sim \boldsymbol{u}^{(0)} + \boldsymbol{u}^{(1)} + \ldots, \quad \boldsymbol{u}^{(0)} = \sum_{j=1}^{n} h_j \boldsymbol{v}^{(j)}. \tag{3.2.5}$$

The vector $\boldsymbol{u}^{(0)}$ does not satisfy the boundary conditions at the ends of the cut $m_{\alpha(1+\varepsilon)}$. As the result, boundary layers are constructed in the vicinity of the points $\mathcal{N}_\alpha^\pm$ (the method is similar to one described in Section 3.1). The correction term $\Lambda_1(\alpha)$ in the expansion

$$\Lambda(\alpha(1+\varepsilon)) \sim \Lambda(\alpha) + \varepsilon\Lambda_1(\alpha) + \ldots, \tag{3.2.6}$$

and the vector $\boldsymbol{u}^{(1)}$ from (3.2.5) correspond to the problem

$$\mathcal{P}(\Lambda(\alpha))\boldsymbol{u}^{(1)} + \Lambda_1(\alpha)\mathcal{P}'(\Lambda(\alpha))\boldsymbol{u}^{(0)} = \mathbf{0} \ \text{ in } G_\alpha, \tag{3.2.7}$$

$= 0$ on ∂G_α. (3.2.8)

...holds:

$(\varphi_\pm) + O(1)$. (3.2.9)

... and $\Psi^{(j)}$ have the form

$^2(-(2\varkappa+1)\cos(3\varphi/2)$
$- 3\sin(\varphi/2)),\ \Psi_z^{(1)} = 0,$
$^2((2\varkappa+1)\sin(3\varphi/2)$
$- \cos(\varphi/2)),\ \Psi_z^{(2)} = 0,$
$-(2\pi)^{-1/2}\sin(\varphi/2).$

... (3.2.7), (3.2.8) are obtained ... (3.2.7) and the vector $\overline{\mathbf{V}}^{(j)}$, ... $r_\pm < \delta\}$ and passing to the ... conditions appear as a set of

... $(\alpha)h_q = 0$. (3.2.10)

The eigenvalues of this $n \times n$ system are the factors at ε in relation (3.2.6).

For $\alpha = \pi$, with the use of the results of [121, Chapter 14] and [102, §6.4], one can obtain that $n = 3$, $K_{1,\pm}^{(1)} = K_{2,\pm}^{(2)} = K_{3,\pm}^{(3)} = 1$, $K_{i,\pm}^{(j)} = 0$,

$$k_{1,\pm}^{(1)} = -2\mu(\varkappa+1)^{-1},\ k_{2,\pm}^{(2)} = -2\mu(4\mu+\lambda)(\varkappa+1)^{-1}(4\mu+3\lambda)^{-1},$$
$$k_{3,\pm}^{(3)} = \mu(4\mu+5\lambda)(4\mu+3\lambda)^{-1},\ k_{i,\pm}^{(j)} = 0,\ i \neq j,\ i,\ j = 1,2,3,$$

and therefore,

$$\Lambda_1^{(1)}(\pi) = 1,\ \Lambda_1^{(2)}(\pi) = (4\mu+\lambda)(4\mu+3\lambda)^{-1},$$
$$\Lambda_1^{(3)}(\pi) = (4\mu+5\lambda)(4\mu+3\lambda)^{-1}. \quad (3.2.11)$$

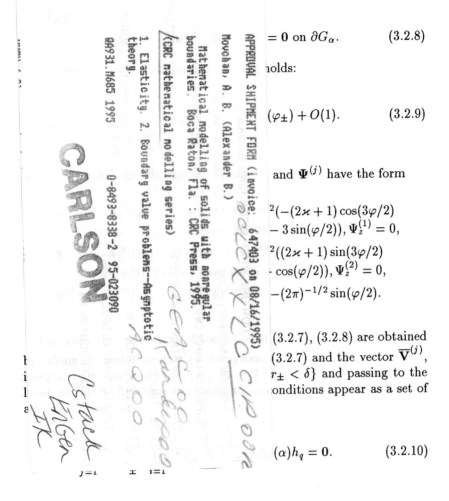

The first term $\Lambda(\alpha)$ in the right–hand side of the relation (3.2.6) for $\alpha = \pi$ is specified by

$$\Lambda(\pi) = 1/2.$$

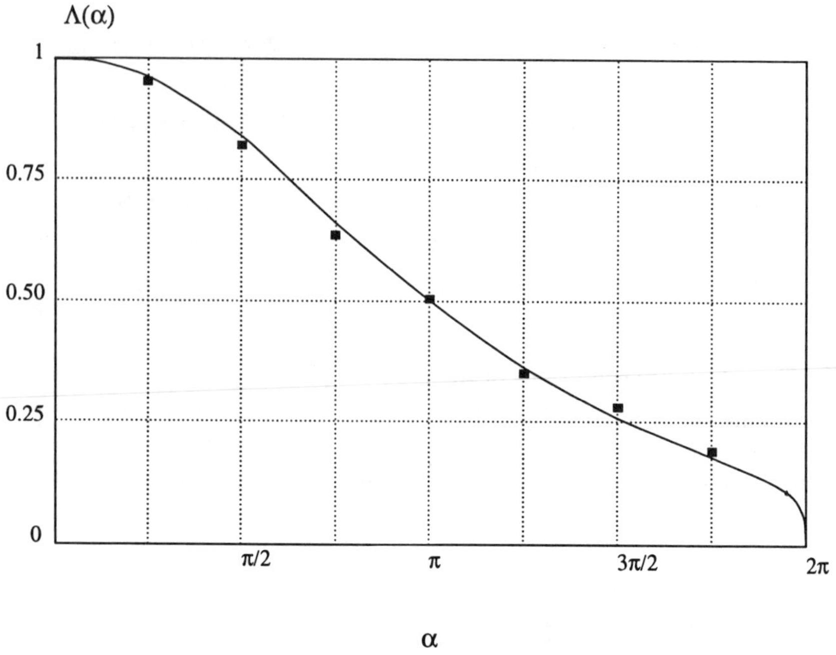

Fig. 3.4: The plot of the exponent Λ versus α.

3.2.3 Numerical results

In [113], [10] (see also [103, p. 324]) the exponent $\Lambda(\alpha)$ was calculated for angular cracks under a Mode-I loading (marked by squares in Fig. 3.4).

Figure 3.5 shows asymptotic curves plotted on the basis of (3.2.2), (3.2.3) and (3.2.11) (1 refers to the normal opening (in this case, there is no correlation with Poisson's ratio ν), 2 and 3 refer to the shear along the axes Ox_1 and Ox_2, with $\nu = 1/3$).

We see from Fig. 3.5 that the smallest eigenvalue, generating a stress singularity, is associated with a shear load. The above asymptotic representations can be applied to small perturbations of the angle α.

However, the range of α that can be treated by these formulae is much wider. We can split the asymptotics obtained in the case of the normal opening for $\alpha \to 0$ and $\alpha \to \pi$ by means of the cut-off functions χ and $1 - \chi$. The function χ should, naturally, be subjected to constraints $\chi(0) = 1$, $\chi(\pi) = 0$, $\chi'(0) = \chi'(\pi) = 0$. Such polynomial of the least degree is equal to $1 - 3\pi^{-2}\alpha^2 + 2\pi^{-3}\alpha^3$. The result is presented in Fig. 3.4: the curve approximates quite well the numerical solution. Since the estimate of the residue in expression (3.2.2) is weak, near the point $\alpha = 2\pi$ we take two-term asymptotic with error $O(|\ln(2 - \alpha\pi^{-1})|^{-3})$.

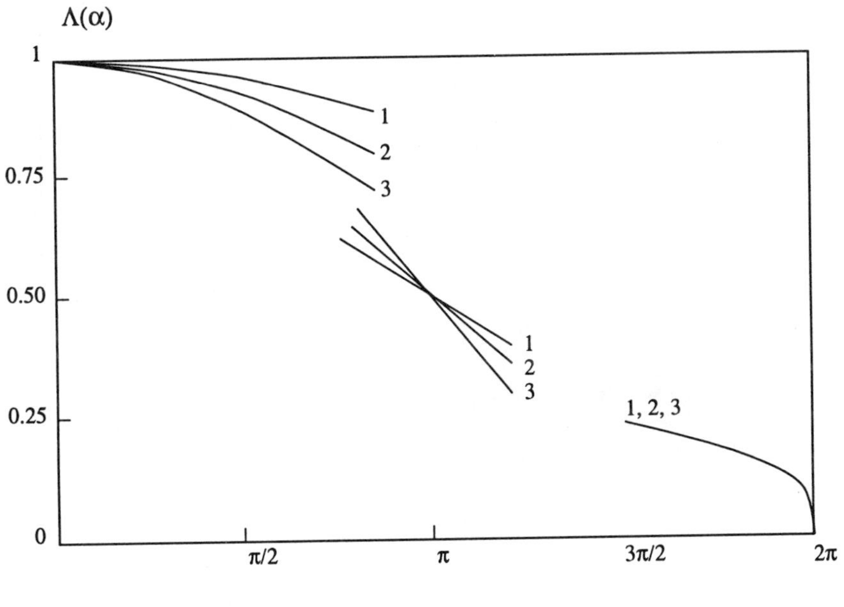

Fig. 3.5: Asymptotic approximations for different intervals of values α.

3.3 Equations of linear elasticity in thin domains

In the present section we discuss the asymptotic algorithm based on the polynomial structure of solutions to the Lamé system in cylindrical domains. It can be used for the derivation of equations of thin plates, shells and elastic beams. This derivation is based on the asymptotic analysis of the Lamé system in a thin domain and provides the justification of the engineering approach. The technique is well known, and we refer to the papers [90], [89], [3], [49]. Simple examples are considered for homogeneous elastic beams and composite elastic beams with a layered structure.

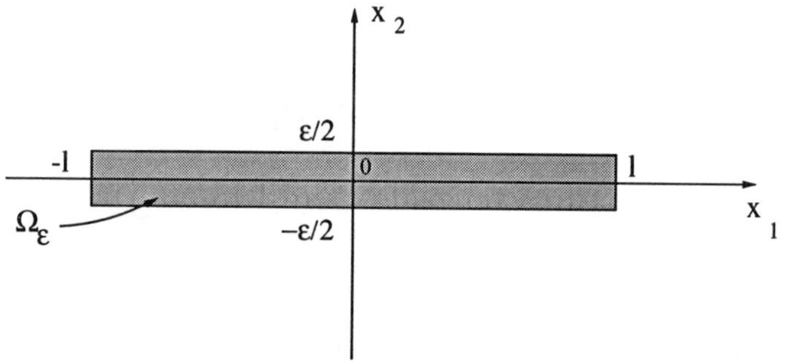

Fig. 3.6: A thin elastic rectangular region.

Here, the elastic material is assumed to be isotropic. However, the same approach works for the case of anisotropic media. We present in the Appendix the REDUCE program which shows how to use the method with the software of symbolic manipulations.

3.3.1 Anti-plane shear of a thin rectangular region with mixed boundary conditions

Consider a rectangular region $\Omega_\varepsilon = \{\boldsymbol{x} \in \mathbb{R}^2 : |x_1| < 1,\ |x_2| < \frac{1}{2}\varepsilon\}$, where $0 < \varepsilon \ll 1$ is a small parameter (see Fig. 3.6). Assume that Ω_ε is occupied by an elastic material with the Lamé moduli λ and μ, and it is subjected to the anti-plane shear. Thus, the first two components of the displacement vector are equal to zero, and the third component (it will be denoted by w) satisfies a mixed boundary value problem in Ω_ε :

$$-\mu\Delta w(\boldsymbol{x}, \varepsilon) = 0, \quad \boldsymbol{x} \in \Omega_\varepsilon, \tag{3.3.1}$$

$$\mu\frac{\partial w}{\partial x_2}(x_1, \pm\tfrac{1}{2}\varepsilon) = \varepsilon p_\pm(x_1), \quad x_1 \in (-1,1), \tag{3.3.2}$$

$$w(\pm 1, x_2) = \phi_\pm(\varepsilon^{-1}x_2), \quad |x_2| < 1/2\varepsilon. \tag{3.3.3}$$

The functions $p_\pm, \phi_\pm$ are assumed to be sufficiently smooth. The leading order differential operators correspond to differentiation in the transversal direction (the Ox_2 axis). We seek the solution of (3.3.1)–(3.3.3) in the form

$$w(\boldsymbol{x}, \varepsilon) \sim w_0(x_1) + \varepsilon^2(W(x_1, t) + w_1(x_1)), \tag{3.3.4}$$

where $t = \varepsilon^{-1}x_2$. Direct substitution of (3.3.4) into the Laplace equation and the Neumann boundary conditions yields

$$\partial_t^2 W(x_1, t) = -\partial_{x_1}^2 w_0(x_1), \quad |t| < 1/2, \tag{3.3.5}$$

$$\partial_t W(x_1, \pm 1/2) = \mu^{-1}p_\pm(x_1), \quad |x_1| < 1. \tag{3.3.6}$$

The problem (3.3.5), (3.3.6) is solvable if and only if

$$-\int_{-1/2}^{1/2} \partial_{x_1}^2 w_0(x_1)dt = (p_+(x_1) - p_-(x_1))/\mu. \tag{3.3.7}$$

This solvability condition yields the second-order differential equation for the leading order term in (3.3.4):

$$\partial_{x_1}^2 w_0(x_1) = -\mu^{-1}\{p_+(x_1) - p_-(x_1)\}. \tag{3.3.8}$$

We also need the boundary conditions at the points $x_1 = \pm 1$ for the function w_0. Direct substitution into (3.3.3) shows that $w_0(x_1)$ does not necessarily satisfy the Dirichlet boundary condition. We construct a boundary layer which should compensate the discrepancy, and the condition of the exponential decay will yield the boundary conditions for (3.3.8).

Consider, for example, the left end region of the thin rectangle. Introduce the local system of scaled coordinates $(\xi_1, \xi_2) = (\frac{x_1+1}{\varepsilon}, \frac{x_2}{\varepsilon})$. As $\varepsilon \to +0$, the limit domain coincides with a semi-infinite rectangular strip. Let $\mathcal{V}$ be the solution of the boundary value problem

$$\Delta_\xi \mathcal{V}(\boldsymbol{\xi}) = 0 \quad \text{in } \Pi = \{\boldsymbol{\xi} : |\xi_1| < 1/2, \ \xi_1 > 0\}, \tag{3.3.9}$$

$$\left.\frac{\partial \mathcal{V}}{\partial \xi_2}\right|_{\xi_2 = \pm 1/2} = 0, \quad \xi_1 > 0, \tag{3.3.10}$$

$$\mathcal{V}(0, \xi_2) = \phi_-(\xi_2) - w_0(-1). \tag{3.3.11}$$

The solution admits the representation

$$\mathcal{V}(\boldsymbol{\xi}) = \int_{-1/2}^{1/2} (\phi_-(\xi_2) - w_0(-1)) d\xi_2 + O(e^{-\gamma \xi_1}), \ \gamma > 0. \tag{3.3.12}$$

The field $\mathcal{V}$ of the boundary layer type vanishes at infinity if and only if

$$w_0(-1) = \int_{-1/2}^{1/2} \phi_-(\xi_2) d\xi_2, \tag{3.3.13}$$

which gives the boundary condition for w_0 on the left end of the interval. Similarly, on the right end one can derive

$$w_0(1) = \int_{-1/2}^{1/2} \phi_+(\xi_2) d\xi_2. \tag{3.3.14}$$

The problem (3.3.8), (3.3.13), (3.3.14) describes completely the leading order term of the displacement field within a thin rectangular region subjected to the state of anti-plane shear.

In a similar manner one can consider next terms of the asymptotic series and derive the equation and boundary conditions for the function w_1 from (3.3.4). These results are well-known. Here, we make an emphasis on the leading order term of the asymptotic expansion, and for more detailed analysis, including the general ansatz, we refer the reader to the work of Nazarov [90], [89].

3.3.2 Plane strain state within a "thin" region

In this section we shall analyze the equations of linear elasticity corresponding to the plane strain of a thin rectangular region Ω_ε. For the leading order terms of the displacement components we shall derive the differential equations and the boundary conditions that correspond to a bending or a longitudinal deformation of a thin elastic beam.

Let the displacement vector $\boldsymbol{u} = (u_1, u_2)^T$ satisfy the Lamé system of equations of equilibrium and mixed boundary conditions,

$$\mu\Delta \boldsymbol{u}(\varepsilon, \boldsymbol{x}) + (\lambda + \mu)\nabla\nabla \cdot \boldsymbol{u}(\varepsilon, \boldsymbol{x}) = \boldsymbol{0}, \quad \boldsymbol{x} \in \Omega_\varepsilon, \tag{3.3.15}$$

$$\boldsymbol{\sigma}^{(2)}(\boldsymbol{u}; x_1, \pm\varepsilon/2) = \varepsilon \boldsymbol{p}^{\pm}(x_1), \quad |x_1| < 1, \tag{3.3.16}$$

$$\boldsymbol{u}(\varepsilon, \pm 1, x_2) = \boldsymbol{\phi}^{\pm}(t), \quad |t| < 1/2. \tag{3.3.17}$$

Here $\boldsymbol{\sigma}^{(2)} = (\sigma_{12}, \sigma_{22})^T$ is the stress vector; the scaled variable t is the same as in the previous section. In this case the structure of the asymptotic expansion is more complicated than for the anti-plane shear. It can be shown (see, for example, [90]) that the displacement vector $\boldsymbol{u}$ admits the asymptotic form (it holds in the middle region of a thin rectangle):

$$\boldsymbol{u}(\boldsymbol{x}) \sim \sum_{k\geq 0} \varepsilon^k \left\{ \sum_{q=0}^{2} \varepsilon^q \boldsymbol{u}^{(I,k,q)}\left(x_1, \frac{x_2}{\varepsilon}\right) \right. \tag{3.3.18}$$

$$\left. + \varepsilon^{-2} \sum_{q=0}^{4} \varepsilon^q \boldsymbol{u}^{(II,k,q)}\left(x_1, \frac{x_2}{\varepsilon}\right) + \mathbf{U}^{(k)}\left(x_1, \frac{x_2}{\varepsilon}\right) \right\}.$$

The first two sums in braces provide the solvability of the problem with respect to $\mathbf{U}^{(k)}$ obtained after substitution into the equations and the traction boundary conditions.

It should be noted that the structure of the asymptotics (3.3.18) is closely related to the form of polynomial (with respect to x_1) solutions of the homogeneous Lamé system in an infinite strip with homogeneous traction boundary conditions prescribed on the upper and lower surfaces.

The approximation (3.3.18) does not work near the ends of Ω_ε where the boundary layer is required to satisfy the Dirichlet-type boundary conditions (3.3.17).

First, we shall discuss the leading ansatz of the expansion (3.3.18) and derive the differential equation for the principal part of the displacement field. This differential equation, in fact, is the same as the equation of an elastic beam subjected to an external load applied on the upper and lower surfaces.

Direct substitution of (3.3.18) into the system (3.3.15) and the traction boundary conditions (3.3.16) yields the following set of boundary value problems on $[-1/2, 1/2]$

$$\partial_t^2 u_1^{(j,0,q)}(x_1,t) = -\mu^{-1}(\lambda+\mu)\partial_{tx_1}^2 u_2^{(j,0,q-1)}(x_1,t)$$

$$-\mu^{-1}(\lambda+2\mu)\partial_{x_1}^2 u_1^{(j,0,q-2)}(x_1,t), \quad |t|<1/2, \tag{3.3.19}$$

$$\partial_t u_1^{(j,0,q)}(x_1,\pm 1/2) = -\partial_{x_1} u_2^{(j,0,q-1)}(x_1,\pm 1/2), \tag{3.3.20}$$

and

$$\mu^{-1}(\lambda+2\mu)\partial_t^2 u_2^{(j,0,q)}(x_1,t) = -\mu^{-1}(\lambda+\mu)\partial_{tx_1}^2 u_1^{(j,0,q-1)}(x_1,t)$$

$$-\partial_{x_1}^2 u_2^{(j,0,q-2)}(x_1,t), \quad |t|<1/2, \tag{3.3.21}$$

$$(\lambda+2\mu)\partial_t u_2^{(j,0,q)}(x_1,\pm 1/2) = -\lambda\partial_{x_1} u_1^{(j,0,q-1)}(x_1,\pm 1/2), \tag{3.3.22}$$

where $j = I, II$, and $q = 0, ..., m_j - 1$; $m_1 = 2$, $m_2 = 4$; for the sake of convenience we use the notations $\partial_{x_1} = \partial/\partial x_1, \partial_t = \partial/\partial t$.

Consequently for the first set of vector functions $\boldsymbol{u}^{(I,0,q)}$, $q = 0, 1$, one has

$$\boldsymbol{u}^{(I,0,0)}(x_1,t) = \begin{pmatrix} u_1^{(0)}(x_1) \\ 0 \end{pmatrix}, \tag{3.3.23}$$

$$\boldsymbol{u}^{(I,0,1)}(x_1,t) = \begin{pmatrix} 0 \\ -\frac{\lambda t}{\lambda+2\mu}\partial_{x_1} u_1^{(0)}(x_1) \end{pmatrix},$$

where $u_1^{(0)}$ is some sufficiently smooth function which depends on x_1 only.

For the second set $u^{(II,0,q)}$, $q = 0,1,2,3$, it can be verified by direct calculations that

$$u^{(II,0,0)}(x_1,t) = \begin{pmatrix} 0 \\ u_2^{(0)}(x_1) \end{pmatrix}, \quad u^{(II,0,1)}(x_1,t) = \begin{pmatrix} -t\partial_{x_1} u_2^{(0)}(x_1) \\ 0 \end{pmatrix},$$

$$u^{(II,0,2)}(x_1,t) = \begin{pmatrix} 0 \\ \frac{\lambda}{2(\lambda+\mu)} t^2 \partial_{x_1}^2 u_2^{(0)}(x_1) \end{pmatrix},$$

$$u^{(II,0,3)}(x_1,t) = \begin{pmatrix} \left\{ \frac{4\mu+3\lambda}{2\mu+\lambda}\frac{t^3}{6} - \frac{\lambda+\mu}{2\mu+\lambda}\frac{t}{2} \right\} \partial_{x_1}^3 u_2^{(0)}(x_1) \\ 0 \end{pmatrix}, \tag{3.3.24}$$

where $u_2^{(0)}(x_1)$ is assumed to be sufficiently smooth. It should be emphasized that the structure of (3.3.23), (3.3.24) is closely related to (3.3.60), (3.3.61).

Next, we make the step which involves components of the vector function $\mathbf{U}^{(0)}$ from the expansion (3.3.18). The components $U_1^{(0)}, U_2^{(0)}$ satisfy the following boundary value problems:

$$\partial_t^2 U_1^{(0)}(x_1,t) = 4(2\mu+\lambda)^{-1}(\lambda+\mu)\partial_{x_1}^2 u_1^{(0)}(x_1), \quad |t| < 1/2, \tag{3.3.25}$$

$$\mu\partial_t U_1^{(0)}(x_1, \pm 1/2) = p_1^{\pm}(x_1), \tag{3.3.26}$$

and

$$\frac{2\mu+\lambda}{\mu}\partial_t^2 U_2^{(0)}(x_1,t) = \frac{1}{3}\frac{\lambda+\mu}{2\mu+\lambda}\partial_{x_1}^4 u_2^{(0)}(x_1), |t| < 1/2, \tag{3.3.27}$$

$$(2\mu+\lambda)\partial_t U_2^{(0)}(x_1, \pm 1/2) = p_2^{\pm}(x_1). \tag{3.3.28}$$

To provide the uniqueness, we subject $U_j^{(0)}$, $j = 1,2$, to the following orthogonality conditions:

$$\int_{-1/2}^{1/2} U_j^{(0)}(x_1,t)d\zeta = 0, \quad j = 1,2, \quad |x_1| < 1. \tag{3.3.29}$$

The solvability conditions of the boundary value problems (3.3.25), (3.3.26) and (3.3.27), (3.3.28) take the form

$$4\mu(\lambda+\mu)(2\mu+\lambda)^{-1}\partial_{x_1}^2 u_1^{(0)}(x_1) = p_1^+(x_1) - p_1^-(x_1), \tag{3.3.30}$$

and

$$[3(2\mu+\lambda)]^{-1}\mu(\lambda+\mu)\partial_{x_1}^4 u_2^{(0)}(x_1) = p_2^+(x_1) - p_2^-(x_1), \qquad (3.3.31)$$

where $|x_1| < 1$.

The expansion (3.3.18) shows that if the vertical component of the external force is not self-balanced, then

$$\boldsymbol{u}(\boldsymbol{x}) \sim \varepsilon^{-2} u_2^{(0)}(x_1)\boldsymbol{e}^{(2)},$$

where the function $u_2^{(0)}$ satisfies the fourth-order ordinary differential equation (3.3.31). One can see that (3.3.31) is the well-known beam equation. It was derived on the basis of the asymptotic analysis of the two-dimensional Lamé system in a thin rectangle without additional physical assumptions.

The ansatz of the asymptotic expansion (3.3.18) was constructed in order to satisfy the system (3.3.15) and the traction boundary conditions (3.3.16). However, the above displacement field does not satisfy the required conditions (3.3.17) at the ends of the thin rectangular domain Ω_ε. Similar to Section 3.3.1 we construct the boundary layer. Introduce the scaled coordinates (ξ_1, ξ_2) in the vicinity of the left end of the rectangle (for the right end the consideration is the same). In our case the expansion of the boundary layer type can be found in the form of the asymptotic series

$$\sum_{k=0}^{\infty} \varepsilon^{k-2} \boldsymbol{\mathcal{V}}^{(k)}(\boldsymbol{\xi}).$$

The coefficients $\boldsymbol{\mathcal{V}}^{(k)}$ satisfy the mixed boundary value problems in the semi-infinite strip Π

$$\mu\Delta_{\boldsymbol{\xi}}\boldsymbol{\mathcal{V}}^{(k)}(\boldsymbol{\xi}) + (\lambda+\mu)\nabla\nabla\cdot\boldsymbol{\mathcal{V}}^{(k)}(\boldsymbol{\xi}) = \boldsymbol{0}, \quad \boldsymbol{\xi}\in\Pi, \qquad (3.3.32)$$

$$\boldsymbol{\sigma}^{(2)}(\boldsymbol{\mathcal{V}}^{(k)}, \xi_1, \pm 1/2) = \boldsymbol{0}, \quad \boldsymbol{\xi}\in\partial\Pi, \quad \xi_1 > 0, \qquad (3.3.33)$$

$$\boldsymbol{\mathcal{V}}^{(k)}(0, \xi_2) = \boldsymbol{\Phi}^{(k)}(\xi_2), \qquad (3.3.34)$$

with some right-hand side $\boldsymbol{\Phi}^{(k)}$. Our objective is to construct the leading part of the boundary layer which decays exponentially at infinity and, hence, contributes no error terms in the equations (3.3.15) and the traction boundary conditions (3.3.16).

We use the basic idea, established by Maz'ya and Plamenevskii [62], for calculation of coefficients in the asymptotic representation at infinity (for problems of cracks the method was published earlier by Bueckner [17]). The boundary value problem is uniquely solvable and its solution admits the asymptotic form

$$\mathcal{V}^{(k)} = \boldsymbol{a}^{(k)} + b^{(k)}(-\xi_2, \xi_1)^T + O(\exp(-\gamma\xi_1)), \quad \text{as } \xi_1 \to +\infty, \quad (3.3.35)$$

with $a_1^{(k)}, a_2^{(k)}, b^{(k)}$ being constant, and $\gamma > 0$ (see, for example, [3]).

The formal algorithm requires certain weight functions $\zeta^{(I)}$, $\zeta^{(II)}, \zeta^{(III)}$, and the coefficients $a_i^{(k)}$, $b^{(k)}$ in (3.3.35) can be written in the form

$$a_1^{(k)} = \mathcal{A}_1(\mathbf{\Phi}^{(k)}; \zeta^{(I)}), \; a_2^{(k)} = \mathcal{A}_2(\mathbf{\Phi}^{(k)}; \zeta^{(II)}),$$

$$b^{(k)} = \mathcal{B}(\mathbf{\Phi}^{(k)}; \zeta^{(III)}), \quad (3.3.36)$$

where $\mathcal{A}_1, \mathcal{A}_2$, $\mathcal{B}$ are linear functionals. We shall present the vector functions $\zeta^{(i)}$, $i = I, II, III$, and construct the linear functionals $\mathcal{A}_1, \mathcal{A}_2, \mathcal{B}$ in the next section (see (3.3.68)).

Now, we consider the discrepancy (up to the order $O(1)$) from the expansion (3.3.18) in the Dirichlet boundary conditions (3.3.17). For the left end one has

$$\varepsilon^{-2}\boldsymbol{u}^{(II,0,0)}(-1) + \varepsilon^{-1}(\boldsymbol{u}^{(II,1,0)}(-1) + \boldsymbol{u}^{(II,0,1)}(-1, \xi_2))$$

$$+ \boldsymbol{u}^{(I,0,0)}(-1) + \boldsymbol{u}^{(II,2,0)}(-1) + \boldsymbol{u}^{(II,1,1)}(-1, \xi_2)$$

$$+ \boldsymbol{u}^{(II,0,2)}(-1, \xi_2). \quad (3.3.37)$$

To compensate the discrepancy of order $O(\varepsilon^{-2})$ and $O(\varepsilon^{-1})$ we set

$$u_2^{(0)}(-1) = \frac{\partial u_2^{(0)}}{\partial x_1}(-1) = 0, \quad (3.3.38)$$

$$u_2^{(1)}(-1) = 0. \quad (3.3.39)$$

In a similar way one can derive for the right end that

$$u_2^{(0)}(1) = \frac{\partial u_2^{(0)}}{\partial x_1}(1) = 0, \tag{3.3.40}$$

$$u_2^{(1)}(1) = 0. \tag{3.3.41}$$

The equalities (3.3.38), (3.3.40) yield the set of boundary conditions for the fourth-order differential equation (3.3.31). These are the clamping conditions for an elastic beam.

The functionals $\mathcal{A}_1, \mathcal{A}_2$ and $\mathcal{B}$ have the following properties:

$$\mathcal{A}_i(\boldsymbol{e}^{(j)}; \boldsymbol{\zeta}^{(k)}) = \delta_{ij}\delta_{ki}, \quad i,j = 1,2; \tag{3.3.42}$$

$$\mathcal{B}(\xi_2 \boldsymbol{e}^{(1)}; \boldsymbol{\zeta}^{(k)}) = \delta_{k3}, \quad k = I, II, III. \tag{3.3.43}$$

If one is interested in the derivation of boundary conditions for the longitudinal component of the displacement vector, then the analysis of the discrepancy of order $O(1)$ is needed. The function $\boldsymbol{\Phi}^{(0)}$ is given by

$$\boldsymbol{\Phi}^{(0)}(\xi_2) = \boldsymbol{\phi}(\xi_2) - \boldsymbol{u}^{(I,0,0)}(-1) - \boldsymbol{u}^{(II,2,0)}(-1) - \boldsymbol{u}^{(II,1,1)}(-1, \xi_2)$$
$$- \boldsymbol{u}^{(II,0,2)}(-1, \xi_2). \tag{3.3.44}$$

Thus, it follows from (3.3.44) and (3.3.42), (3.3.43) that

$$u_1^{(0)}(-1) = \mathcal{A}_1(\boldsymbol{\phi}^-; \boldsymbol{\zeta}^{(I)}) - \mathcal{A}_1(\boldsymbol{u}^{(II,2,0)}(-1) + \boldsymbol{u}^{(II,1,1)}(-1, \xi_2)$$
$$+ \boldsymbol{u}^{(II,0,2)}(-1, \xi_2); \boldsymbol{\zeta}^{(I)}). \tag{3.3.45}$$

In a similar way it can be shown that

$$u_1^{(0)}(+1) = \mathcal{A}_1(\boldsymbol{\phi}^+; \boldsymbol{\zeta}^{(1)}) - \mathcal{A}_1(\boldsymbol{u}^{(II,2,0)}(1)$$
$$+ \boldsymbol{u}^{(II,1,1)}(1, \xi_2) + \boldsymbol{u}^{(II,0,2)}(1, \xi_2); \boldsymbol{\zeta}^{(1)}). \tag{3.3.46}$$

It should be mentioned that, in general, the weight functions should be introduced for each end of the rectangle separately. For the sake of simplicity we have assumed a certain symmetry and, hence, we write the same $\boldsymbol{\zeta}^{(I)}$ in the right-hand sides of (3.3.45) and (3.3.46).

The equalities (3.3.45), (3.3.46), deduced from the assumption of the exponential decay of the boundary layer, give the boundary conditions for the second-order differential equation (3.3.30). These relations are important for the case when the beam is subjected to a longitudinal load, and when the transversal component of the traction vector equals zero.

The objective of the next section is to clarify the definition (3.3.36) and describe the structure of polynomial solutions of the Lamé system in an infinite strip.

3.3.3 The homogeneous Neumann problem in an infinite strip

In this auxiliary section we shall discuss some polynomial solutions of the homogeneous Lamé system that have proved to be useful for the description of the displacement field in a thin rectangle.

Assume that the vector function $\boldsymbol{W}(\xi_1, \xi_2)$ satisfies the homogeneous Lamé system and the homogeneous traction boundary conditions in the infinite strip $\{\boldsymbol{\xi} : \xi_1 \in \mathbb{R}^1, |\xi_2| < 1/2\}$.

$$\mu \left(\frac{\partial^2}{\partial \xi_1^2} + \frac{\partial^2}{\partial \xi_2^2} \right) W_i(\xi_1, \xi_2) + (\lambda + \mu) \frac{\partial}{\partial \xi_i} \left(\frac{\partial W_1}{\partial \xi_1}(\xi_1, \xi_2) \right.$$

$$\left. + \frac{\partial W_2}{\partial \xi_2}(\xi_1, \xi_2) \right) = 0, \quad i = 1, 2, \quad \xi_1 \in \mathbb{R}^1, \quad |\xi_2| < 1/2, \quad (3.3.47)$$

$$\mu \left(\frac{\partial W_1}{\partial \xi_2} + \frac{\partial W_2}{\partial \xi_1} \right) (\xi_1, \pm 1/2) = 0, \qquad (3.3.48)$$

$$(2\mu + \lambda) \frac{\partial W_2}{\partial \xi_2}(\xi_1, \pm 1/2) + \lambda \frac{\partial W_1}{\partial \xi_1}(\xi_1, \pm 1/2) = 0.$$

For the sake of convenience we introduce the matrix differential operators $\boldsymbol{L}$ and $\boldsymbol{T}$ for the Lamé system and traction boundary conditions

$$\boldsymbol{L}(\partial_{\xi_1}, \partial_{\xi_2}) = \begin{pmatrix} (2\mu + \lambda)\partial_{\xi_1}^2 + \mu \partial_{\xi_2}^2 & (\lambda + \mu)\partial_{\xi_1}\partial_{\xi_2} \\ (\lambda + \mu)\partial_{\xi_1}\partial_{\xi_2} & (2\mu + \lambda)\partial_{\xi_2}^2 + \mu \partial_{\xi_1}^2 \end{pmatrix}, \qquad (3.3.49)$$

$$\boldsymbol{T}(\partial_{\xi_1}, \partial_{\xi_2}) = \begin{pmatrix} \mu \partial_{\xi_2} & \mu \partial_{\xi_1} \\ \lambda \partial_{\xi_1} & (2\mu + \lambda)\partial_{\xi_2} \end{pmatrix}. \qquad (3.3.50)$$

Then, (3.3.47), (3.3.48) can be rewritten as

$$\boldsymbol{L}(\partial_{\xi_1}, \partial_{\xi_2})\boldsymbol{W}(\xi_1, \xi_2) = \boldsymbol{0}, \quad \xi_1 \in \mathbb{R}^1, \quad |\xi_2| < 1/2, \tag{3.3.51}$$

$$\boldsymbol{T}(\partial_{\xi_1}, \partial_{\xi_2})\boldsymbol{W}(\xi_1, \xi_2)|_{\xi_2=\pm 1/2} = \boldsymbol{0}, \quad \xi_1 \in \mathbb{R}^1. \tag{3.3.52}$$

Further, we shall use the brief notation

$$\{\mathbf{L}(\partial_{\xi_1}, \partial_{\xi_2}),\ \mathbf{T}(\partial_{\xi_1}, \partial_{\xi_2})\}\boldsymbol{W} = \boldsymbol{0}. \tag{3.3.53}$$

We seek the vector polynomials $\boldsymbol{W}$ in the form

$$\boldsymbol{W}(\xi_1, \xi_2) = \sum_{j=0}^{k} \frac{\xi_1^j}{j!} \boldsymbol{P}^{(k-j)}(\xi_2), \quad k \in \mathbb{N}. \tag{3.3.54}$$

Here $\boldsymbol{P}^{(0)}$ is an eigenvector, and $\boldsymbol{P}^{(1)}, \ldots, \boldsymbol{P}^{(n)}$ are generalized eigenvectors corresponding to the eigenvalue $\Lambda = 0$ of the operator $\{\boldsymbol{L}(\Lambda, \partial_{\xi_2}),\ \boldsymbol{T}(\Lambda, \partial_{\xi_2})\}$. Namely,

$$\{\boldsymbol{L}(0, \partial_{\xi_2}),\ \boldsymbol{T}(0, \partial_{\xi_2})\}\boldsymbol{P}^{(0)} = \boldsymbol{0}, \tag{3.3.55}$$

and

$$\{\boldsymbol{L}(0, \partial_{\xi_2}), \boldsymbol{T}(0, \partial_{\xi_2})\}\boldsymbol{P}^{(k)} \tag{3.3.56}$$

$$= -\sum_{j=1}^{k} \frac{1}{j!} \partial_\Lambda^j \{\boldsymbol{L}(\Lambda, \partial_{\xi_2}), \boldsymbol{T}(\Lambda, \partial_{\xi_2})\}\boldsymbol{P}^{(k-j)}|_{\Lambda=0}.$$

It should be mentioned that the number $n+1$ of elements of the set $\{\boldsymbol{P}^{(0)}, .., \boldsymbol{P}^{(n)}\}$ is always finite. In other words, for some k the problem (3.3.56) is unsolvable, and $n = k - 1$.

The system (3.3.55) has two eigenvectors

$$\boldsymbol{P}^{(I,0)} = (1, 0)^T \text{ and } \boldsymbol{P}^{(II,0)} = (0, 1)^T. \tag{3.3.57}$$

It can be verified by direct calculations that the generalized eigenvectors are given by

$$\boldsymbol{P}^{(I,1)} = \begin{pmatrix} 0 \\ -\frac{\lambda}{2\mu+\lambda}\xi_2 \end{pmatrix}, \tag{3.3.58}$$

and

$$\boldsymbol{P}^{(II,1)} = \begin{pmatrix} -\xi_2 \\ 0 \end{pmatrix}, \ \boldsymbol{P}^{(II,2)} = \begin{pmatrix} 0 \\ \frac{\lambda}{2(2\mu+\lambda)}\xi_2^2 \end{pmatrix},$$

$$\boldsymbol{P}^{(II,3)} = \begin{pmatrix} \frac{4\mu+3\lambda}{6(2\mu+\lambda)}\xi_2^3 - \frac{\lambda+\mu}{(2\mu+\lambda)}\xi_2 \\ 0 \end{pmatrix}. \tag{3.3.59}$$

The first eigenvector $\mathbf{P}^{(I,0)}$ has the only generalized one ($K_I = 1$), and the number K_{II} of generalized eigenvectors, corresponding to $\mathbf{P}^{(II,0)}$, is equal to 3. The detailed analysis of spectral problems in cylindrical domains is given in [90], [89]. In the present case the structure of polynomial solutions seems to be very simple. The polynomials (3.3.53) are specified by

$$\begin{pmatrix} 1 \\ 0 \end{pmatrix}, \ \begin{pmatrix} \xi_1 \\ -\frac{\lambda}{2\mu+\lambda}\xi_2 \end{pmatrix}, \tag{3.3.60}$$

and

$$\begin{pmatrix} 0 \\ 1 \end{pmatrix}, \ \begin{pmatrix} -\xi_2 \\ \xi_1 \end{pmatrix}, \ \begin{pmatrix} -\xi_1\xi_2 \\ \frac{1}{2}\left(\xi_1^2 + \frac{\lambda}{2\mu+\lambda}\xi_2^2\right) \end{pmatrix}, \tag{3.3.61}$$

$$\begin{pmatrix} \frac{4\mu+3\lambda}{2\mu+\lambda}\frac{\xi_2^3}{3} - \xi_1^2\xi_2 - \frac{\lambda+\mu}{(2\mu+\lambda)}\xi_2 \\ \frac{\xi_1^3}{3} + \frac{\lambda}{(2\mu+\lambda)}\xi_1\xi_2^2 \end{pmatrix}.$$

Let us mention that the vectors $\boldsymbol{e}^{(j)}, j = 1, 2$, and $-\xi_2\boldsymbol{e}^{(1)} + \xi_1\boldsymbol{e}^{(2)}$ represent rigid body translations and the axial rotation; these fields produce no contribution to the elastic energy. The elastic energy, calculated on the remaining three vector fields (3.3.60), (3.3.61), is infinite. The vectors (3.3.60), (3.3.61) span the set of all polynomial solutions of the homogeneous boundary value problem (3.3.47), (3.3.48).

Now, introduce the following vector functions

$$\boldsymbol{\zeta}^{(I)} = \xi_1\boldsymbol{e}^{(1)} - \frac{\lambda}{2\mu+\lambda}\xi_2\boldsymbol{e}^{(2)} + \boldsymbol{R}^{(1)}(\boldsymbol{\xi}),$$

$$\boldsymbol{\zeta}^{(II)} = \left\{(2\mu+\lambda)^{-1}\left[\frac{1}{3}(4\mu+3\lambda)\xi_2^3 - (\lambda+\mu)\xi_2\right] - \xi_1^2\xi_2\right\}\boldsymbol{e}^{(1)}$$

$$+\left\{\frac{1}{3}\xi_1^3 + (2\mu+\lambda)^{-1}\lambda\xi_1\xi_2^2\right\}\boldsymbol{e}^{(2)} + \boldsymbol{R}^{(2)}(\boldsymbol{\xi}),$$

$$\zeta^{(III)} = -\xi_1\xi_2 e^{(1)} + \frac{1}{2}\left(\xi_1^2 + \frac{\lambda}{2\mu+\lambda}\xi_2^2\right) e^{(2)} + \boldsymbol{R}^{(3)}(\boldsymbol{\xi}), \qquad (3.3.62)$$

which satisfy the homogeneous boundary value problem (3.3.32)–(3.3.34). Note that the remainder terms $\boldsymbol{R}^{(i)}$, $i = 1,2,3$, produce the stress fields that decay exponentially at infinity, and the vectors $\boldsymbol{R}^{(i)}$, $i = 1,2,3$, admit the form

$$\boldsymbol{R}^{(i)}(\boldsymbol{\xi}) = \boldsymbol{d}^{(i)} + c^{(i)}(-\xi_2, \xi_1)^T + O(e^{-\gamma\xi_1}), \quad \gamma > 0,$$

as $\xi \to +\infty$, with $d_1^{(i)}, d_2^{(i)}, c^{(i)}$ being constant.

To obtain the form of the linear functionals (3.3.36) we apply the Betti formula to the "weight functions" $\zeta^{(i)}$ and the boundary layer (3.3.35). We consider a bounded rectangular domain $\mathcal{R}_q = \{\boldsymbol{\xi} : 0 < \xi_1 < q, |\xi_2| < 1/2\}$, where $q \gg 1$. First, we take into account the fields $\zeta^{(1)}$ and $\mathcal{V}^{(k)}$ (for some fixed k). The following relation holds:

$$\begin{aligned} 0 &= \int_{\mathcal{R}_q} \left\{\zeta^{(I)} \cdot (\boldsymbol{L}(\partial_{\xi_1}, \partial_{\xi_2})\mathcal{V}^{(k)}) - \mathcal{V}^{(k)} \cdot (\boldsymbol{L}(\partial_{\xi_1}, \partial_{\xi_2})\zeta^{(I)})\right\} d\xi \\ &= \int_{-1/2}^{1/2} \left\{\zeta_1^{(I)}\sigma_{11}(\mathcal{V}^{(1)}; q, \xi_2) + \zeta_2^{(I)}\sigma_{12}(\mathcal{V}^{(k)}; q, \xi_2)\right. \\ &- \left. \mathcal{V}_1^{(k)}\sigma_{11}(\zeta^{(I)}; q, \xi_2) - \mathcal{V}_2^{(k)}\sigma_{12}(\zeta^{(I)}; q, \xi_2)\right\} d\xi_2 \\ &+ \int_{-1/2}^{1/2} \left\{\Phi_1^{(k)}\sigma_{11}(\zeta^{(I)}; 0, \xi_1) + \Phi_2^{(k)}\sigma_{12}(\zeta^{(I)}; 0, \xi_2)\right\} d\xi_2. \end{aligned} \qquad (3.3.63)$$

Here we used the fact that the boundary layer $\mathcal{V}^{(k)}$ satisfies the homogeneous Lamé system and traction boundary conditions (3.3.33).

It can be verified by direct calculations that the first integral in (3.3.63) is approximated by

$$-\frac{4\mu(\lambda+\mu)}{2\mu+\lambda} a_1^{(k)} + O(e^{-\gamma q}), \quad \gamma > 0, \qquad (3.3.64)$$

where $a_1^{(k)}$ is the coefficient from the asymptotics (3.3.35). Taking the limit $q \to +\infty$, we obtain

$$a_1^{(k)} = \frac{2\mu + \lambda}{4\mu(\lambda + \mu)} \int_{-1/2}^{1/2} \{\Phi_1^{(k)} \sigma_{11}(\zeta^{(I)}, 0, \xi_2) \tag{3.3.65}$$

$$+ \Phi_2^{(k)} \sigma_{12}(\zeta^{(I)}; 0, \xi_2)\} d\xi_2.$$

In a similar way, applying the Betti formula to the vector functions $\zeta^{(II)}, \mathcal{V}^{(k)}$, and then to $\zeta^{(III)}, \mathcal{V}^{(k)}$, we obtain in the limit $q \to +\infty$ that

$$a_2^{(k)} = -\frac{3(2\mu + \lambda)}{\mu(\lambda + \mu)} \int_{-1/2}^{1/2} \{\Phi_1^{(k)} \sigma_{11}(\zeta^{(II)}; 0, \xi_2) \tag{3.3.66}$$

$$+ \Phi_2^{(k)} \sigma_{12}(\zeta^{(II)}; 0, \xi_2)\} d\xi_2,$$

$$b^{(k)} = \frac{3(2\mu + \lambda)}{\mu(\lambda + \mu)} \int_{-1/2}^{1/2} \{\Phi_1^{(k)} \sigma_{11}(\zeta^{(III)}; 0, \xi_2) \tag{3.3.67}$$

$$+ \Phi_2^{(k)} \sigma_{12}(\zeta^{(III)}; 0, \xi_2)\} d\xi_2.$$

Thus, we can see that the linear functionals from (3.3.36) have a very simple form:

$$\mathcal{A}_1(\mathbf{\Phi}; \zeta) = \frac{2\mu + \lambda}{4\mu(\lambda + \mu)} \mathcal{F}(\mathbf{\Phi}; \zeta),$$

$$\mathcal{A}_2(\mathbf{\Phi}; \zeta) = -\frac{3(2\mu + \lambda)}{\mu(\lambda + \mu)} \mathcal{F}(\mathbf{\Phi}; \zeta),$$

$$\mathcal{B}(\mathbf{\Phi}, \zeta) = \frac{3(2\mu + \lambda)}{\mu(\lambda + \mu)} \mathcal{F}(\mathbf{\Phi}; \zeta), \tag{3.3.68}$$

where

$$\mathcal{F}(\mathbf{\Phi}; \zeta) = \int_{-1/2}^{1/2} \{\Phi_1 \sigma_{11}(\zeta; 0, \xi_2) + \Phi_2 \sigma_{12}(\zeta; 0, \xi_2)\} d\xi_2.$$

Now, the orthogonality conditions (3.3.42), (3.3.43) seem to be obvious and can be easily verified by direct substitution.

3.4 Asymptotic analysis of the longitudinal jump in adhesively bonded joints

We shall use the asymptotic algorithm presented in the previous section to study the displacement field within a thin elastic region with the layered structure. Here we analyze the displacement field within a thin compound region which consists of three layers. The middle layer, modelling an adhesive material, is assumed to be relatively thin and relatively soft in comparison with two other layers. The ends of a thin rectangle are fixed, and on the interface surfaces the ideal contact conditions hold. On the upper and lower parts of the boundary the traction boundary conditions are prescribed. The variational approach for the analysis of the adhesively bonded joints was developed in [37]. The asymptotic procedure given here is based on the analysis of equations of linear elasticity in thin regions. The basic elements of this asymptotic algorithm are presented in [59]. Some particular cases were studied in [71], [3].

Two cases of external load will be considered. First, the anti-plane shear of a layered structure is analyzed. Second, we consider the state of plane strain. The systems of ordinary differential equations are derived for the leading order components of the displacement field within a thin compound domain.

3.4.1 Formulation of boundary value problems

In contrast with the previous part we suppose here that a thin rectangular domain Ω_ε consists of three parts:

$$\Omega_0 = \{\boldsymbol{x} \in \mathbb{R}^2 : |x_1| < l, |x_2| < \varepsilon^2 h_0/2\}, \tag{3.4.1}$$

and Ω_i, $i = 1, 2$, such that

$$\Omega_1 = \{\boldsymbol{x} \in \mathbb{R}^2 : |x_1| < l, \varepsilon^2 h_0/2 < x_2 < \varepsilon^2 h_0/2 + \varepsilon h\}, \tag{3.4.2}$$

$$\Omega_2 = \{\boldsymbol{x} \in \mathbb{R}^2 : |x_1| < l, -\varepsilon h - \varepsilon^2 h_0/2 < x_2 < -\varepsilon^2 h_0/2\}, \tag{3.4.3}$$

where the quantities l and h, h_0 have the same order of magnitude. Thus, the interface boundary includes S_+ and S_- specified by

$$S_\pm = \{\boldsymbol{x} : |x_1| < l, x_2 = \pm\varepsilon^2 h_0/2\}.$$

The upper and lower surfaces of the compound region are

$$\Gamma_\pm = \{\boldsymbol{x} : |x_1| < l, x_2 = \pm(\varepsilon^2 h_0/2 + \varepsilon h)\}.$$

The region Ω_ε, described above, is shown in Fig. 3.7.

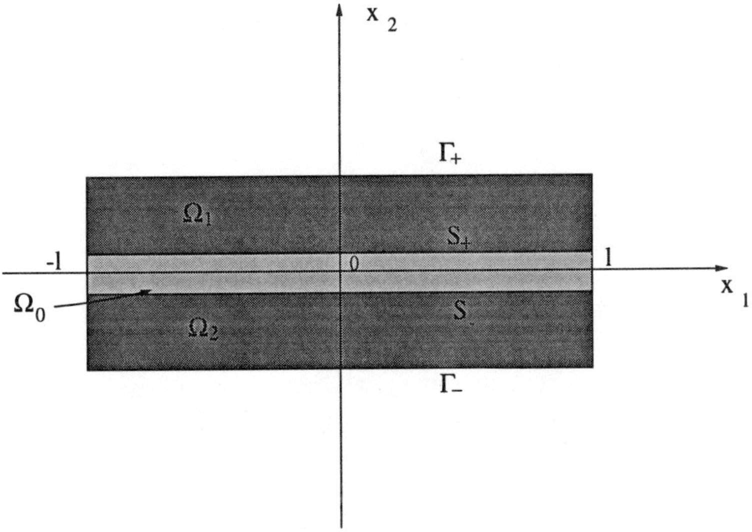

Fig. 3.7: A composite layered structure.

The elastic materials of the regions Ω_i, $i = 0, 1, 2$, are characterized by the Young moduli E_i, $i = 0, 1, 2$, and by the values ν_i, $i = 0, 1, 2$, of the Poisson ratio. Assume that ν_i, $i = 0, 1, 2$, have the same order of magnitude for all three materials, and

$$E_0 = \varepsilon^3 E, \tag{3.4.4}$$

where E is comparable with E_1 and E_2. Here $\varepsilon > 0$ is a small nondimensional parameter. By λ_i, μ_i, $i = 0, 1, 2$, we denote the Lamé constants of the elastic materials

$$\lambda_i = \frac{E_i \nu_i}{(1+\nu_i)(1-2\nu_i)}, \quad \mu_i = \frac{E_i}{2(1+\nu_i)}.$$

Also, we use the notations λ, μ for the normalized Lamé constants related to the middle layer:

$$\lambda = \frac{E\nu_0}{(1+\nu_0)(1-2\nu_0)}, \quad \mu = \frac{E}{2(1+\nu_0)}.$$

Consider the following two formulations:

1. Antiplane shear. Assume that the third component $U^{(i)}$, $i = 0, 1, 2$, of the displacement vector satisfies the equations

$$\nabla^2 U^{(i)}(\boldsymbol{x}) = 0, \ \boldsymbol{x} \in \Omega_i, \ i = 0, 1, 2, \tag{3.4.5}$$

$$\mu_1 \frac{\partial}{\partial x_2} U^{(1)}(x_1, \varepsilon h + \frac{1}{2}\varepsilon^2 h_0) = \varepsilon p^{(1)}, \ |x_1| < l,$$

$$\mu_2 \frac{\partial}{\partial x_2} U^{(2)}(x_1, -\varepsilon h - \frac{1}{2}\varepsilon^2 h_0) = \varepsilon p^{(2)}, \ |x_1| < l, \tag{3.4.6}$$

The conditions of ideal physical contact are prescribed on the interface boundary

$$\mu_1 \frac{\partial}{\partial x_2} U^{(1)}(\boldsymbol{x}) = \mu_0 \frac{\partial}{\partial x_2} U^{(0)}(\boldsymbol{x}), U^{(1)}(\boldsymbol{x}) = U^{(0)}(\boldsymbol{x}), \boldsymbol{x} \in S_+,$$

$$\mu_2 \frac{\partial}{\partial x_2} U^{(2)}(\boldsymbol{x}) = \mu_0 \frac{\partial}{\partial x_2} U^{(0)}(\boldsymbol{x}), U^{(2)}(\boldsymbol{x}) = U^{(0)}(\boldsymbol{x}), \boldsymbol{x} \in S_-. \tag{3.4.7}$$

The ends of the thin rectangle are fixed

$$U^{(i)}(\pm l, x_2) = 0, \ \ \boldsymbol{x} \in \overline{\Omega_i}, \ i = 0, 1, 2. \tag{3.4.8}$$

2. Plane strain state. The displacement vector has the form

$$\boldsymbol{u}^{(i)} = (u_1(x_1, x_2), u_2(x_1, x_2), 0)^T, \ \ i = 0, 1, 2, \tag{3.4.9}$$

and the vectors $\boldsymbol{u}^{(i)}$ satisfy the homogeneous Lamé systems

$$\mu_i \nabla^2 \boldsymbol{u}^{(i)}(\boldsymbol{x}) + (\lambda_i + \mu_i)\nabla\nabla \cdot \boldsymbol{u}^{(i)}(\boldsymbol{x}) = \boldsymbol{0}, \ \ \boldsymbol{x} \in \Omega_i, \ \ i = 0, 1, 2. \tag{3.4.10}$$

On the upper and lower surfaces of the compound region Ω_ε we prescribe tractions:

$$\mu_1(\frac{\partial}{\partial x_1}u_2^{(1)} + \frac{\partial}{\partial x_2}u_1^{(1)}) = \varepsilon p_1^{(1)} \text{ on } \Gamma_+, \tag{3.4.11}$$

$$(2\mu_1 + \lambda_1)\frac{\partial}{\partial x_2}u_2^{(1)} + \lambda_1\frac{\partial}{\partial x_1}u_1^{(1)} = \varepsilon^3 p_2^{(1)} \text{ on } \Gamma_+, \tag{3.4.12}$$

$$\Gamma_+ = \{\boldsymbol{x} : x_2 = \varepsilon^2 h_0/2 + \varepsilon h, |x_1| < l\},$$

and

$$\mu_2(\frac{\partial}{\partial x_1}u_2^{(2)} + \frac{\partial}{\partial x_2}u_1^{(2)}) = \varepsilon p_1^{(2)} \text{ on } \Gamma_-, \tag{3.4.13}$$

$$(2\mu_2 + \lambda_2)\frac{\partial}{\partial x_2}u_2^{(2)} + \lambda_2\frac{\partial}{\partial x_1}u_1^{(2)} = \varepsilon^3 p_2^{(2)} \text{ on } \Gamma_-, \tag{3.4.14}$$

$$\Gamma_- = \{\boldsymbol{x} : x_2 = -\varepsilon^2 h_0/2 - \varepsilon h, |x_1| < l\}.$$

It should be emphasized that we make the preference to the special kind of external load where the longitudinal external force is much greater than the transversal force. We shall consider the case of a bending load separately.

The conditions of the ideal interface contact have the form

$$\boldsymbol{u}^{(1)}(\boldsymbol{x}) = \boldsymbol{u}^{(0)}(\boldsymbol{x}),$$

$$\mu_1(\frac{\partial}{\partial x_1}u_2^{(1)} + \frac{\partial}{\partial x_2}u_1^{(1)}) = \mu_0(\frac{\partial}{\partial x_1}u_2^{(0)} + \frac{\partial}{\partial x_2}u_1^{(0)}),$$

$$(2\mu_1 + \lambda_1)\frac{\partial}{\partial x_2}u_2^{(1)} + \lambda_1\frac{\partial}{\partial x_1}u_1^{(1)} = (2\mu_0 + \lambda_0)\frac{\partial}{\partial x_2}u_2^{(0)} + \lambda_0\frac{\partial}{\partial x_1}u_1^{(0)},$$

$$\boldsymbol{x} \in S_+, \tag{3.4.15}$$

and

$$\boldsymbol{u}^{(2)}(\boldsymbol{x}) = \boldsymbol{u}^{(0)}(\boldsymbol{x}),$$

$$\mu_2(\frac{\partial}{\partial x_1}u_2^{(2)} + \frac{\partial}{\partial x_2}u_1^{(2)}) = \mu_0(\frac{\partial}{\partial x_1}u_2^{(0)} + \frac{\partial}{\partial x_2}u_1^{(0)}),$$

$$(2\mu_2 + \lambda_2)\frac{\partial}{\partial x_2}u_2^{(2)} + \lambda_2\frac{\partial}{\partial x_1}u_1^{(2)} = (2\mu_0 + \lambda_0)\frac{\partial}{\partial x_2}u_2^{(0)} + \lambda_0\frac{\partial}{\partial x_1}u_1^{(0)},$$

$$\boldsymbol{x} \in S_-. \tag{3.4.16}$$

The ends of a thin compound beam are fixed

$$\boldsymbol{u}^{(i)}(\pm l, x_2) = \mathbf{0}, \ \boldsymbol{x} \in \overline{\Omega_i}, \ i = 0, 1, 2. \tag{3.4.17}$$

In the next section we shall introduce scaled variables in the direction of the Ox_2 axis, which would allow us to split the differential operators into the groups related to longitudinal and transversal displacements.

3.4.2 Splitting up procedure

The structure of the asymptotic expansion for the displacement components follows from the asymptotic algorithm presented in Section 3.3.

Introduce the variables

$$t_0 = \varepsilon^{-2} x_2, \ t_1 = \varepsilon^{-1}(x_2 - (\varepsilon^2 h_0 + \varepsilon h)/2),$$

and

$$t_2 = \varepsilon^{-1}(x_2 + (\varepsilon^2 h_0 + \varepsilon h)/2).$$

In this case

$$t_i \in [-h/2, h/2], \ i = 1, 2; \ \ t_0 \in [-h_0/2, h_0/2]$$

within $\overline{\Omega_i}$, $i = 0, 1, 2$, and one can see that

$$\frac{\partial}{\partial x_2} = \varepsilon^{-2} \frac{\partial}{\partial t_0}, \ \frac{\partial}{\partial x_2} = \varepsilon^{-1} \frac{\partial}{\partial t_i}, \ i = 1, 2. \tag{3.4.18}$$

Then, the differential operators of equations within Ω_i and of the boundary conditions, including those prescribed on the interface surface, can be split up into the longitudinal and transversal parts. Also, assume that for the anti–plane shear the displacement is approximated by

$$U^{(i)}(\boldsymbol{x}) \sim U^{(i,0)}(x_1, t_i) + \varepsilon U^{(i,1)}(x_1, t_i) + \varepsilon^2 U^{(i,2)}(x_1, t_i),$$

$$i = 0, 1, 2. \tag{3.4.19}$$

In contrast, in the case of plane strain we use the following asymptotic formulae

$$\boldsymbol{u}^{(i)}(\boldsymbol{x}) \sim \sum_{k=0}^{\infty} \varepsilon^k (\mathbf{U}^{(i,k)}(\varepsilon, x_1, t_i) + \mathbf{V}^{(i,k)}(\varepsilon, x_1, t_i)), \ i = 0, 1, 2, \tag{3.4.20}$$

here $\mathbf{U}^{(i,k)}$ and $\mathbf{V}^{(i,k)}$ are "related" to the bending mode and to the longitudinal deformation, respectively. We consider the first ansatz of the above series and use the notations

$$\mathcal{U}^{(i)} := \mathbf{U}^{(i,0)}, \quad \mathcal{V}^{(i)} := \mathbf{V}^{(i,0)}. \tag{3.4.21}$$

Following the idea described in Section 3.3, we take the vector-valued functions (3.4.21) in the form

$$\mathcal{U}^{(i)} = \sum_{j=0}^{2} \varepsilon^j \mathcal{U}^{(i,j)}(x_1, t_i),$$

$$\mathcal{V}^{(i)} = \sum_{j=0}^{4} \varepsilon^j \mathcal{V}^{(i,j)}(x_1, t_i), \tag{3.4.22}$$

where

$$\mathcal{U}^{(i,0)} = \mathcal{U}_1^{(i,0)} \mathbf{e}^{(1)}, \quad \mathcal{V}^{(i,0)} = \mathcal{V}_2^{(i,0)} \mathbf{e}^{(2)}, \tag{3.4.23}$$

with $\mathbf{e}^{(i)}, i = 1, 2$, being the unit basis vectors of the Cartesian system of coordinates.

Direct substitution of (3.4.19) and (3.4.20) into the equations and boundary conditions yields the following systems of recurrent relations in Ω_i, $i = 0, 1, 2$, and on the parts of the boundary (including the interface surface) which are parallel to the Ox_1 axis.

1. Antiplane shear. The coefficients of the expansion (3.4.19) satisfy the system of equations

$$\frac{\partial^2}{\partial t_i^2} U^{(i,k)} = -\frac{\partial^2}{\partial x_1^2} U^{(i,k-2)} \text{ in } \Omega_i, \; i = 1, 2, \tag{3.4.24}$$

$$\frac{\partial^2}{\partial t_0^2} U^{(0,k)} = -\frac{\partial^2}{\partial x_1^2} U^{(0,k-4)} \text{ in } \Omega_0. \tag{3.4.25}$$

The boundary conditions on the upper and lower parts of $\partial\Omega$ yield

$$\frac{\partial}{\partial t_i} U^{(1,k)} = \delta_{k2} p^{(1)} \text{ on } \Gamma_+,$$

$$\frac{\partial}{\partial t_i} U^{(2,k)} = \delta_{k2} p^{(2)} \text{ on } \Gamma_-, \tag{3.4.26}$$

and, due to the contact conditions on the interface boundary

$$U^{(1,k)}(x_1, -h/2) = U^{(0,k)}(x_1, h_0/2),$$

$$\mu_1 \frac{\partial}{\partial t_1} U^{(1,k)}|_{t_1=-h/2} = \mu \frac{\partial}{\partial t_0} U^{(0,k-2)}|_{t_0=h_0/2}, \tag{3.4.27}$$

$$U^{(2,k)}(x_1, h/2) = U^{(0,k)}(x_1, -h_0/2),$$

$$\mu_2 \frac{\partial}{\partial t_2} U^{(2,k)}|_{t_2=h/2} = \mu \frac{\partial}{\partial t_0} U^{(0,k-2)}|_{t_0=-h_0/2}. \tag{3.4.28}$$

It is assumed here and further in this section that all terms with negative indices are equal to zero.

2. The state of plane strain. If one substitutes the series

$$\sum_j \varepsilon^j \boldsymbol{u}^{(i,j)}(x_1, t_i), \quad i = 0, 1, 2,$$

into equations (3.4.10) and boundary conditions (3.4.11)–(3.4.16), and equates the coefficients near like powers of ε, then it yields the recurrent system of relations (for the sake of convenience we use the notations $\partial_{t_i} = \partial/\partial t_i$, $\partial_1 = \partial/\partial x_1$):

$$\mu_i \partial_{t_i}^2 u_1^{(i,k)} + (\lambda_i + \mu_i)\partial_{t_i}\partial_1 u_2^{(i,k-1)} + (\lambda_i + 2\mu_i)\partial_1^2 u_1^{(i,k-2)} = 0, \tag{3.4.29}$$

$$(2\mu_i + \lambda_i)\partial_{t_i}^2 u_2^{(i,k)} + (\lambda_i + \mu_i)\partial_{t_i}\partial_1 u_1^{(i,k-1)} + \mu_i \partial_1^2 u_2^{(i,k-2)} = 0 \text{ in } \Omega_i, \ i = 1, 2, \tag{3.4.30}$$

for the upper and lower layers, and

$$\mu \partial_{t_0}^2 u_1^{(0,k)} + (\lambda + \mu)\partial_{t_0}\partial_1 u_2^{(0,k-2)} + (\lambda + 2\mu)\partial_1^2 u_1^{(0,k-4)} = 0, \tag{3.4.31}$$

$$(2\mu + \lambda)\partial_{t_0}^2 u_2^{(0,k)} + (\lambda + \mu)\partial_{t_0}\partial_1 u_1^{(0,k-2)} + \mu \partial_1^2 u_2^{(0,k-4)} = 0, \text{ in } \Omega_0, \tag{3.4.32}$$

for the middle layer. On the interface boundary

$$\mu_1(\partial_{t_1} u_1^{(1,k)} + \partial_1 u_2^{(1,k-1)}) = \mu(\partial_{t_0} u_1^{(0,k-2)} + \partial_1 u_2^{(0,k-4)}), \tag{3.4.33}$$

$$(2\mu_1 + \lambda_1)\partial_{t_1} u_2^{(1,k)} + \lambda_1 \partial_1 u_1^{(1,k-1)}$$

$$= (2\mu + \lambda)\partial_{t_0} u_2^{(0,k-2)} + \lambda \partial_1 u_1^{(0,k-4)}, \qquad (3.4.34)$$

$$\boldsymbol{u}^{(0,k)} = \boldsymbol{u}^{(1,k)} \text{ on } S_+, \qquad (3.4.35)$$

and

$$\mu_2(\partial_{t_2} u_1^{(2,k)} + \partial_1 u_2^{(2,k-1)}) = \mu(\partial_{t_0} u_1^{(0,k-2)} + \partial_1 u_2^{(0,k-4)}), \qquad (3.4.36)$$

$$(2\mu_2 + \lambda_2)\partial_{t_2} u_2^{(2,k)} + \lambda_2 \partial_1 u_1^{(2,k-1)}$$

$$= (2\mu + \lambda)\partial_{t_0} u_2^{(0,k-2)} + \lambda \partial_1 u_1^{(0,k-4)}, \qquad (3.4.37)$$

$$\boldsymbol{u}^{(0,k)} = \boldsymbol{u}^{(2,k)} \text{ on } S_-. \qquad (3.4.38)$$

On the upper and lower surface we have

$$\mu_1(\partial_{t_1} u_1^{(1,k)} + \partial_1 u_2^{(1,k-1)}) = \delta_{k2} p_1^{(1)},$$

$$(2\mu_1 + \lambda_1)\partial_{t_1} u_2^{(1,k)} + \lambda_1 \partial_1 u_1^{(1,k-1)} = \delta_{k4} p_2^{(1)} \text{ on } \Gamma_+, \qquad (3.4.39)$$

$$\mu_2(\partial_{t_2} u_1^{(2,k)} + \partial_1 u_2^{(2,k-1)}) = \delta_{k2} p_1^{(2)},$$

$$(2\mu_2 + \lambda_2)\partial_{t_2} u_2^{(2,k)} + \lambda_2 \partial_1 u_1^{(2,k-1)} = \delta_{k4} p_2^{(2)} \text{ on } \Gamma_-. \qquad (3.4.40)$$

In the next two sections we show that the leading parts of the displacements depend on x_1 only, and we derive the ordinary differential equations with respect to these functions and the displacement jump across the thin middle layer of the compound rectangle.

3.4.3 Leading part of the displacement in the case of anti-plane shear

1. First, we consider the case $k = 0$. One can see that

$$\partial_{t_i}^2 U^{(i,0)} = 0, \ t_i \in (-h/2, h/2), \qquad (3.4.41)$$

$$\partial_{t_i} U^{(i,0)}(x_1, \pm h/2) = 0, \ \ i = 1, 2, \qquad (3.4.42)$$

and, clearly, the functions $U^{(i,0)}$, $i = 1, 2$, depend on x_1 only.

From the displacement contact conditions on the interface boundary we deduce

$$U^{(0,0)} = U^{(2,0)}(x_1) + \frac{t_0 + h_0/2}{h_0} d^{(0)}(x_1), \tag{3.4.43}$$

where

$$d^{(0)}(x_1) = U^{(1,0)}(x_1) - U^{(2,0)}(x_1)$$

is the leading order term of the displacement jump across the thin middle layer.

2. We make the next step and consider $k = 1$. The functions $U^{(i,1)}$ are subjected to the equations

$$\partial_{t_i}^2 U^{(i,1)}(x_1, t_i) = 0, \quad i = 0, 1, 2;$$

$$t_0 \in (-h_0/2, h_0/2), \quad t_i \in (-h/2, h/2), \quad i = 1, 2. \tag{3.4.44}$$

On the outer boundary one has

$$\partial_{t_1} U^{(1,1)}|_{\Gamma_+} = 0, \quad \partial_{t_2} U^{(2,1)}|_{\Gamma_-} = 0. \tag{3.4.45}$$

The interface conditions yield

$$\partial_{t_1} U^{(1,1)}|_{S_+} = 0, \quad U^{(0,1)}|_{S_+} = U^{(1,1)}|_{S_+}, \tag{3.4.46}$$

$$\partial_{t_2} U^{(2,1)}|_{S_-} = 0, \quad U^{(0,1)}|_{S_-} = U^{(2,1)}|_{S_-}. \tag{3.4.47}$$

Therefore, the functions $U^{(i,1)}$, $i = 1, 2$, do not depend on t_i, and $U^{(0,1)}$ can be represented in the form

$$U^{(0,1)} = U^{(2,1)}(x_1) + \frac{t_0 + h_0/2}{h_0} d^{(1)}(x_1), \tag{3.4.48}$$

$$d^{(1)}(x_1) = U^{(1,1)}(x_1) - U^{(2,1)}(x_1).$$

3. This is the last step (k = 2), where we derive the ordinary differential equations with respect to $U^{(i,0)}(x_1)$. The equations are

$$\partial_{t_i}^2 U^{(i,2)}(x_1, t_i) = -\partial_1^2 U^{(i,0)}(x_1, t_i), t_i \in (-h/2, h/2), i = 1, 2, \tag{3.4.49}$$

$$\partial_{t_0}^2 U^{(0,2)}(x_1, t_0) = 0, \quad t_0 \in (-h_0/2, h_0/2). \tag{3.4.50}$$

Similar to (3.4.45), we have the boundary conditions

$$\partial_{t_1} U^{(1,2)}|_{\Gamma_+} = \frac{p^{(1)}(x_1)}{\mu_1}, \quad \partial_{t_2} U^{(2,2)}|_{\Gamma_-} = \frac{p^{(2)}(x_1)}{\mu_2}. \tag{3.4.51}$$

The conditions on the interface boundary yield

$$\mu_1 \partial_{t_1} U^{(1,2)}|_{t_1=-h/2} = \frac{\mu}{h_0} d^{(0)}(x_1),$$

$$\mu_2 \partial_{t_2} U^{(2,2)}|_{t_2=h/2} = \frac{\mu}{h_0} d^{(0)}(x_1). \tag{3.4.52}$$

The solvability conditions of the boundary value problems (3.4.49)–(3.4.52) give

$$\begin{cases} \partial_1^2 U^{(1,0)}(x_1) = \frac{\mu}{\mu_1 h h_0}(U^{(1,0)}(x_1) - U^{(2,0)}(x_1)) - \frac{p^{(1)}(x_1)}{h\mu_1}, \\ \partial_1^2 U^{(2,0)}(x_1) = -\frac{\mu}{\mu_2 h h_0}(U^{(1,0)}(x_1) - U^{(2,0)}(x_1)) + \frac{p^{(2)}(x_1)}{h\mu_2}. \end{cases} \tag{3.4.53}$$

Thus, we have obtained a system of differential equations with respect to $U^{(i,0)}$ which includes the displacement jump across the thin middle layer and the normalized shear modulus of the bonding material.

4. The boundary conditions at the ends $x_1 = \pm l$ of the interval are required for the system (3.4.53). These relations are derived as conditions of the exponential decay of the boundary layer near the ends of the thin rectangle. We have presented this analysis in Section 3.3, and these results will be used here. One can see immediately that the discrepancy of order $O(1)$ in the boundary conditions can be removed provided

$$\mathcal{U}^{(i,0)}(\pm l) = 0, \quad i = 1, 2, \tag{3.4.54}$$

and it gives a set of boundary conditions for the system (3.4.53).

In the next section the idea presented above will be illustrated on the example related to the plane-strain state of a thin compound rectangle.

3.4.4 The case of plane strain

Perhaps the reader noticed that the principal component of external load is the longitudinal one (see (3.4.11)–(3.4.14)). We need this restriction to provide the same order of magnitude for transversal and longitudinal components of the displacement vector (see (3.4.20)–(3.4.23)). Our intention is to consider the coupling effect for the transversal and the longitudinal modes (it does not occur for the case of a homogeneous beam). To make the presentation simpler, we shall use the symmetry assumption made at the beginning of Section 3.4.1.

1. Longitudinal deformation. Substitute the terms $\mathcal{U}^{(i)}$, $i = 0, 1, 2$, into the equations (3.4.29)–(3.4.32) and the boundary conditions (3.4.33)–(3.4.40).

Taking $k = 0$ we can see (the consideration is the same as in the previous section) that the vector functions $\mathcal{U}^{(i,0)}$, $i = 1, 2$, has zero second component, and the first component depends on x_1 only. (The second component is zero by assumption related to the structure of the asymptotic expansion; see (3.4.20)–(3.4.23).) For the middle layer

$$\mathcal{U}_2^{(0,0)} = 0, \tag{3.4.55}$$

$$\mathcal{U}_1^{(0,0)} = \frac{t_0 - h_0/2}{h_0}\left(\mathcal{U}_1^{(1,0)}(x_1) - \mathcal{U}_1^{(2,0)}(x_1)\right) + \mathcal{U}_1^{(1,0)}(x_1).$$

Now make the next step and consider $k = 1$. The equations in Ω_i, $i = 0, 1, 2$, yield that the vector valued functions $\mathcal{U}^{(i,1)}$, $i = 0, 1, 2$, are linear with respect t_i. The traction boundary conditions on $\Gamma_\pm$ give

$$\partial_{t_1}\mathcal{U}_1^{(1,1)}|_{t_1=h/2} = 0,$$

$$(2\mu_1 + \lambda_1)\partial_{t_1}\mathcal{U}_2^{(1,1)}|_{t_1=h/2} + \lambda_1\partial_1\mathcal{U}_1^{(1,0)} = 0, \tag{3.4.56}$$

and

$$\partial_{t_2}\mathcal{U}_1^{(2,1)}|_{t_2=-h/2} = 0,$$

$$(2\mu_2 + \lambda_2)\partial_{t_2}\mathcal{U}_2^{(2,1)}|_{t_2=-h/2} + \lambda_2\partial_1\mathcal{U}_1^{(2,0)} = 0. \tag{3.4.57}$$

From the interface traction boundary conditions we deduce

$$\partial_{t_1}\mathcal{U}_1^{(1,1)}|_{t_1=-h/2} = 0,$$

$$(2\mu_1 + \lambda_1)\partial_{t_1}\mathcal{U}_2^{(1,1)}|_{t_1=-h/2} + \lambda_1\partial_1\mathcal{U}_1^{(1,0)} = 0, \tag{3.4.58}$$

$$\partial_{t_2}\mathcal{U}_1^{(2,1)}|_{t_2=h/2} = 0,$$

$$(2\mu_2 + \lambda_2)\partial_{t_2}\mathcal{U}_2^{(2,1)}|_{t_2=h/2} + \lambda_2\partial_1\mathcal{U}_1^{(2,0)} = 0. \tag{3.4.59}$$

The displacement interface conditions yield

$$\mathcal{U}_j^{(0,1)}(x_1, h_0/2) = \mathcal{U}_j^{(1,1)}(x_1, -h/2),$$

$$\mathcal{U}_j^{(0,1)}(x_1, -h_0/2) = \mathcal{U}_j^{(2,1)}(x_1, h/2). \tag{3.4.60}$$

Consequently, the components of vectors $\mathcal{U}^{(i,1)}$, $i = 0, 1, 2$, can be represented in the form

$$\mathcal{U}_1^{(i,1)} = 0, \ i = 1, 2, \tag{3.4.61}$$

$$\mathcal{U}_2^{(i,1)} = -\frac{\lambda_i}{2\mu_i + \lambda_i} t_i \partial_1 \mathcal{U}_1^{(i,0)}(x_1), \ i = 1, 2, \tag{3.4.62}$$

$$\mathcal{U}_j^{(0,1)} = \frac{t_0 - h_0/2}{h_0}\left(\mathcal{U}_j^{(1,1)}(x_1, -h/2) - \mathcal{U}_j^{(2,1)}(x_1, h/2)\right)$$

$$+\mathcal{U}_j^{(1,1)}(x_1, -h/2). \tag{3.4.63}$$

Now, we are going to take the last step for the longitudinal mode and consider $k = 2$.

The following equations hold in Ω_i, $i = 1, 2$,

$$\mu_i \partial_{t_i}^2 \mathcal{U}_1^{(i,2)} + (\lambda_i + \mu_i)\partial_{t_i}\partial_1 \mathcal{U}_2^{(i,1)} + (\lambda_i + 2\mu_i)\partial_1^2 \mathcal{U}_1^{(i,0)} = 0, \tag{3.4.64}$$

$$(2\mu_i + \lambda_i)\partial_{t_i}^2 \mathcal{U}_2^{(i,2)} + (\lambda_i + \mu_i)\partial_{t_i}\partial_1 \mathcal{U}_1^{(i,1)} + \mu_i \partial_1^2 \mathcal{U}_2^{(i,0)} = 0, \tag{3.4.65}$$

and in Ω_0

$$\mu \partial_{t_0}^2 \mathcal{U}_1^{(0,2)} + (\lambda + \mu)\partial_{t_0}\partial_1 \mathcal{U}_2^{(0,0)} = 0, \tag{3.4.66}$$

$$(2\mu + \lambda)\partial_{t_0}^2 \mathcal{U}_2^{(0,2)} + (\lambda + \mu)\partial_{t_0}\partial_1 \mathcal{U}_1^{(0,0)} = 0. \tag{3.4.67}$$

On the interface boundary we have

$$\mu_1\left(\partial_{t_1}\mathcal{U}_1^{(1,2)} + \partial_1 \mathcal{U}_2^{(1,1)}\right)|_{t_1=-h/2} = \mu \partial_{t_0}\mathcal{U}_1^{(0,0)}|_{t_0=h_0/2}, \tag{3.4.68}$$

$$\left((2\mu_1 + \lambda_1)\partial_{t_1}\mathcal{U}_2^{(1,2)} + \lambda_1 \partial_1 \mathcal{U}_1^{(1,1)}\right)|_{t_1=-h/2}$$

$$= (2\mu + \lambda)\partial_{t_0}\mathcal{U}_2^{(0,0)}|_{t_0=h_0/2}, \tag{3.4.69}$$

$$\mathcal{U}_j^{(0,2)}(x_1, h_0/2) = \mathcal{U}_j^{(1,2)}(x_1, -h/2), \tag{3.4.70}$$

and

$$\mu_2\left(\partial_{t_2}\mathcal{U}_1^{(2,2)}+\partial_1\mathcal{U}_2^{(2,1)}\right)|_{t_2=h/2}=\mu\partial_{t_0}\mathcal{U}_1^{(0,0)}|_{t_0=-h_0/2}, \quad (3.4.71)$$

$$\left((2\mu_2+\lambda_2)\partial_{t_2}\mathcal{U}_2^{(2,2)}+\lambda_2\partial_1\mathcal{U}_1^{(2,1)}\right)|_{t_2=h/2}$$

$$=(2\mu+\lambda)\partial_{t_0}\mathcal{U}_2^{(0,0)}|_{t_0=-h_0/2}, \quad (3.4.72)$$

$$\mathcal{U}_j^{(0,2)}(x_1,-h_0/2)=\mathcal{U}_j^{(2,2)}(x_1,h/2). \quad (3.4.73)$$

From the traction boundary conditions on the upper and lower parts of $\partial\Omega$ one can deduce

$$\mu_1\left(\partial_{t_1}\mathcal{U}_1^{(1,2)}+\partial_1\mathcal{U}_2^{(1,1)}\right)|_{t_1=h/2}=p_1^{(1)}, \quad (3.4.74)$$

$$\left((2\mu_1+\lambda_1)\partial_{t_1}\mathcal{U}_2^{(1,2)}+\lambda_1\partial_1\mathcal{U}_1^{(1,1)}\right)|_{t_1=h/2}=0, \quad (3.4.75)$$

$$\mu_2\left(\partial_{t_2}\mathcal{U}_1^{(2,2)}+\partial_1\mathcal{U}_2^{(2,1)}\right)|_{t_2=-h/2}=p_1^{(2)}, \quad (3.4.76)$$

$$\left((2\mu_2+\lambda_2)\partial_{t_2}\mathcal{U}_2^{(2,2)}+\lambda_2\partial_1\mathcal{U}_1^{(2,1)}\right)|_{t_2=-h/2}=0. \quad (3.4.77)$$

Consider the solvability conditions of the boundary value problems (3.4.64), (3.4.65), (3.4.74)–(3.4.77) with respect to $\mathcal{U}_j^{(i,2)}(t_i,\cdot)$, $i,j=1,2$. In $\overline{\Omega_1}$ one has

$$\mu_1\partial_{t_1}^2\mathcal{U}_1^{(1,2)}=-\mu_1\frac{3\lambda_1+4\mu_1}{2\mu_1+\lambda_1}\partial_1^2\mathcal{U}_1^{(1,0)}, \quad t_1\in(-h/2,h/2), \quad (3.4.78)$$

$$\partial_{t_1}\mathcal{U}_1^{(1,2)}|_{t_1=-h/2}=\frac{\mu}{h_0\mu_1}d_1^{(0)}(x_1)-\frac{\lambda_1}{2\mu_1+\lambda_1}\frac{h}{2}\partial_1^2\mathcal{U}_1^{(1,0)}, \quad (3.4.79)$$

$$\partial_{t_1}\mathcal{U}_1^{(1,2)}|_{t_1=h/2}=\frac{p_1^{(1)}}{\mu_1}+\frac{\lambda_1}{2\mu_1+\lambda_1}\frac{h}{2}\partial_1^2\mathcal{U}_1^{(1,0)}. \quad (3.4.80)$$

where $d_1^{(0)} = \mathcal{U}_1^{(1,0)} - \mathcal{U}_1^{(2,0)}$.

The boundary value problem (3.4.78)–(3.4.80) is solvable if and only if

$$-\frac{4(\lambda_1+\mu_1)}{2\mu_1+\lambda_1}\partial_1^2\mathcal{U}_1^{(1,0)} = \frac{p_1^{(1)}}{h\mu_1} - \frac{\mu}{hh_0\mu_1}d_1^{(0)}(x_1), |x_1| < l. \tag{3.4.81}$$

Similarly, we can analyze the lower layer Ω_2, and we can deduce that $\mathcal{U}_1^{(2,0)}$ satisfies the differential equation

$$-\frac{4(\lambda_2+\mu_2)}{2\mu_2+\lambda_2}\partial_1^2\mathcal{U}_1^{(2,0)} = -\frac{p_1^{(2)}}{h\mu_2} + \frac{\mu}{hh_0\mu_2}d_1^{(0)}(x_1), |x_1| < l. \tag{3.4.82}$$

Thus, we have derived that the leading order terms of the displacement in the upper and lower layers satisfy the following system of differential equations on the interval $(-l,\ l)$

$$\begin{cases} \frac{4(\lambda_1+\mu_1)}{2\mu_1+\lambda_1}\partial_1^2\mathcal{U}_1^{(1,0)} &= -\frac{p_1^{(1)}}{h\mu_1} + \frac{\mu}{hh_0\mu_1}(\mathcal{U}_1^{(1,0)}(x_1) - \mathcal{U}_1^{(2,0)}(x_1)), \\ \frac{4(\lambda_2+\mu_2)}{2\mu_2+\lambda_2}\partial_1^2\mathcal{U}_1^{(2,0)} &= \frac{p_1^{(2)}}{h\mu_2} - \frac{\mu}{hh_0\mu_2}(\mathcal{U}_1^{(1,0)}(x_1) - \mathcal{U}_1^{(2,0)}(x_1)), \end{cases} \tag{3.4.83}$$

and the displacement components in the middle layer are specified by (3.4.54).

2. Bending deformation. Now, consider the terms $\mathcal{V}^{(i)}$, $i = 0,1,2$, from (3.4.20)–(3.4.23) that correspond to the transversal displacement. For the sake of simplicity we assume in this section that $\lambda_1 = \lambda_2$, $\mu_1 = \mu_2$. Direct substitution of (3.4.20) into equations (3.4.29)–(3.4.32) and boundary conditions (3.4.33)–(3.4.40) ($k = 0$) yields that $\mathcal{V}_2^{(i,0)}$, $i = 1,2$, depend on x_1 only. Note that for the case of a compound beam the longitudinal and transversal modes may be coupled.

In a way, similar to one described in the previous section, we obtain for $k = 1$

$$\mathcal{V}_1^{(i,1)} = -t_i\partial_1\mathcal{V}_2^{(i,0)} + \mathfrak{V}_1^{(i,1)}(x_1),\ i = 1,2, \tag{3.4.84}$$

$$\mathcal{V}_2^{(i,1)} = 0,\ i = 1,2, \tag{3.4.85}$$

$$\mathcal{V}_j^{(0,1)} = \frac{t_0 - h_0/2}{h_0}\left(\mathcal{V}_j^{(1,1)}(x_1,-h/2) - \mathcal{V}_j^{(2,1)}(x_1,h/2)\right) + \mathcal{V}_j^{(1,1)}(x_1,-h/2). \tag{3.4.86}$$

Here $\mathfrak{V}_1^{(i,1)}$, $i = 1, 2$, are some unknown functions of x_1. Due to the symmetry assumption

$$\mathfrak{V}_1^{(1,1)}(x_1) = -\mathfrak{V}_1^{(2,1)}(x_1). \tag{3.4.87}$$

Note that the relation (3.4.87) does not hold if the layers Ω_1, Ω_2 have different thickness or different elastic moduli. The nonsymmetric case involves some routine calculations and will be considered separately. The idea of the coupling can be demonstrated for the symmetric structure, and we follow this way in order to make the presentation clear and understandable.

On the third step ($k = 2$) one can check that the second component $d_2^{(0)}$ of the displacement jump across the soft layer must be zero. To see that, we take the region Ω_1 (for Ω_2 the analysis is exactly the same) and derive the equations and boundary conditions with respect to $\mathcal{V}_2^{(1,2)}$

$$\partial_{t_1}^2 \mathcal{V}_2^{(1,2)} = \frac{\lambda_1}{2\mu_1 + \lambda_1} \partial_1^2 \mathcal{V}_2^{(1,0)}, \; t_1 \in (-h/2, h/2), \tag{3.4.88}$$

$$\partial_{t_1} \mathcal{V}_2^{(1,2)}|_{t_1=-h/2} = \frac{2\mu + \lambda}{2\mu_1 + \lambda_1} d_2^{(0)}(x_1)$$
$$- \frac{\lambda_1}{2\mu_1 + \lambda_1} \frac{h}{2} \partial_1^2 \mathcal{V}_2^{(1,0)} - \frac{\lambda_1}{2\mu_1 + \lambda_1} \partial_1 \mathfrak{V}_1^{(1,1)}(x_1), \tag{3.4.89}$$

$$\partial_{t_1} \mathcal{V}_2^{(1,2)}|_{t_1=h/2} = \frac{\lambda_1}{2\mu_1 + \lambda_1} \frac{h}{2} \partial_1^2 \mathcal{V}_2^{(1,0)}$$
$$- \frac{\lambda_1}{2\mu_1 + \lambda_1} \partial_1 \mathfrak{V}_1^{(1,1)}(x_1). \tag{3.4.90}$$

The boundary value problem (3.4.88)–(3.4.90) is solvable if and only if

$$d_2^{(0)}(x_1) := \mathcal{V}_2^{(1,0)} - \mathcal{V}_2^{(2,0)} = 0. \tag{3.4.91}$$

Further we shall study the upper layer Ω_1 (the leading part of the transversal displacement in the lower part is the same).

For our particular case, the components $\mathcal{V}_k^{(1,2)}$, $k = 1, 2$, can be chosen in the form

$$\mathcal{V}_2^{(1,2)} = \frac{\lambda_1}{2\mu_1 + \lambda_1} \left(\frac{t_1^2}{2} \partial_1^2 \mathcal{V}_2^{(1,0)} - t_1 \partial_1 \mathfrak{V}_1^{(1,1)} \right), \tag{3.4.92}$$

$$\mathcal{V}_1^{(1,2)} = 0. \tag{3.4.93}$$

The fourth step corresponds to $k = 3$. The solvability condition for the boundary value problem with respect to $\mathcal{V}_2^{(1,3)}$ takes the form

$$d_2^{(1)}(x_1) = 0, \tag{3.4.94}$$

of the continuity relation for the second-order transversal component of displacement. The explicit representation for the function $\mathcal{V}_2^{(1,3)}$ is not required.

In the upper layer Ω_1 the following equation is valid for the function $\mathcal{V}_1^{(1,3)}$:

$$\partial_{t_1}^2 \mathcal{V}_1^{(1,3)} = \frac{3\lambda_1 + 4\mu_1}{2\mu_1 + \lambda_1}\{t_1 \partial_1^3 \mathcal{V}_2^{(1,0)} - \partial_1^2 \mathfrak{V}_1^{(1,1)}\}. \tag{3.4.95}$$

The following boundary conditions hold on the boundary of the layer (they follow directly from the interface conditions (3.4.15))

$$\partial_{t_1} \mathcal{V}_1^{(1,3)}|_{t_1=-h/2} = -\frac{\lambda_1}{2\mu_1 + \lambda_1}\left\{\frac{h^2}{8}\partial_1^3 \mathcal{V}_2^{(1,0)} + \frac{h}{2}\partial_1^2 \mathfrak{V}_1^{(1,1)}\right\}$$

$$+ \frac{\mu}{\mu_1 h_0}\{h\partial_1 \mathcal{V}_2^{(1,0)} + 2\mathfrak{V}_1^{(1,1)}\} \tag{3.4.96}$$

and

$$\partial_{t_1} \mathcal{V}_1^{(1,3)}|_{t_1=h/2} = -\frac{\lambda_1}{2\mu_1 + \lambda_1}\left(\frac{h^2}{8}\partial_1^3 \mathcal{V}_2^{(1,0)} - \frac{h}{2}\partial_1^2 \mathfrak{V}_1^{(1,1)}\right). \tag{3.4.97}$$

The solvability condition of the boundary value problem (3.4.95)–(3.4.97) takes the form of the second-order differential equation with respect to $\mathfrak{V}_1^{(1,1)}$,

$$\partial_1^2 \mathfrak{V}_1^{(1,1)} = \frac{(2\mu_1 + \lambda_1)\mu}{4\mu_1(\lambda_1 + \mu_1)}\frac{1}{hh_0}\{2\mathfrak{V}_1^{(1,1)} + h\partial_1 \mathcal{V}_2^{(1,0)}\}. \tag{3.4.98}$$

Then, the solution of (3.4.95)–(3.4.97) admits the representation

$$\mathcal{V}_1^{(1,3)} = \frac{3\lambda_1 + 4\mu_1}{2\mu_1 + \lambda_1}\left\{\frac{t_1^3}{6}\partial_1^3 \mathcal{V}_2^{(1,0)} - \frac{t_1^2}{2}\partial_1^2 \mathfrak{V}_1^{(1,1)}\right\}$$

$$+\frac{\lambda_1+\mu_1}{2\mu_1+\lambda_1}t_1\left\{-\frac{h^2}{2}\partial_1^3\mathcal{V}_2^{(1,0)}+2h\partial_1^2\mathfrak{W}_1^{(1,1)}\right\}. \tag{3.4.99}$$

It is verified by direct calculations that the function $\mathcal{V}_2^{(1,3)}$ satisfies the homogeneous boundary value problem (3.4.95)–(3.4.97), and in our analysis it can be chosen to be zero.

Now, we are ready to make the last step and consider the case $k = 4$, where the boundary value problem for $\mathcal{V}_2^{(1,4)}$ will be posed. The solvability condition of this problem will lead to the fourth-order ordinary differential equation with respect to $\mathcal{V}_2^{(1,0)}$.

The function $\mathcal{V}_2^{(1,4)}$ is subjected to the following equation,

$$\partial_{t_1}^2\mathcal{V}_2^{(1,4)}=\frac{1}{2}\left\{\left(\frac{\lambda_1+\mu_1}{2\mu_1+\lambda_1}\right)^2h^2-\frac{3\lambda_1+2\mu_1}{2\mu_1+\lambda_1}t_1^2\right\}\partial_1^4\mathcal{V}_2^{(1,0)}$$

$$+\left\{\frac{2\mu_1+3\lambda_1}{2\mu_1+\lambda_1}t_1-2h\left(\frac{\lambda_1+\mu_1}{2\mu_1+\lambda_1}\right)^2\right\}\partial_1^3\mathfrak{W}_1^{(1,1)}, \tag{3.4.100}$$

and the boundary conditions

$$\partial_{t_1}\mathcal{V}_2^{(1,4)}|_{t_1=h/2}=\frac{\lambda_1}{(2\mu_1+\lambda_1)^2}\left[\frac{h^3}{48}(9\lambda_1+8\mu_1)\partial_1^4\mathcal{V}_2^{(1,0)}\right.$$

$$\left.-\frac{h^2}{8}(4\mu_1+5\lambda_1)\partial_1^3\mathfrak{W}_1^{(1,1)}\right]+\frac{p_2^{(1)}}{2\mu_1+\lambda_1}, \tag{3.4.101}$$

$$\partial_{t_1}\mathcal{V}_2^{(1,4)}|_{t_1=-h/2}=-\frac{\mu}{2\mu_1+\lambda_1}\partial_1\mathcal{U}_1^{(1,0)}(x_1)-\frac{\lambda_1h^2}{8(2\mu_1+\lambda_1)^2}$$

$$\times\left\{\frac{1}{6}h(9\lambda_1+8\mu_1)\partial_1^4\mathcal{V}_2^{(1,0)}-(11\lambda_1+12\mu_1)\partial_1^3\mathfrak{W}_1^{(1,1)}\right\}. \tag{3.4.102}$$

Here we used the symmetry assumption (in general, the boundary condition (3.4.102) looks more complicated) which allows us to say that the analysis for the function $\mathcal{V}_2^{(2,4)}$ follows the same pattern.

The boundary value problem (3.4.100)–(3.4.102) is solvable if and only if the function $\mathcal{V}_2^{(0)} := \mathcal{V}_2^{(1,0)} = \mathcal{V}_2^{(2,0)}$ satisfies the differential equation

$$\frac{2\mu_1(\mu_1+\lambda_1)}{3(2\mu_1+\lambda_1)}h^3\partial_1^4\mathcal{V}_2^{(0)}-\mu\frac{h}{h_0}\{2\partial_1\mathfrak{W}_1^{(1,1)}+h\partial_1^2\mathcal{V}_2^{(0)}\}$$

$$-\frac{1}{2}\mu\partial_1 d_1^{(0)}(x_1) = p_2^{(1)} - p_2^{(2)}. \qquad (3.4.103)$$

Thus, we have obtained the fourth-order differential equation for the principal part of the transversal component of the displacement vector. We said nothing about the functions $\mathcal{V}_1^{(i,k)}, i = 1, 2$; in fact the solvability conditions for the corresponding boundary value problems will require certain correction terms of order $O(\varepsilon^2)$ in the asymptotic representation of the longitudinal displacement. These quantities are not important here if we talk about the principal part of the displacement vector. Nevertheless, in the next section we pay special attention to this effect.

It should be emphasized that the second and third terms in the left–hand side (3.4.103) characterize the soft middle layer and show the coupling effect between the longitudinal and transversal components. The vector function $\mathfrak{V}_1^{(1,1)}$ is specified as a solution of the differential equation (3.4.98) which vanishes at $x_1 = \pm l$.

3. Boundary conditions. Expanding the displacement field in the form of the power series in the vicinity of the ends of a thin rectangle, we can see that the discrepancy of order $O(1)$ in the boundary conditions for longitudinal displacements will be removed provided

$$\mathcal{U}_1^{(i,0)}(\pm l) = 0, \quad i = 1, 2, \qquad (3.4.104)$$

and it follows from the analysis of the discrepancy in the boundary conditions for transversal displacement (up to $O(\varepsilon)$) that

$$\mathcal{V}_2^{(1,0)}(\pm l) = 0, \quad \partial_1 \mathcal{V}_2^{(1,0)}|_{x_1=\pm l} = 0, \quad \mathfrak{V}_1^{(1,1)}(\pm l) = 0. \qquad (3.4.105)$$

Also, it should be mentioned that the leading part of the transversal displacement jump across the soft middle layer is equal to zero (see (3.4.91)). Relations (3.4.104), (3.4.105) deliver the boundary conditions required for the system of differential equations (3.4.83), (3.4.103), (3.4.98).

3.4.5 Nonsymmetric layered structure. Bending

Here we consider the asymptotic representation of the displacement components which satisfy the homogeneous system (3.4.10), the contact conditions (3.4.15), (3.4.16) and the clamping conditions (3.4.17). However, we assume that the structure is not symmetric any more, and that the traction components have the same order of magnitude. Namely, the regions $\Omega_i, i = 1, 2$, are defined by

$$\Omega_1 = \{\boldsymbol{x} \in \mathbb{R}^2 : |x_1| < l, \varepsilon^2 h_0/2 < x_2 < \varepsilon^2 h_0/2 + \varepsilon h_1\},$$

$$\Omega_2 = \{\boldsymbol{x} \in \mathbb{R}^2 : |x_1| < l, -\varepsilon h_2 - \varepsilon^2 h_0/2 < x_2 < -\varepsilon^2 h_0/2\}, \quad (3.4.106)$$

where h_1 and h_2 may be different; the definition of Ω_0 is the same as in (3.4.1). Also, we assume that the elastic moduli of the layers Ω_1, Ω_2 may be different.

The traction boundary conditions on the upper and lower surfaces take the form

$$\mu_1(\frac{\partial u_2^{(1)}}{\partial x_1} + \frac{\partial u_1^{(1)}}{\partial x_2}) = \varepsilon p_1^{(1)},$$

$$(2\mu_1 + \lambda_1)\frac{\partial u_2^{(1)}}{\partial x_2} + \lambda_1 \frac{\partial u_1^{(1)}}{\partial x_1} = \varepsilon p_2^{(1)} \text{ on } \Gamma_+, \quad (3.4.107)$$

$$\mu_2(\frac{\partial u_2^{(2)}}{\partial x_1} + \frac{\partial u_1^{(2)}}{\partial x_2}) = \varepsilon p_1^{(2)},$$

$$(2\mu_2 + \lambda_2)\frac{\partial u_2^{(2)}}{\partial x_2} + \lambda_2 \frac{\partial u_1^{(2)}}{\partial x_1} = \varepsilon p_2^{(2)} \text{ on } \Gamma_-. \quad (3.4.108)$$

In this case the asymptotic approximation (3.4.20) is to be replaced by the following formula:

$$\boldsymbol{u}_\varepsilon^{(i)}(\boldsymbol{x}) \sim \sum_{k=0}^{\infty} \varepsilon^k \{\varepsilon^{-2}\mathcal{V}^{(i,0,k)}(x_1) + \varepsilon^{-1}\mathcal{V}^{(i,1,k)}(x_1, t_i)$$

$$+ \mathcal{V}^{(i,2,k)}(x_1, t_i) + \varepsilon\mathcal{V}^{(i,3,k)}(x_1, t_i) + \mathcal{U}^{(i,0,k)}(x_1, t_i)$$

$$+ \varepsilon\mathcal{U}^{(i,1,k)}(x_1, t_i) + \varepsilon^2\mathcal{W}^{(i,k)}(x_1, t_i)\}, \quad i = 1, 2, \quad (3.4.109)$$

where $\mathcal{V}_1^{(i,0,k)} = \mathcal{U}_2^{(i,0,k)} = 0$ for $k \in \mathbb{Z}_+$, $i = 1, 2$. We shall consider just the first ansatz of the expansion (3.4.109) which shows clearly that the leading order part of the displacement corresponds to the bending phenomena. To describe the longitudinal displacement, it is useful to know the components $\mathcal{U}_1^{(i,0,0)}(x_1)$. Basically, the procedure is very much similar to one described in previous sections. However, it requires considering of each layer separately. We present the final formulae and leave the standard derivation to the reader as an exercise. The function $\mathcal{V}_2^{(0)}(x_1) := \mathcal{V}_2^{(1,0,0)}(x_1) = \mathcal{V}_2^{(2,0,0)}(x_1)$ satisfies the fourth-order differential equation

$$\frac{1}{3}\left\{\frac{\mu_1(\mu_1 + \lambda_1)}{2\mu_1 + \lambda_1} h_1^3 + \frac{\mu_2(\mu_2 + \lambda_2)}{2\mu_2 + \lambda_2} h_2^3\right\} \partial_1^4 \mathcal{V}_2^{(0)}$$

$$+ 2\left\{ \frac{\mu_2(\lambda_2+\mu_2)}{2\mu_2+\lambda_2} h_2^2 \partial_1^3 v^{(2)} - \frac{\mu_1(\mu_1+\lambda_1)}{2\mu_1+\lambda_1} h_1^2 \partial_1^3 v^{(1)} \right\}$$

$$= p_2^{(1)} - p_2^{(2)}, \tag{3.4.110}$$

where

$$\partial_1^2 v^{(1)} = \frac{2\mu_1+\lambda_1}{4(\lambda_1+\mu_1)h_1} \frac{\mu}{\mu_1 h_0} \{ v^{(1)} - v^{(2)} + \frac{1}{2}(h_1+h_2)\partial_1 \mathcal{V}_2^{(0)} \}, \tag{3.4.111}$$

$$\partial_1^2 v^{(2)} = -\frac{2\mu_2+\lambda_2}{4(\lambda_2+\mu_2)h_2} \frac{\mu}{\mu_2 h_0} \{ v^{(1)} - v^{(2)} + \frac{1}{2}(h_1+h_2)\partial_1 \mathcal{V}_2^{(0)} \}. \tag{3.4.112}$$

The functions $v^{(i)}$, $i = 1, 2$, and $\mathcal{V}_2^{(0)}$ satisfy the homogeneous boundary conditions at $x_1 = \pm l$,

$$\mathcal{V}_2^{(0)}(\pm l) = \partial_1 \mathcal{V}_2^{(0)}(\pm l) = 0, \tag{3.4.113}$$

$$v^{(i)}(\pm l) = 0, \quad i = 1, 2. \tag{3.4.114}$$

The functions $\mathcal{U}_1^{(i,0)}(x_1), i = 1, 2$, characterizing the longitudinal displacement in Ω_i, $i = 1, 2$, are defined by the following system of second-order differential equations:

$$-\frac{4(\mu_1+\lambda_1)}{2\mu_1+\lambda_1} h_1 \partial_1^2 \mathcal{U}_1^{(1,0)} + \frac{\mu}{\mu_1 h_0}(\mathcal{U}_1^{(1,0)} - \mathcal{U}_1^{(2,0)})$$

$$+ \frac{\mu}{8\mu_1 h_0}\left(\frac{\lambda_1}{2\mu_1+\lambda_1} h_1^2 - \frac{\lambda_2}{2\mu_2+\lambda_2} h_2^2 \right) \partial_1^2 \mathcal{V}_2^{(0)} + \frac{\mu}{\mu_1} \partial_1 \mathcal{V}_2^{(0)}$$

$$+ \frac{\mu}{2\mu_1 h_0}\left(\frac{\lambda_1}{2\mu_1+\lambda_1} h_1 \partial_1 v^{(1)} + \frac{\lambda_2}{2\mu_2+\lambda_2} h_2 \partial_1 v^{(2)} \right) = \frac{p_1^{(1)}}{\mu_1}, \tag{3.4.115}$$

$$\frac{4(\lambda_2+\mu_2)}{2\mu_2+\lambda_2} h_2 \partial_1^2 \mathcal{U}_1^{(2,0)} + \frac{\mu}{\mu_2 h_0}(\mathcal{U}_1^{(1,0)} - \mathcal{U}_1^{(2,0)})$$

$$+ \frac{\mu}{8\mu_2 h_0}\left(\frac{\lambda_1}{2\mu_1+\lambda_1} h_1^2 - \frac{\lambda_2}{2\mu_2+\lambda_2} h_2^2 \right) \partial_1^2 \mathcal{V}_2^{(0)} + \frac{\mu}{\mu_2} \partial_1 \mathcal{V}_2^{(0)}$$

$$+ \frac{\mu}{2h_0\mu_2}\left(\frac{\lambda_1}{2\mu_1+\lambda_1} h_1 \partial_1 v^{(1)} + \frac{\lambda_2}{2\mu_2+\lambda_2} h_2 \partial_1 v^{(2)} \right) = \frac{p_1^{(2)}}{\mu_2}. \tag{3.4.116}$$

As before, the clamp conditions at the ends $x_1 = \pm l$ yield

$$\mathcal{U}_1^{(i,0)}(\pm l) = 0, \quad i = 1, 2. \tag{3.4.117}$$

In this section we have shown the example of the asymptotic derivation of equations of thin composite beams. We should say that the boundary layer, which occurs near the clamped ends, decays exponentially. This phenomena is of great importance; however, we have no intention to discuss this in detail here.

The present algorithm can be extended to the case of elastic layered plates of nonconstant thickness. Also, the extensions to the cases of anisotropic materials are possible. One can see that this algorithm is good for programming with REDUCE (see Appendix) or another package of symbolic manipulations which enables one to tackle very complicated structures in thin domains.

3.5 Lekhnitskii's problems

In this section we present the analysis of classical problem (see Lekhnitskii [46], [47]) on the deformation of an elastic half–space, subjected to the the gravity forces, with an infinite cylindrical mine perpendicular to the surface. First, we analyze polynomial solutions of the Lamé system in a half–space and show that the solution of the Lekhnitskii problem is not unique. We discuss the correct formulations, where physically appropriate conditions at infinity provide uniqueness.

The Lekhnitskii problem posed in an infinite region can be treated as a model problem of the boundary layer type that could occur in certain examples of the singular perturbation theory. We discuss some interesting asymptotic interpretation of the Lekhnitskii problems.

3.5.1 Particular solutions of the Lekhnitskii problem

First, we formulate the boundary value problem. Let g be a two–dimensional domain with a smooth boundary ∂g. For the sake of convenience assume that g contains the origin. By G denote a semi–infinite cylinder

$$\left\{ \boldsymbol{x} \in \mathbb{R}^3 : \ (x_1, x_2) \in g, \ x_3 > 0 \right\},$$

and introduce the following notation:

$$\Omega = \mathbb{R}^3_+ \setminus \overline{G},$$

where $\mathbb{R}^3_+$ is a half-space

$$\mathbb{R}^3_+ = \{\boldsymbol{x} : x_3 > 0\}.$$

It will be convenient to work in cylindrical system of coordinates (r, φ, z) such that

$$x_1 = r\cos\varphi,\ x_2 = r\sin\varphi, x_3 = z.$$

The displacement vector $\boldsymbol{u}$ is assumed to satisfy the following boundary value problem:

$$\mu\Delta\boldsymbol{u}(\boldsymbol{x}) + (\lambda+\mu)\nabla\nabla\cdot\boldsymbol{u}(\boldsymbol{x}) + \gamma e = \boldsymbol{0},\ \boldsymbol{x}\in\Omega, \tag{3.5.1}$$

$$\boldsymbol{\sigma}^{(n)}(\boldsymbol{u};\boldsymbol{x}) = \boldsymbol{0},\ \boldsymbol{x}\in\partial\Omega. \tag{3.5.2}$$

As usual, λ, μ are the Lamé constant of elastic material; further we shall also use ν for the Poisson ratio and E for the Young modulus of the elastic medium; the constant γ is the intensity of the gravity field, $\boldsymbol{e}$ is a unit vector.

For the case where g is a disk $\{(x_1,x_2) : x_1^2 + x_2^2 < R^2\}$ and $\boldsymbol{e} = \boldsymbol{e}^{(3)}$, the Ox_3 basis vector, the problem (3.5.1), (3.5.2) was posed by Lekhnitskii [46]. The axial cross-section of the domain is shown in Fig. 3.8. A particular axisymmetric solution was presented in [46]:

$$\begin{aligned} u_r(\boldsymbol{x}) &= -\nu\frac{1+\nu}{1-\nu}\frac{R^2}{Er}\gamma z, \\ u_z(\boldsymbol{x}) &= \frac{2\nu^2+\nu-1}{2(1-\nu)}\frac{\gamma z^2}{E} + \nu\frac{1+\nu}{1-\nu}\frac{\gamma R^2}{E}\ln\frac{r}{R}, \end{aligned} \tag{3.5.3}$$

$$\begin{aligned} \sigma_{rr}(\boldsymbol{u};\boldsymbol{x}) &= -\frac{\nu\gamma z}{1-\nu}\left(1-\frac{R^2}{r^2}\right), \\ \sigma_{\varphi\varphi}(\boldsymbol{u};\boldsymbol{x}) &= -\frac{\nu\gamma z}{1-\nu}\left(1+\frac{R^2}{r^2}\right), \end{aligned}$$

$$\sigma_{zz}(\boldsymbol{u};\boldsymbol{x}) = -\gamma z. \tag{3.5.4}$$

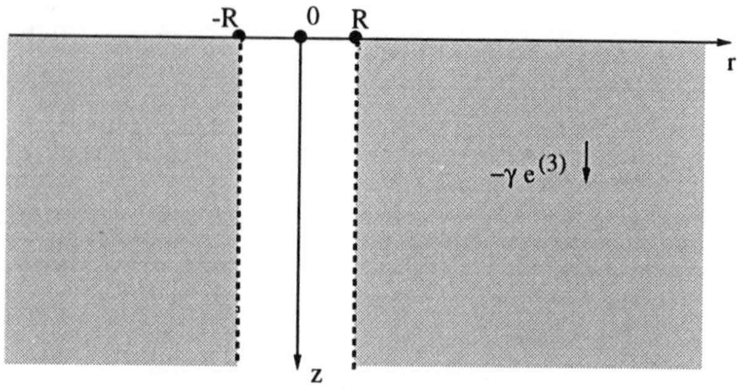

Fig. 3.8: An elastic half-space with a cylindrical shaft.

The remaining stress and displacement components are equal to zero. It should be noted that for the axisymmetric region Ω a solution of problem (3.5.1), (3.5.2) for an arbitrary vector $\boldsymbol{e}$ can be obtained by a superposition of the Lekhnitskii solution and the solution corresponding to $\boldsymbol{e} = \boldsymbol{e}^{(1)}$ which has been found by Geogdzhaev; in this case the displacement components can be written as follows:

$$u_1(\boldsymbol{x}) = \frac{\gamma}{4E}\Bigg\{(3+2\nu)(1+\nu)\frac{R^4}{4r^2}\cos 2\varphi$$

$$-\left[(3+2\nu)(1+\nu)(\cos\varphi)^2 + (17+15\nu-2\nu^2)(\sin\varphi)^2\right]\frac{r^2}{4}$$

$$+(3-2\nu)(1+\nu)R^2\ln(R/r) + (1+\nu)R^2(\sin\varphi)^2\Bigg\},$$

$$u_2(\boldsymbol{x}) = \frac{\gamma}{4E}\Big\{(3+2\nu)(1+\nu)\frac{R^4}{2r^2} + (14+10\nu-4\nu^2)\frac{r^2}{4}$$

$$- (1+\nu)R^2\Big\}\sin\varphi\cos\varphi,$$

$$u_3(\boldsymbol{x}) = -\nu(1+\nu)\frac{\gamma z}{2E}(\frac{R^2}{r}+r)\cos\varphi. \tag{3.5.5}$$

The stress components are given by

$$\sigma_{rr}(\boldsymbol{u};\boldsymbol{x}) = -\frac{\gamma}{4}\left[\frac{1}{2}(3+2\nu)(\frac{R^4}{r^3}-r)+3(r-\frac{R^2}{r})\right]\cos\varphi,$$

$$\sigma_{\varphi\varphi}(\boldsymbol{u};\boldsymbol{x}) = \frac{\gamma}{4}\left[\frac{1}{4}(3+2\nu)(\frac{R^4}{r^3}-r)+(1+2\nu)r-(1-2\nu)\frac{R^2}{r}\right]\cos\varphi,$$

$$\sigma_{r\varphi}(\boldsymbol{u};\boldsymbol{x}) = -\frac{\gamma}{4}\left[\frac{1}{4}(3+2\nu)(\frac{R^4}{r^2}-r)-(1-\nu)(r-\frac{R^2}{r})\right]\sin\varphi,$$

$$\sigma_{rz}(\boldsymbol{u};\boldsymbol{x}) = \nu\gamma\frac{z}{4}(\frac{R^2}{r^2}-1)\cos\varphi,$$

$$\sigma_{\varphi z}(\boldsymbol{u};\boldsymbol{x}) = \nu\gamma\frac{z}{4}(\frac{R^2}{r^2}+1)\sin\varphi,$$

$$\sigma_{zz}(\boldsymbol{u};\boldsymbol{x}) = 0. \tag{3.5.6}$$

One can easily see that both fields (3.5.3) and (3.5.5) are characterized by the quadratic growth at infinity. This gives a motivation to study solution of the homogeneous Neumann boundary value problem for the Lamé system in a half-space which is $O(||\boldsymbol{x}||^2)$, as $||\boldsymbol{x}|| \to \infty$.

3.5.2 Polynomial solution of the homogeneous equations in a half-space

Clearly, at infinity the points of the cylinder G belong to any conical neighbourhood of the Ox_3 axis. Then, it follows from Maz'ya, Nazarov and Plamenevskii [57] and Arutyunyan, Movchan and Nazarov [3] that the principal part of the asymptotic expansion of the solution of (3.5.1), (3.5.2) at infinity should agree with the appropriate solution of the homogeneous Neumann problem in a half-space. Here, we present those polynomial solutions which do not decrease but have no more than quadratic growth at infinity.

Let $\mathbf{V}^{(m,q)}$ denote the vector polynomials of degree $m = 0, 1, 2$, which satisfy the homogeneous Lamé system in the half-space and the homogeneous traction boundary conditions on the surface. The number of such linearly independent polynomials is equal to $3(m+1)$.

First, take $m = 0$. The vectors $\mathbf{V}^{(0,q)}$, $q = 1, 2, 3$, correspond to the rigid-body translations

$$\boldsymbol{e}^{(1)}, \ \boldsymbol{e}^{(2)}, \ \text{and} \ \boldsymbol{e}^{(3)}. \tag{3.5.7}$$

For the case $m = 1$ three of the polynomials $\mathbf{V}^{(1,q)}$, $q = 1, 2, \ldots, 6$, give rigid-body rotations

$$\boldsymbol{x} \times \boldsymbol{e}^{(q)}, \ q = 1, 2, 3,$$

and the remaining three polynomials are specified by

$$(x_1, -x_2, 0)^T, \ (x_1, x_2, -\frac{2\lambda}{\lambda + \mu} x_3)^T, \ (x_2, x_1, 0)^T \tag{3.5.8}$$

The last group $\mathbf{V}^{(2,q)}$, $q = 1, 2, \ldots, 9$, of quadratic polynomials can be written in the form

$$(x_2 x_3, x_3 x_1, -x_1 x_2)^T, \ (-2x_1 x_3, 0, x_1^2 + \lambda(\lambda + 2\mu)^{-1} x_3^2)^T,$$

$$(0, -2x_2 x_3, x_2^2 + \lambda(\lambda + 2\mu)^{-1} x_3^2)^T, \ (0, x_1^2 - x_3^2, 0)^T, \ (x_2^2 - x_3^2, 0, 0)^T,$$

$$((\lambda^{-1}\mu + 1)x_1^2, \ -(4\lambda^{-1}\mu + 3)x_1 x_2, x_1 x_3)^T, \ (x_2^2 - x_1^2, 2x_1 x_2, 0)^T,$$

$$(2x_1 x_2, x_1^2 - x_2^2, 0)^T, \ (-(4\lambda^{-1}\mu + 3)x_1 x_2, (\lambda^{-1}\mu + 1)x_2^2, x_2 x_3)^T. \tag{3.5.9}$$

The corresponding stress tensor components $\sigma_{jk}^{(m,q)} := \sigma_{jk}(\mathbf{V}^{(m,q)}; \boldsymbol{x})$ are

$$\sigma_{11}^{(1,1)} = -\sigma_{22}^{(1,1)} = \sigma_{12}^{(1,2)} = 2\mu,$$

$$\sigma_{11}^{(1,3)} = \sigma_{22}^{(1,3)} = 2\mu(2\mu + 3\lambda)(2\mu + \lambda)^{-1},$$

$$\sigma_{22}^{(2,2)} = \sigma_{11}^{(2,1)} = -8\mu(\lambda + \mu)(\lambda + 2\mu)^{-1} x_3,$$

$$\sigma_{22}^{(2,1)} = \sigma_{11}^{(2,2)} = -4\mu\lambda(\lambda + 2\mu)^{-1} x_3,$$

$$\sigma_{12}^{(2,2+j)} = 2\mu x_j,\ \sigma_{3,3-j}^{(2,2+j)} = -2\mu x_3,\ \sigma_{jj}^{(2,4+j)} = -4\mu x_j,$$

$$\sigma_{3-j,3-j}^{(2,4+j)} = 4\mu x_j,\ \sigma_{12}^{(2,4+j)} = 4\mu x_{3-j},\ \sigma_{jj}^{(2,6+j)} = 2\mu(\lambda+2\mu)(\lambda+\mu)^{-1}x_2,$$

$$\sigma_{3-j,3-j}^{(2,6+j)} = -8\mu x_2,\ \sigma_{3j}^{(2,6+j)} = \lambda\mu(\lambda+\mu)^{-1}x_3,$$

$$\sigma_{12}^{(2,6+j)} = -\mu(3\lambda+4\mu)(\lambda+\mu)^{-1}x_{3-j},\ j=1,2,$$

$$\sigma_{12}^{(2,9)} = 2\mu x_3. \tag{3.5.10}$$

The remaining stress tensor components are assumed to be zero.

3.5.3 Auxiliary two–dimensional problems

It is important to mention that the set of two–dimensional model problems is required to describe the displacement and stress field near a cylindrical cavity. It can be easily seen that a solution of the homogeneous Neumann boundary value problem for the Lamé system in a half–space leaves a discrepancy in the boundary conditions on ∂G. This discrepancy can be compensated by solutions of certain boundary value problems posed in $\mathbb{R}^2\backslash g$. The appropriate differential operators can be obtained from the system (3.5.1), (3.5.2) by eliminating derivatives with respect to z. These operators correspond to the plane-strain and anti–plane shear problems on $\mathbb{R}^2 \setminus \overline{g}$,

$$\mu\Delta\mathbf{W}(\boldsymbol{y}) + (\lambda+\mu)\nabla\nabla\cdot\mathbf{W}(\boldsymbol{y}) = \mathbf{0},\ \mu\Delta w_3(\boldsymbol{y}) = 0,\ \boldsymbol{y}\in\mathbb{R}^2\backslash\overline{g}, \tag{3.5.11}$$

$$\boldsymbol{\sigma}^{(n)}(\mathbf{W};\boldsymbol{y}) = \mathbf{P}(\boldsymbol{y}), \mu\frac{\partial w_3}{\partial \boldsymbol{n}}(\boldsymbol{y}) = p_3(\boldsymbol{y}),\ \boldsymbol{y}\in\partial g. \tag{3.5.12}$$

Here, $\mathbf{W} = (w_1,\ w_2)$; in the boundary conditions the two–dimensional traction operator is assumed; $\boldsymbol{n}$ is a unit outward normal vector with respect to ∂g.

Let $\boldsymbol{w}$ denote the vector $(w_1,\ w_2,\ w_3)$. Then, the solution $\boldsymbol{w}$ of the problem (3.5.11), (3.5.12) exists for any smooth loads $\mathbf{P} := (p_1, p_2)^T$ and p_3, and the following asymptotic expansion holds at infinity:

$$\boldsymbol{w}(\boldsymbol{y}) = \sum_{j=1}^{3}\left\{c_j\mathbf{T}^{(j)}(\boldsymbol{y}) + \sum_{k=1}^{2} c_{jk}\frac{\partial}{\partial y_k}\mathbf{T}^{(j)}(\boldsymbol{y})\right\}$$

$$+ O(\|\boldsymbol{y}\|^{-2}),\ \text{as } \|\boldsymbol{y}\|\to\infty, \tag{3.5.13}$$

where c_j, c_{jk} are constants, $\mathbf{T}^{(j)}$ are columns of the 3×3 matrix with the block–diagonal structure. The first diagonal block has the size 2×2

and it corresponds to the two–dimensional Somigliana tensor; the (33)–element is

$$-\frac{1}{2\pi\mu}\ln\|\boldsymbol{y}\|,$$

which is the fundamental solution for the operator $-\mu\Delta$ in $\mathbb{R}^2$.

Displacement field (3.5.13) has a logarithmic growth at infinity. It decays if and only if the principal force vector of **P** and the mean value of p_3 on ∂g are equal to zero, and, therefore, the coefficients c_j vanish. Assume that in (3.5.12)

$$p_j(\boldsymbol{y}) = \sum_{k=1}^{3} n_k(\boldsymbol{y})(a_0^{(j,k)} + a_1^{(j,k)}y_1 + a_2^{(j,k)}y_2), \tag{3.5.14}$$

with constant $a_p^{(j,k)}$, then with the use of the Green's theorem in a finite but large region and by taking the limit as the diameter of this region tends to infinity, we obtain the following formula for the coefficients in the asymptotic expansion (3.5.13):

$$c_j = -\text{mes}g(a_1^{(j,1)} + a_2^{(j,2)}). \tag{3.5.15}$$

3.5.4 Solutions of the homogeneous boundary value problem in a half-space with the vertical cylindrical cavity

We shall construct the solution which is different from one presented by Lekhnitskii, namely, the displacement field that is not decay at infinity. In this case the Lekhnitskii problem requires specification of the conditions at infinity for its correct formulation.

Consider displacement fields that solve the homogeneous problem (3.5.1), (3.5.2) and do not decay at infinity. One can select immediately six rigid–body displacements. Also, there are solutions of the homogeneous equations which have the polynomial asymptotics (3.5.8), (3.5.9) at infinity. Using the formulae (3.5.10) for the stress tensor components one can calculate the discrepancy left by these polynomial fields in the traction boundary conditions.

Introduce the cut–off function $\chi \in \mathbb{C}^\infty(\mathbb{R}^1)$ such that $\chi(t) = 1$ for $t \le 2R$ and $\chi(t) = 0$ for $t \ge 3R$, for some quantity R which is selected in such a way that the domain g belongs to the circle with the centre at the origin and the radius R. Next, define the field $\boldsymbol{v}^{(1,q)}$ by the following equality:

$$\boldsymbol{v}^{(1,q)}(\boldsymbol{x}) = \mathbf{V}^{(1,q)}(\boldsymbol{x}) + \chi(\|\boldsymbol{y}\|/z)\boldsymbol{w}^{(1,q)}(\boldsymbol{y}). \tag{3.5.16}$$

We look at the discrepancy left by the field (3.5.16) in the homogeneous equations and tractions boundary conditions (3.5.1), (3.5.2). We are interested in the asymptotic behaviour of the solution at infinity (as $||\boldsymbol{x}|| \to \infty$). The boundary conditions are satisfied, and the discrepancy (let us call it $\mathbf{F}^{(1,q)}$) in the Lamé system is concentrated within the cone

$$K_R = \left\{ \boldsymbol{x} : z^{-1}||\boldsymbol{y}|| \in (2R, 3R) \right\}.$$

The inequalities

$$c_1 z < ||\boldsymbol{x}|| < C_1 z, \; c_2||\boldsymbol{y}|| < ||\boldsymbol{x}|| < C_2||\boldsymbol{y}|| \tag{3.5.17}$$

are valid for all points of this cone, where c_k and C_k are certain positive constant. The following estimate holds for large values of $||\boldsymbol{x}||$:

$$||\mathbf{F}^{(1,q)}(\boldsymbol{x})|| \leq \text{const } ||\boldsymbol{x}||^{-3}.$$

There exists a field $\boldsymbol{u}^{(1,q)}$ with the finite elastic energy such that it cancels the discrepancy, and the sum

$$\mathbf{U}^{(1,q)}(\boldsymbol{x}) = \boldsymbol{v}^{(1,q)}(\boldsymbol{x}) + \boldsymbol{u}^{(1,q)}(\boldsymbol{x}), \; q = 1, 2, 3, \tag{3.5.18}$$

satisfies the homogeneous boundary value problem (3.5.1), (3.5.2).

Now, consider the polynomial fields (3.5.9). Due to (3.5.10), the residual of the vector field $\boldsymbol{v}^{(2,q)}$ in the boundary conditions (3.5.2) has the form (3.5.14) with constant coefficients $a_1^{(j,k)}$, $a_2^{(j,k)}$ and $a_0^{(j,k)} = c_{jk}z$, $c_{jk} = \text{const}$. Then, there exists a vector function

$$\boldsymbol{w}^{(2,q)}(\boldsymbol{y}, z) = \boldsymbol{w}^{(2,q,0)}(\boldsymbol{y}) + z\boldsymbol{w}^{(2,q,1)}(\boldsymbol{y}), \tag{3.5.19}$$

subjected to the homogeneous Lamé system. It cancels the discrepancy in the boundary conditions on ∂G. The terms in the right–hand side of (3.5.19) admit the following estimate for large values of $\boldsymbol{y}$:

$$\boldsymbol{w}^{(2,q,1)}(\boldsymbol{y}) = O(||\boldsymbol{y}||^{-1}), \;\; \boldsymbol{w}^{(2,q,0)}(\boldsymbol{y}) = O(|\ln||\boldsymbol{y}|||).$$

The vector function

$$\mathbf{V}^{(2,q)}(\boldsymbol{x}) + \chi(\frac{||\boldsymbol{y}||}{z})\boldsymbol{w}^{(2,q)}(\boldsymbol{y})$$

satisfies the boundary conditions on $\partial\Omega$ outside a spherical neighbourhood of the origin. The discrepancy in the Lamé system is concentrated within the cone K_R and admits the representation

$$\mathbf{F}^{(2,q)}(\boldsymbol{x}) + O(\|\boldsymbol{x}\|^{-3}|\ln\|\boldsymbol{x}\||). \tag{3.5.20}$$

The first term in (3.5.20) has the form

$$\mathbf{F}^{(2,q)}(\boldsymbol{x}) = \frac{\ln\|\boldsymbol{x}\|}{\|\boldsymbol{x}\|^2}\boldsymbol{f}^{(q,0)}(\frac{\boldsymbol{x}}{\|\boldsymbol{x}\|}) + \frac{1}{\|\boldsymbol{x}\|^2}\boldsymbol{f}^{(q,1)}(\frac{\boldsymbol{x}}{\|\boldsymbol{x}\|}). \tag{3.5.21}$$

The inequalities (3.5.17) are used in order to estimate the remainder. Here $\boldsymbol{f}^{(q,j)}$, $j = 0, 1$, are smooth functions defined on the semi–sphere

$$\mathbb{S}_+ = \{\boldsymbol{x} : \|\boldsymbol{x}\| = 1, x_3 > 0\},$$

with the supports in the intersection $\mathbb{S}_+ \cap K_R$.

The boundary value problem for the Lamé system in $\mathbb{R}^3_+$ with the right–hand side (3.5.21) and the homogeneous traction boundary conditions has the particular solution of the form

$$\boldsymbol{\psi}^{(2,q)}(\boldsymbol{x}) = (\ln\|\boldsymbol{x}\|)^2\boldsymbol{\psi}^{(2,q,2)}(\frac{\boldsymbol{x}}{\|\boldsymbol{x}\|}) + \ln\|\boldsymbol{x}\|\boldsymbol{\psi}^{(2,q,1)}(\frac{\boldsymbol{x}}{\|\boldsymbol{x}\|})$$

$$+\,\boldsymbol{\psi}^{(2,q,0)}(\frac{\boldsymbol{x}}{\|\boldsymbol{x}\|}), \tag{3.5.22}$$

with $\boldsymbol{\psi}^{(2,q,j)}$ being smooth functions in $\mathbb{S}_+$.

The vector field (3.5.22) cancels the principal part of the residual (3.5.20) in the Lamé system; however, it leaves the discrepancy $\mathbf{P}$ in the traction boundary conditions on ∂G,

$$\mathbf{P}(\boldsymbol{x}) = z^{-1}[(\ln z)^2\boldsymbol{p}^{(q,2)}(\boldsymbol{y}) + (\ln z)\boldsymbol{p}^{(q,1)}(\boldsymbol{y}) + \boldsymbol{p}^{(q,0)}(\boldsymbol{y})]$$

$$+\,O(z^{-2}|\ln z|^2), \tag{3.5.23}$$

where the components of the vectors $\boldsymbol{p}^{(q,j)}$, $j = 0, 1, 2$, have zero mean value over ∂g. In this case there exist solutions $\boldsymbol{\Psi}^{(q,j)}$ of the boundary value problem (3.5.1), (3.5.2) with the right–hand sides $\boldsymbol{p}^{(q,j)}$ such that

$$\boldsymbol{\Psi}^{(q,j)} = O(\|\boldsymbol{y}\|^{-1}).$$

On the basis of the above analysis we set

$$v^{(2,q)}(x) = \mathbf{V}^{(2,q)}(x) + \chi(\frac{||y||}{z})w^{(2,q)}(y,z) + \psi^{(2,q)}(x) \qquad (3.5.24)$$

$$+\frac{1}{z}\chi(\frac{||y||}{z})\left[(\ln z)^2\Psi^{(q,2)}(y) + (\ln z)\Psi^{(q,1)}(y) + \Psi^{(q,0)}(y)\right].$$

Due to (3.5.21), (3.5.23), the residuals $\mathbf{F}$ and $\mathbf{S}$, left by the displacement field (3.5.24) in the Lamé system (3.5.1) and the traction boundary conditions (3.5.2), admit the following estimates:

$$||\mathbf{S}(x)|| \leq \text{const } |\frac{\ln||x||}{||x||}|^2, \; ||\mathbf{F}(x)|| \leq \text{const } |\frac{\ln||x||}{||x||^3}|, \text{ for } ||y|| > 3Rz,$$

and

$$||\mathbf{F}(x)|| \leq \text{const } \frac{|\ln||x|||}{||x||^2}\left(\frac{1}{||x||} + \frac{|\ln||x|||}{||y||^2}\right), \text{ for } ||y|| < 3Rz.$$

It follows that the integrals

$$\int_\Omega ||x||^2||\mathbf{F}(x)||^2 dx, \; \int_{\partial\Omega} ||x||^2||\mathbf{S}(x)||^2 ds,$$

are finite, and, consequently, there exist vector functions $u^{(2,q)}$ that cancel the above residuals and vanish at infinity.

Finally, the displacement fields

$$\mathbf{U}^{(2,q)}(x) = v^{(2,q)}(x) + u^{(2,q)}(x), \; q = 1, \ldots, 9, \qquad (3.5.25)$$

satisfy the homogeneous boundary value problem (3.5.1), (3.5.2).

We should mention that the vectors (3.5.18) and (3.5.25) are linearly independent and span the set of all solutions $\mathbf{U}$ of the homogeneous boundary value problem (3.5.1), (3.5.2) that satisfy the inequality

$$||\mathbf{U}(x)|| \leq \text{const } ||x||^2. \qquad (3.5.26)$$

3.5.5 Correct formulations of the Lekhnitskii problem

Here we consider the axisymmetric boundary value problem in Ω, where the right–hand sides of the equations and traction boundary conditions were posed by Lekhnitskii. The particular solution, obtained by Lekhnitskii, has the form (3.5.3), (3.5.4). Among the linear combinations of the vectors (3.5.8), (3.5.9) one can find just one vector which is invariant with respect to a rotation about the Oz axis. It is given by

$$\mathbf{V}(\boldsymbol{x}) = \mathbf{V}^{(2,1)}(\boldsymbol{x}) + \mathbf{V}^{(2,2)}(\boldsymbol{x})$$

$$= (-2y_1 z, -2y_2 z, r^2 + \frac{2\lambda}{\lambda + 2\mu} z^2)^T. \tag{3.5.27}$$

The solution

$$\mathbf{U}(\boldsymbol{x}) = \mathbf{U}^{(2,1)}(\boldsymbol{x}) + \mathbf{U}^{(2,2)}(\boldsymbol{x}) \tag{3.5.28}$$

of the homogeneous problem (3.5.1), (3.5.2), corresponding to (3.5.27), has the form

$$U_r(r,z) = -2zr + 2z\frac{1+\nu}{1-\nu}\frac{R^2}{r}, \quad U_\varphi(r,z) = 0,$$

$$U_z(r,z) = \frac{2\nu z^2}{1-\nu} + r^2 + 2\frac{1+\nu}{1-\nu}R^2 \ln\frac{r}{R}. \tag{3.5.29}$$

To obtain this solution one can follow the way described above. It should be emphasized that the algorithm provides the asymptotic forms only. However, in our particular axisymmetric case the solution has been found in the closed form.

The stress tensor components admit the representation

$$\sigma_{rr}(\mathbf{U}; r, z) = -2\mu z\frac{1+\nu}{1-\nu}(1 - \frac{R^2}{r^2}), \quad \sigma_{\varphi\varphi}(\mathbf{U}; r, z) = -2\mu z\frac{1+\nu}{1-\nu}(1 + \frac{R^2}{r^2}),$$

$$\sigma_{zz}(\mathbf{U}; r, z) = \sigma_{rz}(\mathbf{U}; r, z) = 0. \tag{3.5.30}$$

It is interesting to compare (3.5.29) with (3.5.3). One can see that in (3.5.3) the components u_1 and u_2 decay in the layer

$$Q_d = \{\boldsymbol{x} \in \Omega : z < d\}, \; d \in \mathbb{R}^1,$$

as $||\boldsymbol{x}|| \to \infty$. In contrast, the displacement components (3.5.29) do not have similar properties. Thus, the additional conditions that would yield the uniqueness for the solution (3.5.3), (3.5.4) have the form

$$u_j(\boldsymbol{x}) = o(1), \text{ as } ||\boldsymbol{x}|| \to \infty, \; \boldsymbol{x} \in Q_d, \; j = 1, 2. \tag{3.5.31}$$

Clearly, in order to satisfy (3.5.31), one has to assume that the coefficient C in the representation

$$\boldsymbol{u}^{(1)} = \boldsymbol{u} + C\mathbf{U}, \tag{3.5.32}$$

equals zero. On the other hand, the validity of the conditions

$$\sigma_{jk}(\boldsymbol{u}^{(1)}; \boldsymbol{x}) = o(1), \text{ as } \|\boldsymbol{x}\| \to \infty, \ \boldsymbol{x} \in Q_d, \ j,k = 1,2, \tag{3.5.33}$$

can be achieved for the vector (3.5.32).

Direct calculations show that if

$$C = -\frac{\nu\gamma}{2\mu(1+\nu)},$$

then the components of the vector (3.5.32) take the form

$$u_r^{(1)}(\boldsymbol{x}) = \frac{\nu\gamma z r}{E}, \ u_z^{(1)}(\boldsymbol{x}) = -\gamma\frac{z^2 + \nu r^2}{2E}, \ u_\varphi^{(1)}(\boldsymbol{x}) = 0, \tag{3.5.34}$$

and the corresponding stress tensor components are specified by

$$\sigma_{zz}(\boldsymbol{u}^{(1)}; \boldsymbol{x}) = -\gamma z,$$

$$\sigma_{rr}(\boldsymbol{u}^{(1)}; \boldsymbol{x}) = \sigma_{\varphi\varphi}(\mathbf{u}^{(1)}; \boldsymbol{x}) = \sigma_{rz}(\boldsymbol{u}^{(1)}; \boldsymbol{x}) = 0. \tag{3.5.35}$$

We can summarize that the Lekhnitskii problem admits two formulations in the class of vector functions which satisfy the estimate (3.5.26). The first formulation involves the additional requirement (3.5.31) of vanishing of the displacement component u_r in any layer of a finite thickness. The second formulation is related to the condition (3.5.33) of decay of the stress tensor components σ_{rr} and $\sigma_{\varphi\varphi}$. The unique solutions, corresponding to these formulations, are given by (3.5.3), (3.5.4) and (3.5.34), (3.5.35).

3.5.6 Vertical half-space with a cylindrical cavity

The formulae (3.5.5) show the solution constructed by Geogdzhaev for $e = e^{(1)}$. This field increases at infinity like $O(\|\boldsymbol{x}\|^2)$ in the layer Q_d. The solution of this problem is determined up to a linear combination of the 12 vector functions (3.5.18), (3.5.25). The coefficients are chosen in such a way that the sum of the linear combination mentioned and the field (3.5.5) has the least possible growth in the layer Q_d. The result is

$$\boldsymbol{u}^{(2)} = \boldsymbol{u} + \frac{\gamma(1+\nu)}{16E}\left\{16\mathbf{U}^{(2,4)} + (1-2\nu)\mathbf{U}^{(2,5)} + 4\mathbf{U}^{(2,7)}\right\}. \quad (3.5.36)$$

Now, following the procedure stated above, we describe the asymptotic form of the field (3.5.36) which solves the boundary value problem (3.5.1), (3.5.2), with $\boldsymbol{e} = \boldsymbol{e}^{(1)}$.

First, consider the following solution of the Lamé system (3.5.1)

$$\mathbf{V}(\boldsymbol{x}) = -\frac{\gamma}{2\mu}(z^2, 0, 0)^T. \quad (3.5.37)$$

The only non-zero component of the stress tensor is

$$\sigma_{rz}(\mathbf{V}; \boldsymbol{x}) = -\gamma z \cos\varphi.$$

The approximation of the vector function $\boldsymbol{u}^{(1)}$ is given by the sum

$$z\boldsymbol{w}^{(1)}(\boldsymbol{y}) + \boldsymbol{w}^{(0)}(\boldsymbol{y}). \quad (3.5.38)$$

The vector function $\boldsymbol{w}^{(1)}$ from (3.5.38) is

$$\boldsymbol{w}^{(1)} = (0, 0, w_3^{(1)})^T,$$

where $w_3^{(1)}$ satisfies the Neumann boundary value problem in the exterior of the domain g,

$$\mu\Delta w_3^{(1)}(\boldsymbol{y}) = 0, \ \boldsymbol{y} \in \mathbb{R}^2 \setminus \overline{g}, \ \mu\frac{\partial w_3^{(1)}}{\partial r}(\boldsymbol{y}) = \gamma\cos\varphi, \boldsymbol{y} \in \partial g, \quad (3.5.39)$$

and is given by the equality

$$w_3^{(1)} = -\frac{\gamma R^2}{\mu r}\cos\varphi. \quad (3.5.40)$$

Let us recall that in our case the domain g is the disk of the radius R.

A two–dimensional vector $\mathbf{W}^{(0)}$ with components which coincide with the first two components of the vector

$$\boldsymbol{w}^{(0)} = (w_1^{(0)}, \ w_2^{(0)}, \ 0)^T$$

from (3.5.38), satisfies the following plane strain problem in $\mathbb{R}^2 \setminus g$:

$$\mu\Delta \mathbf{W}^{(0)}(\boldsymbol{y}) + (\lambda+\mu)\nabla\nabla\cdot\mathbf{W}^{(0)}(\boldsymbol{y})$$

$$-\gamma R^2 \frac{\lambda+\mu}{\mu}\nabla\left(\frac{\cos\varphi}{r}\right) = \mathbf{0},\ \boldsymbol{y}\in\mathbb{R}^2\setminus\overline{g},$$

$$\sigma_{rr}(\mathbf{W}^{(0)}; R,\varphi) = \gamma R\lambda\mu^{-1}\cos\varphi, \sigma_{r\varphi}(\mathbf{W}^{(0)}; R,\varphi) = 0.$$

The solution of this problem is given by the formula

$$\mathbf{W}^{(0)}(r,\varphi) = \frac{\gamma R^2}{2\mu\nu}\begin{pmatrix} (3-2\nu)\ln(rR^{-1})\cos\varphi \\ -((3-2\nu)\ln(rR^{-1})+1)\sin\varphi \end{pmatrix}$$

$$-\frac{\gamma R^4(1+2\nu)}{16\mu r^2}\begin{pmatrix}\cos\varphi\\ \sin\varphi\end{pmatrix}$$

$$-\frac{\gamma R^2(4-\nu)(3-4\nu)}{8\mu(1-\nu)}\begin{pmatrix} \ln(rR^{-1})\cos\varphi \\ -(\ln(rR^{-1})+(3-4\nu)^{-1})\sin\varphi\end{pmatrix}. \quad (3.5.41)$$

The approximation required has the form

$$\boldsymbol{v}(\boldsymbol{x}) = \mathbf{V}(\boldsymbol{x}) + z\boldsymbol{w}^{(1)}(\boldsymbol{y}) + \boldsymbol{w}^{(0)}(\boldsymbol{y}). \quad (3.5.42)$$

Here the functions (3.5.40), (3.5.41) are defined everywhere in Ω, and we do not use the multiplication by the cut–off function (compare with (3.5.16), (3.5.24)). The sum (3.5.42) satisfies the homogeneous boundary conditions (3.5.2) on the boundary of the half–space, provided

$$\sigma_{33}(z\boldsymbol{w}^{(1)} + \boldsymbol{w}^{(0)}; \boldsymbol{y}, z) := (2\mu+\lambda)w_3^{(1)}(\boldsymbol{y}) + \nabla\cdot\mathbf{W}^{(0)}(\boldsymbol{y}) = 0. \quad (3.5.43)$$

Using (3.5.40), (3.5.41) one can verify that (3.5.43) does not hold. Therefore, the representation (3.5.42) gives only the asymptotic approximation

$$\boldsymbol{u}^{(2)}(\boldsymbol{x}) = \boldsymbol{v}(\boldsymbol{x}) + O(|\ln\|\boldsymbol{x}\||^2), \quad (3.5.44)$$

but not the exact solution.

The solution (3.5.36) is not specified explicitly in terms of elementary functions, and we can only state its existence and asymptotic properties at infinity. The solution mentioned is characterized by a growth of the displacement at infinity within the layer Q_d (see (3.5.44) and (3.5.38)–(3.5.42)). Since all the solutions (3.5.18), (3.5.25) of the homogeneous boundary value problem grow like $O(r)$ or $O(r^2)$ (compare with (3.5.8),

(3.5.9)) within this layer, the solution of the nonhomogeneous problem, which admits the growth

$$u^{(2)} = O(|\ln \|x\||^2), \text{ as } \|x\| \to \infty, \ x \in Q_d, \tag{3.5.45}$$

is unique (up to a rigid body displacement).

Next, we shall examine the sum of the vector field (3.5.5) and a linear combination of (3.5.18) and (3.5.25). The components of stress (3.5.6) grow like $O(r^2)$ within the layer Q_d. A certain choice of coefficients of the linear combination mentioned provides the least possible growth of the stress in this layer. The result is

$$u^{(3)} = u + c_1 \mathbf{V}^{(2,4)} + c_2 \mathbf{V}^{(2,5)} + c_3 \mathbf{V}^{(2,7)}, \tag{3.5.46}$$

where

$$c_1 = -2c(2\nu^2 + 9\nu + 2), \ c_2 = c(2\nu^2 + 9\nu - 13), \ c_3 = -10c, \ c = \gamma/(16E).$$

The stress component σ_{13}, corresponding to (3.5.46), equals

$$\gamma z/4 + o(1),$$

and is bounded within the layer Q_d. The remaining components of the stress tensor decrease in Q_d, as $\|x\| \to \infty$.

It follows from the formulae (3.5.10) that the functions $\sigma_{ik}(\mathrm{U}^{(j,q)}; x)$, where $j = 1, 2$; $q = 1, \ldots, 3^n$; $i, k = 1, 2$, do not vanish at infinity. Consequently, the solution (3.5.46) of the problem (3.5.1), (3.5.2) for $e = e^{(1)}$, subjected to the additional condition (3.5.33), is unique (up to a rigid body displacement).

Thus, (3.5.46) and (3.5.46) are two different solutions of the problem (3.5.1), (3.5.2) for $e = e^{(1)}$ that satisfy the conditions (3.5.45) and (3.5.33), respectively.

3.5.7 Remarks on a different orientation of the gravity field

Due to the linearity of the original formulation, the solution of (3.5.1), (3.5.2) can be obtained for an arbitrary orientation of the vector e by a superposition of solutions of the Lekhnitskii problem and of the problem analyzed in the previous subsection. Although we considered the case where g is a disk, the asymptotic algorithm holds for the nonsymmetric case as well. Two correct formulations exist.

The restriction of growth of the displacement components, given by (3.5.45), yields the first correct formulation of the problem. The solution of (3.5.1), (3.5.2) and (3.5.45) is unique up to a rigid body displacement. Moreover, the constraint (3.5.45) can be made weaker and replaced by

$$\boldsymbol{u}(\boldsymbol{x}) = o(\|\boldsymbol{x}\|), \text{ as } \|\boldsymbol{x}\| \to \infty, \ \boldsymbol{x} \in Q_d. \tag{3.5.47}$$

The second correct formulation is obtained by means of constraints on the asymptotics of the stress components in the layer Q_d. The corresponding problem (3.5.1), (3.5.2), (3.5.33) has the unique solution (specified up to a rigid body displacement). Moreover, we can make the stronger assumption and prescribe

$$\sigma_{jk}(\boldsymbol{u};\boldsymbol{x}) = o(1), \ j,k = 1,2,$$

$$\boldsymbol{\sigma}^{(3)}(\boldsymbol{u};\boldsymbol{x}) = O(1), \text{ as } \|\boldsymbol{x}\| \to \infty, \ \boldsymbol{x} \in Q_d. \tag{3.5.48}$$

The boundary value problems (3.5.1), (3.5.2), (3.5.48) and (3.5.1), (3.5.2), (3.5.33) are equivalent.

3.5.8 Corollaries

Note that both groups of fields (3.5.3), (3.5.4) and (3.5.34), (3.5.35), corresponding to correct formulations of the Lekhnitskii problem can be obtained from the solution of the problem on a semi–infinite cylinder of a large diameter D in the limit $D \to \infty$. Suppose that

$$C_D = \{\boldsymbol{x} \in \mathbb{R}^3 : z > 0, \ r < D/2\}, \ \Gamma_D = \{\boldsymbol{x} \in \mathbb{R}^3 : z > 0, r = D/2\},$$

and

$$\Omega_D = \Omega \cap C_D.$$

Then the solution (3.5.3), (3.5.4) of the boundary value problem (3.5.1), (3.5.2), (3.5.31) for $\boldsymbol{e} = \boldsymbol{e}^{(3)}$ is the limit of the solution $\boldsymbol{u}_D$ of the problem posed in Ω_D with the conditions of rigid clamping on Γ_D, as $D \to \infty$.

On the other hand, the solution (3.5.34), (3.5.35) of the problem (3.5.1), (3.5.2), (3.5.33) can be treated as a limit of solution of the boundary value problem posed in a large cylinder where the lateral surface Γ_D is free of tractions.

3.6 Asymptotic interpretation of solutions to Lekhnitskii's problem

One can see from Section 3.5 that there exists more than one correct mathematical formulation of the Lekhnitskii problem on the deformation under gravity of the elastic half–space with a cylindrical shaft. The choice of some particular conditions at infinity should admit a certain physical interpretation. Clearly, a half–space is an idealization of a large but finite elastic body. Thus, we shall look at the deformation of finite bodies with relatively long shafts subjected to body forces. Namely, we consider

(i) a gravitating sphere;

(ii) a thick layer with the lower base subjected to one of the following conditions:

(a) it lays on a smooth rigid surface,

(b) it is bonded to a perfectly rigid half–space,

(c) it does not have a support.

The classical Lekhnitskii solution is given by

$$u_r^{(1)} = \frac{\nu h^2 \gamma z}{2(1-\nu)\mu r}, \quad u_\varphi^{(1)} = 0,$$

$$u_z^{(1)} = \frac{(1-2\nu)\gamma z^2}{4(1-\nu)\mu} + \frac{\nu\gamma h^2}{2(1-\nu)\mu} \ln\frac{r}{h}, \tag{3.6.1}$$

with the stress components

$$\sigma_{rr}(\boldsymbol{u}^{(1)};\boldsymbol{x}) = \frac{\nu\gamma z}{1-\nu}(1-\frac{h^2}{r^2}), \quad \sigma_{\varphi\varphi}(\boldsymbol{u}^{(1)};\boldsymbol{x}) = \frac{\nu\gamma z}{1-\nu}(1+\frac{h^2}{r^2}), \tag{3.6.2}$$

$$\sigma_{zz}(\boldsymbol{u}^{(1)};\boldsymbol{x}) = \gamma z, \quad \sigma_{rz}(\boldsymbol{u}^{(1)};\boldsymbol{x}) = 0,$$

where h is the radius of the cylinder, ν is the Poisson ratio, (r, φ, z) are cylindrical coordinates.

Another solution has the form

$$u_r^{(2)} = -\frac{\nu}{2\mu(1+\nu)}\gamma z r, \quad u_\varphi^{(2)} = 0, \quad u_z^{(2)} = \frac{\gamma}{4(1+\nu)\mu}(z^2+\nu r^2), \tag{3.6.3}$$

and the stress components can be written as

$$\sigma_{zz}(\boldsymbol{u}^{(2)};\boldsymbol{x}) = \gamma z,$$

$$\sigma_{rr}(\boldsymbol{u}^{(2)};\boldsymbol{x}) = \sigma_{\varphi\varphi}(\boldsymbol{u}^{(2)};\boldsymbol{x}) = \sigma_{rz}(\boldsymbol{u}^{(2)};\boldsymbol{x}) = 0. \tag{3.6.4}$$

It can be seen that the component $u_r^{(2)}$ increases at infinity. The solution (3.6.1) does not have this property. However, in (3.6.2) the stresses σ_{rr} and $\sigma_{\varphi\varphi}$ do not decay, as $r \to \infty$. Also, as $|z| > 2\gamma^{-1}\mu(1+\nu^{-1})$, one can observe the interpretation of the walls of the shaft. In contrast, in equations (3.6.3),(3.6.4) an increase in depth causes an increase in the cross–section of the shaft. These simple observations show that the range of applicability of these fields, obtained for an infinite region, is limited. Further, we shall look at a certain neighbourhood of the boundary of the half–space.

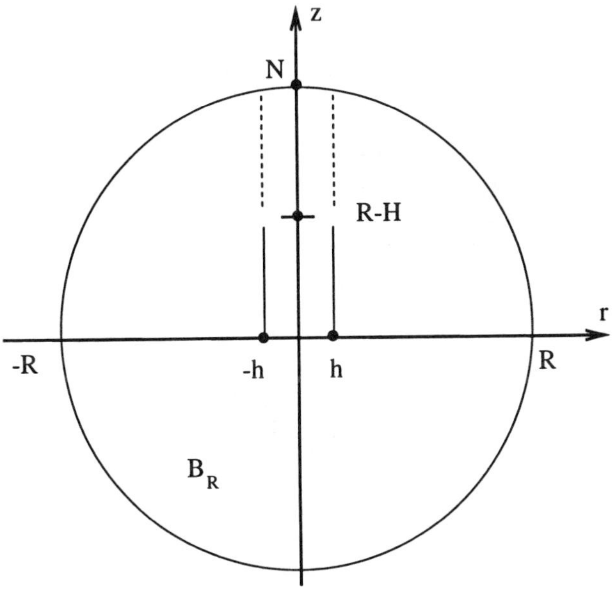

Fig. 3.9: An elastic ball with a cylindrical shaft.

In the examples (a), (b) and (c), mentioned at the beginning of the section, the principal part of the asymptotic solution consists of the following displacement and stress fields:

$$u_r = \frac{\sigma_\infty}{2\mu}\{\frac{1-\nu}{1+\nu}r + \frac{h^2}{r}\}, \quad u_\varphi = 0, \quad u_z = -\frac{\sigma_\infty \nu z}{\mu(1+\nu)}, \tag{3.6.5}$$

$$\sigma_{rr} = \sigma_\infty(1 - \frac{h^2}{r^2}), \quad \sigma_{\varphi\varphi} = \sigma_\infty(1 + \frac{h^2}{r^2}). \tag{3.6.6}$$

The remaining components of stresses are equal to zero. These fields correspond to the axisymmetric tension of intensity σ_∞ (compression, if $\sigma_\infty < 0$) of the half–space with the cylindrical shaft.

It should be noticed that the contribution of the fields (3.6.5), (3.6.6) to the asymptotic solution increases with the increase of the radius R of the sphere and of the radius d and the thickness R of the circular layer. It is well known that the stresses in horizontal planes predominate in mechanics of rock fracture, and the Lekhnitskii solution is not usually used for designing of shafts.

In the example (c) below we show that the Lekhnitskii solution may be used for the description of the asymptotics of the stress–strain state.

Below, we analyze the four examples mentioned at the beginning of the section.

3.6.1 Gravitating ball with a shaft

Let S_R be a ball of the radius R with the centre at the origin. The region S_R is occupied by an elastic material with the Lamé constants λ and μ. Here, $C_h = \{\boldsymbol{x} : ||\boldsymbol{x}|| < h\}$, and we also use the notations

$$\Omega = S_R \setminus \overline{G}, \quad G = \{\boldsymbol{x} \in \overline{C_h} : z > R - H\},$$

where H denotes the depth of the shaft (see Fig. 3.9). It will be assumed that $R \gg H \gg h$.

The displacement vector $\boldsymbol{u}$ satisfies the Lamé system and the homogeneous traction boundary conditions

$$\mathbf{L}(\frac{\partial}{\partial \boldsymbol{x}})\boldsymbol{u}(\boldsymbol{x}) + \mathbf{F}(\boldsymbol{x}) = \mathbf{0}, \quad \boldsymbol{x} \in \Omega; \tag{3.6.7}$$

$$\boldsymbol{\sigma}^{(n)}(\boldsymbol{u}; \boldsymbol{x}) = \mathbf{0}, \quad \boldsymbol{x} \in \partial\Omega, \tag{3.6.8}$$

where

$$\mathbf{F}(\boldsymbol{x}) = -\frac{\gamma\rho}{R}\boldsymbol{e}^{(\rho)} \tag{3.6.9}$$

is the gravity force with $\boldsymbol{e}^{(\rho)}$ being the unit basis vector in the radial direction in the spherical coordinate system (ρ, θ, φ). In this example one should use the singular perturbation approach, and outside some neighbourhood of the shaft G the field $\boldsymbol{u}$ is described (to leading order) by the solution $\boldsymbol{u}^{(0)}$ of the problem for the homogeneous gravitating ball, given by

$$u_\rho^{(0)} = \frac{\gamma\rho}{10(\lambda + 2\mu)R}(\rho^2 - \frac{R^2(6\mu + 5\lambda)}{2\mu + 3\lambda}), \tag{3.6.10}$$

with the stress tensor components

$$\sigma_{\rho\rho} = -\frac{\gamma}{10(\lambda+2\mu)R}(6\mu+5\lambda)(R^2-\rho^2),$$

$$\sigma_{\theta\theta} = \sigma_{\varphi\varphi} = -\frac{\gamma}{10(\lambda+2\mu)R}\{(6\mu+5\lambda)(R^2-\rho^2)+4\mu\rho^2\}. \quad (3.6.11)$$

(Here we do not indicate components which are equal to zero.) Clearly, the vector $\boldsymbol{u}^{(0)}$ does not satisfy the boundary condition (3.6.8) on the shaft surface, and the boundary layer should be constructed in order to compensate the discrepancy. We would like to mention that two boundary layers are involved. Namely, a two-dimensional layer for the middle region of the shaft, and the three-dimensional boundary layer near the surface ∂S. We intend to compare the solution of the problem (3.6.7), (3.6.8) with the Lekhnitskii solution, and, hence, we shall consider the second boundary layer only.

As usual, we introduce the scaled coordinates

$$\boldsymbol{\xi} = \frac{\boldsymbol{x}-\mathcal{N}}{h},$$

where $\boldsymbol{x}$ represents Cartesian coordinates. In the limit $h \to 0$ we obtain the half-space $\mathbb{R}^3_- = \{\boldsymbol{\xi} : \xi_3 < 0\}$ with the semi-infinite shaft C_1. It follows from (3.6.11) that the error, produced by $\boldsymbol{u}^{(0)}$ in the boundary condition on the lateral surface, is given by

$$\sigma_{rr}(\boldsymbol{u}^{(0)};\boldsymbol{x}) = -\frac{2\mu\gamma R}{5(\lambda+2\mu)}\{1+O(h\|\boldsymbol{\xi}\|/R)\}; \quad (3.6.12)$$

it corresponds to the lateral compression.

Consequently, the leading part of the boundary layer is given by $h\boldsymbol{w}(\boldsymbol{\xi})$, where

$$w_r(\boldsymbol{\xi}) = -\frac{\gamma R}{5(\lambda+2\mu)(\xi_1^2+\xi_2^2)^{1/2}}, \quad w_\varphi = w_z = 0, \quad (3.6.13)$$

and it represents the solution of the homogeneous Lamé system in $\mathbb{R}^3_- \setminus C_1$, with given constant normal tractions on the lateral surface of the shaft.

3.6.2 Circular layer with a shaft

(a) Contact with a smooth rigid surface. We consider the deformation of the elastic region $\Omega = \Pi \setminus \overline{G}$, where $\Pi = \{\boldsymbol{x} \in C_d : 0 < x_3 < R\}$ is a circular layer. The axial cross-section of Ω is shown in Fig. 3.10.

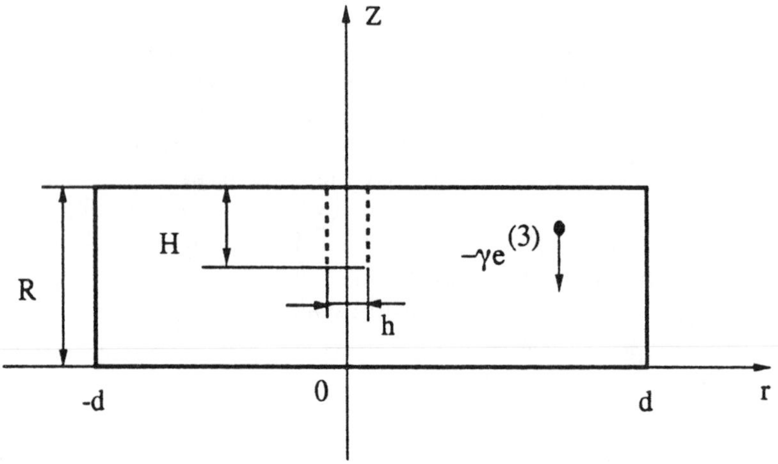

Fig. 3.10: An axisymmetric layer with a shaft.

The contact surface corresponding to the bottom part of the layer is supposed to be absolutely rigid and smooth. The upper part of the boundary ($x_3 = R$) is assumed to be free of tractions. The displacement vector satisfies the following boundary value problem:

$$-\mathbf{L}(\frac{\partial}{\partial \boldsymbol{x}})\boldsymbol{u}(\boldsymbol{x}) + \gamma \boldsymbol{e}^{(3)} = \mathbf{0}, \quad \boldsymbol{x} \in \Omega, \tag{3.6.14}$$

$$\boldsymbol{\sigma}^{(3)}(\boldsymbol{u}; \boldsymbol{x}', R) = \mathbf{0}, \quad h < \|\boldsymbol{x}'\| < d; \tag{3.6.15}$$

$$u_3(\boldsymbol{x}', 0) = 0, \quad \sigma_{i3}(\boldsymbol{u}; \boldsymbol{x}', 0) = 0, \ i = 1, 2, \ \|\boldsymbol{x}'\| < d; \tag{3.6.16}$$

$$\boldsymbol{\sigma}^{(r)}(\boldsymbol{u}; \boldsymbol{x}', z) = -p\boldsymbol{e}^{(r)}, \ \|\boldsymbol{x}'\| = d, \ 0 < z < R; \tag{3.6.17}$$

$$\boldsymbol{\sigma}^{(n)}(\boldsymbol{u}; \boldsymbol{x}) = \mathbf{0}, \boldsymbol{x} \in \partial G \bigcap \Omega, \tag{3.6.18}$$

where $\boldsymbol{x}' = (x_1, x_2)$.

The solution of the problem (3.6.14)–(3.6.18) can be written in the form

$$\boldsymbol{u} = \boldsymbol{u}^{(0)} + \boldsymbol{u}^{(1)}, \tag{3.6.19}$$

where $\boldsymbol{u}^{(0)}$ satisfies (3.6.14), (3.6.15), (3.6.16), but leaves an error in the traction boundary conditions (3.6.17), (3.6.18), and $\boldsymbol{u}^{(1)}$ compensates the above discrepancy. The field $\boldsymbol{u}^{(0)}$ has the components

$$u_r^{(0)} = u_\varphi^{(0)} = 0, \quad u_z^{(0)} = \frac{\gamma z(z - 2R)}{2(2\mu + \lambda)}. \tag{3.6.20}$$

The components of the field $\boldsymbol{u}^{(1)}$ are given by

$$u_r^{(1)} = \Psi_r, \quad u_\varphi^{(1)} = 0, \quad u_z^{(1)} = -\frac{\nu z}{1-\nu}\left(\frac{\partial \Psi_r}{\partial r} + \frac{\Psi_r}{r}\right), \tag{3.6.21}$$

where the auxiliary two–dimensional field $\boldsymbol{\Psi} = (\Psi_r, 0)$ corresponds to an axisymmetric generalized plane-stress state of the disk $\{\boldsymbol{x}' \in \mathbb{R}^2 : ||\boldsymbol{x}'|| < d\}$ with the boundary condition

$$\frac{4\mu(\lambda+\mu)}{2\mu+\lambda}\frac{\partial \Psi_r}{\partial r} + \frac{2\mu\lambda}{2\mu+\lambda}\frac{\Psi_r}{r} = \frac{\lambda R \gamma}{2(2\mu+\lambda)} - p \text{ at } r = d. \tag{3.6.22}$$

The function Ψ_r admits the following explicit representation:

$$\Psi_r = \frac{\lambda R\gamma - 2(2\mu+\lambda)p}{4\mu(2\mu+3\lambda)} r. \tag{3.6.23}$$

It follows from (3.6.22) that the discrepancy of $\boldsymbol{u}^{(0)} + \boldsymbol{u}^{(1)}$ in the boundary condition (3.6.17) is self–balanced (it is characterized by the zero values of the principal force and moment vectors). Due to this fact, the boundary layer, which occurs near the surface of the layer, decays exponentially with the increase of the distance from the surface. In the vicinity of G the field $\boldsymbol{u}^{(0)} + \boldsymbol{u}^{(1)}$ approximates the solution of the problem in the homogeneous region Π up to terms of order $O(R/d)$.

The boundary layer should be constructed in the vicinity of G. It compensates the discrepancy of the field $\boldsymbol{u}^{(0)} + \boldsymbol{u}^{(1)}$ in the boundary condition (3.6.18). The leading order part of the boundary layer, associated with the entrance into the shaft, has the form

$$h\boldsymbol{w}(\boldsymbol{\xi}).$$

Here, we use the scaled coordinates $\boldsymbol{\xi} = h^{-1}(\boldsymbol{x} - \mathcal{N})$. The field $\boldsymbol{w}(\boldsymbol{\xi})$ is given by

$$\boldsymbol{w}(\boldsymbol{\xi}) = \frac{\lambda R\gamma - 2(2\mu + \lambda)p}{4\mu(2\mu + \lambda)(\xi_1^2 + \xi_2^2)^{1/2}} \boldsymbol{e}^{(r)}, \tag{3.6.24}$$

and it compensates the discrepancy

$$\frac{\lambda R\gamma - 2(2\mu + \lambda)p}{2(2\mu + \lambda)}$$

in the homogeneous traction boundary conditions on the lateral surface of the shaft. Clearly, if

$$\lambda R\gamma - 2(2\mu + \lambda)p = 0,$$

then the second term $h^2\boldsymbol{w}^{(1)}(\boldsymbol{\xi})$ of the boundary layer is required. It has the form

$$\boldsymbol{w}^{(1)}(\boldsymbol{\xi}) = \frac{\nu}{1 - \nu}\left\{\frac{\gamma\xi_3}{2\mu(\xi_1^2 + \xi_2^2)^{1/2}} \boldsymbol{e}^{(r)} - \frac{\gamma}{4\mu}\ln(\xi_1^2 + \xi_2^2)\boldsymbol{e}^{(z)}\right\}, \tag{3.6.25}$$

and compensates the discrepancy left by the field $\boldsymbol{u}^{(0)}$ in the homogeneous traction boundary condition on ∂G. It is verified by direct calculations that the sum

$$\boldsymbol{u}^{(0)} + h^2\boldsymbol{w}^{(1)},$$

describing the displacement field near the entrance into the shaft, coincides (up to the rigid body displacement) with the Lekhnitskii solution.

(b) Circular layer with the fixed lower base. We consider the deformation of the circular layer with the shaft and assume that the lower base is bonded to the rigid surface $x_3 = 0$. Then, the displacement vector $\boldsymbol{u}$ satisfies the system (3.6.14) with the boundary conditions (3.6.15), (3.6.17), (3.6.18), and

$$\boldsymbol{u}(\boldsymbol{x}', 0) = \boldsymbol{0}, \quad \|\boldsymbol{x}'\| < d. \tag{3.6.26}$$

The description of this problem is essentially the same as in part (a). Moreover, it is more simple, because we do not need the field $\boldsymbol{u}^{(1)}$.

The field (3.6.20) leaves an error in the boundary condition (3.6.17). This error is compensated by the two–dimensional boundary layer that occurs in the vicinity of S. This boundary layer is the solution of the two–dimensional problem on the deformation of a half–strip where one of the sides is fixed (see (3.6.26)). It decays exponentially while the field (3.6.21) and the boundary layer $h\boldsymbol{w}^{(0)}(\boldsymbol{\xi})$ are absent. Consequently, the three–dimensional boundary layer is determined by $h^2\boldsymbol{w}^{(1)}$, where the vector $\boldsymbol{w}^{(1)}$ is the same as in (3.6.25). As before, the sum $\boldsymbol{u}^{(0)} + h^2\boldsymbol{w}^{(1)}$ gives the Lekhnitskii solution.

(c) The unsupported layer. The same geometry as in (a), (b) will be considered. However, we shall assume that the lower base is free of tractions, while the lateral surface is fixed

$$\boldsymbol{\sigma}^{(3)}(\boldsymbol{u};\boldsymbol{x}',0) = \mathbf{0}, \ \ \|\boldsymbol{x}'\| < d, \tag{3.6.27}$$

$$\boldsymbol{u}(\boldsymbol{x}',z) = \mathbf{0}, \ \ \|\boldsymbol{x}'\| = d, \ \ z \in (0,R). \tag{3.6.28}$$

The leading part of the solution of the problem (3.6.14), (3.6.15), (3.6.18), (3.6.27), (3.6.28) is represented by the displacement field $\boldsymbol{u}^{(0)}$ corresponding to the problem on the bending of a plate:

$$u_j^{(0)} = -R\zeta\frac{\partial\Phi}{\partial x_j}(\boldsymbol{x}') + R^3\left(-\frac{\zeta^3}{6}\frac{\nu-2}{1-\nu} + \zeta\frac{\nu-6}{24(1-\nu)}\right)\Delta\frac{\partial\Phi}{\partial x_j}(\boldsymbol{x}'),$$

$$j = 1,2; \tag{3.6.29}$$

$$u_3^{(0)} = \Phi(\boldsymbol{x}') + R^2\frac{\nu}{1-\nu}\left(\frac{\zeta^2}{2} - \frac{1}{24}\right)\Delta\Phi(\boldsymbol{x}'), \tag{3.6.30}$$

where

$$\zeta = \frac{2z-R}{2R}.$$

The function Φ is represented in the form

$$\Phi(\boldsymbol{x}') = -\frac{3\gamma(1-\nu)}{32\mu}\left(\frac{d}{R}\right)^4\left(1-\frac{r^2}{d^2}\right)^2, \tag{3.6.31}$$

and it solves the bending problem in the Kirchhoff theory of elastic plates,

$$\Delta^2\Phi(\boldsymbol{x}') = -\frac{6\gamma(1-\nu)}{\mu R^4}, \ \text{ as } \|\boldsymbol{x}'\| < d, \tag{3.6.32}$$

$$\Phi(\boldsymbol{x}') = 0, \ \ \frac{\partial\Phi}{\partial r}(\boldsymbol{x}') = 0, \ \text{ as } \|\boldsymbol{x}'\| = d. \tag{3.6.33}$$

The above field $\boldsymbol{u}^{(0)}$ leaves an error in the boundary condition (3.6.18). This error is written as

$$\sigma_{rr} = \frac{3\gamma(1+\nu)d^2}{8R^4}(R - 2\xi_3 h),$$

where

$$\xi_1 = \frac{x_2}{h}, \quad \xi_2 = \frac{x_1}{h}, \quad \xi_3 = \frac{R}{2h}(1 - 2\zeta).$$

This discrepancy is compensated for by the boundary layer

$$h\boldsymbol{w}^{(0)}(\boldsymbol{\xi}) + h^2\boldsymbol{w}^{(1)}(\boldsymbol{\xi}),$$

where

$$\boldsymbol{w}^{(0)}(\boldsymbol{\xi}) = \frac{3\gamma}{16\mu}(1+\nu)\frac{d^2}{R^3}\frac{1}{(\xi_1^2+\xi_2^2)^{1/2}}\boldsymbol{e}^{(r)}, \tag{3.6.34}$$

and

$$\boldsymbol{w}^{(1)}(\boldsymbol{\xi}) = \frac{3(1+\nu)}{4}\frac{d^2}{R^3}\left\{\frac{\xi_3\gamma}{2\mu(\xi_1^2+\xi_2^2)^{1/2}}\boldsymbol{e}^{(r)} - \frac{\gamma}{4\mu} \times \ln(\xi_1^2+\xi_2^2)\boldsymbol{e}^{(z)}\right\}. \tag{3.6.35}$$

As in part (b) the main contribution goes from the field corresponding to the compression of the half–space with the shaft.

3.7 Exercises

We advise the reader to use the REDUCE program, presented in the Appendix, in order to work with the the following exercises.

1. Generalize the algorithm of Section 3.3.2 for the three–dimensional case and derive the fourth-order differential equation with respect to a leading term of the transversal component of displacement within a thin elastic plate.

2. Follow the algorithm of Section 3.3.2 and derive the equation of a thin beam of a variable thickness.

4

Stress–Strain State in the Vicinity of Sharp Inclusions

Two–dimensional and three–dimensional cusp–shaped inclusions are considered in this chapter. In Section 4.1 we analyze the asymptotics of the displacement and stress fields in the vicinity of a vertex of the cusp–shaped interface boundary. For a two–dimensional cusp with the power geometry, the stress components are bounded in a neighbourhood of the vertex. The case of a logarithmic cusp may admit a "weak" logarithmic singularity for the gradient of the displacement field. The cases of three–dimensional elastic and rigid cusp–shaped inclusions are analyzed. Finally, Section 4.2 deals with the stress concentration in the vicinity of soft and relatively rigid cusp–shaped two–dimensional inclusions. The boundary layer occurs in a neighbourhood of the vertex, and its description is very similar to one presented in Section 3.3. It explains the formation of the stress singularity at the vertex of the plane cusp when the Young modulus of the inclusion vanishes or tends to infinity.

4.1 Asymptotic behaviour of the stress-strain state in the vicinity of sharp defects in an elastic body

The present section deals with asymptotic representations of solutions to plane and three-dimensional problems of theory of elasticity in a neighbourhood of a tip of a sharp inclusion. In the two-dimensional situation the cases under consideration correspond to cusp-shaped elastic inclusions and to a thin angular elastic inclusion. It is shown that there is no stress singularity at the tip of the cusp-shaped elastic inclusion with the geometry in the vicinity of the tip satisfying a power law. This type of geometry of the inclusion boundary is compared with the logarithmic behaviour of a boundary of a narrow elastic inclusion. Moreover, the asymptotic analysis of the stress-strain state near cusp-shaped and narrow wedge-shaped elastic inclusions helps to describe the appearance of the stress singularity as the shape of an inclusion changes from that of a peak to that of a wedge. With regard to the equations of three-dimensional elasticity, the cases of cusp-shaped elastic and rigid inclusions are considered. It is proved that the stress components are bounded near the tip of a three-dimensional elastic inclusion, but approaching the tip from different directions which are tangent to the surface of the inclusion gives different limiting values of the stresses. On the other hand, the case of a rigid inclusion corresponds to stress singularity, studied in the vicinity of the irregular point of the boundary. The main results of the present chapter are based on [70], [69], [72].

4.1.1 Two-dimensional cusp-shaped elastic inclusions

Let Ω be a doubly connected plane domain with the boundary $\partial\Omega$ which is smooth, except for the point O, which belongs to the inner part Γ of the boundary (see Fig. 4.1). Let us denote the simply connected domain, which is bounded by Γ, as ω and suppose that, in a neighbourhood of O, the set ω is described by the inequality

$$-A(r) < \varphi < A(r), \tag{4.1.1}$$

where (r, φ) are polar coordinates with origin at O; A is a function, which admits the representation

$$A(r) = ar^{\gamma} + O(r^{\gamma+\delta}), \tag{4.1.2}$$

with positive constants a, γ and δ.

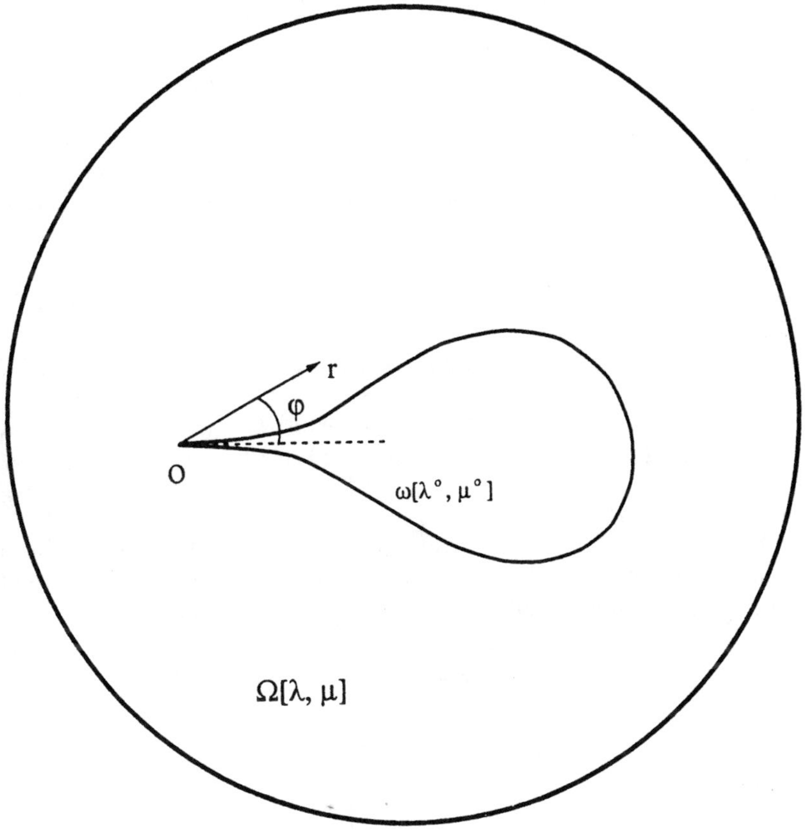

Fig. 4.1: A cusp-shaped elastic inclusion in two dimensions.

Let the domain Ω contain elastic material with Lamé constants λ and μ, and let ω correspond to another material with constants λ°, μ°. Suppose that there are no body forces, and that the loading conditions correspond to forces, distributed on the outer part of the boundary $\partial\Omega$.

Let us consider the displacement vector $\boldsymbol{u} = (u_1, u_2)^T$ and the stress vector

$$\boldsymbol{\sigma}^{(n)}(\boldsymbol{u};\boldsymbol{x}) = n_1((2\mu+\lambda)\frac{\partial u_1}{\partial x_1}(\boldsymbol{x}) + \lambda\frac{\partial u_2}{\partial x_2}(\boldsymbol{x}), \mu(\frac{\partial u_1}{\partial x_2}(\boldsymbol{x}) + \frac{\partial u_2}{\partial x_1}(\boldsymbol{x})))^T$$

$$+ n_2(\mu(\frac{\partial u_1}{\partial x_2}(\boldsymbol{x}) + \frac{\partial u_2}{\partial x_1}(\boldsymbol{x})), (2\mu+\lambda)\frac{\partial u_2}{\partial x_2}(\boldsymbol{x}) + \lambda\frac{\partial u_1}{\partial x_1}(\boldsymbol{x}))^T,$$

where $\boldsymbol{n} = (n_1, n_2)^T$ is the unit outer normal vector to $\partial\Omega$, and suppose that $\boldsymbol{u}$ satisfies the following system of equations:

$$\frac{1}{r^2}\mathbf{L}(r\frac{\partial}{\partial r}, \frac{\partial}{\partial \varphi})\boldsymbol{u}(r,\varphi) = \mathbf{0}, \boldsymbol{x} \in \Omega, \tag{4.1.3}$$

$$\frac{1}{r^2}\mathbf{L}^\circ(r\frac{\partial}{\partial r}, \frac{\partial}{\partial \varphi})\boldsymbol{u}^\circ(r,\varphi) = \mathbf{0}, \boldsymbol{x} \in \omega, \tag{4.1.4}$$

$$\boldsymbol{u}(r,\varphi) = \boldsymbol{u}^\circ(r,\varphi), \boldsymbol{\sigma}^{(n)}(\boldsymbol{u}; r,\varphi) = \boldsymbol{\sigma}^{\circ,(n)}(\boldsymbol{u}^\circ; r,\varphi), \boldsymbol{x} \in \Gamma, \tag{4.1.5}$$

where

$$\mathbf{L}(\xi_1,\xi_2) = \tag{4.1.6}$$

$$\begin{pmatrix} (\lambda+2\mu)(\xi_1^2-1)+\mu\xi_2^2 & \{(\lambda+2\mu)(\xi_1-1)-\mu(\xi_1+1)\}\xi_2 \\ \{(\lambda+2\mu)(\xi_1+1)-\mu(\xi_1-1)\}\xi_2 & (\lambda+2\mu)\xi_2^2+\mu(\xi_1^2-1) \end{pmatrix}$$

and $\boldsymbol{x} = (x_1, x_2) = (r\cos\varphi, r\sin\varphi)$.

Here $\mathbf{L}, \mathbf{L}^\circ$ are the matrix differential operators, corresponding to the Lamé equations with the elastic constants λ, μ and λ°, μ°. Let us note that the condition (4.1.5) can be written in the matrix form

$$r^{-1}\{\mathbf{B}(r\partial/\partial r, \partial/\partial\varphi) + \mathbf{D}^{\pm}(r, r\partial/\partial r, \partial/\partial\varphi)\}\boldsymbol{u}(r,\varphi) \tag{4.1.7}$$

$$= r^{-1}\{\mathbf{B}^\circ(r\partial/\partial r, \partial/\partial\varphi) + \mathbf{D}^{\circ,\pm}(r, r\partial/\partial r, \partial/\partial\varphi)\}\boldsymbol{u}^\circ(r,\varphi);$$

$$\varphi = A(r),$$

where

$$\mathbf{B}(\xi_1,\xi_2) = \begin{pmatrix} \mu\xi_2 & \mu(\xi_1-1) \\ 2\mu+\lambda(\xi_1+1) & (2\mu+\lambda)\xi_2 \end{pmatrix}, \tag{4.1.8}$$

$$\mathbf{D}^{\pm}(r,\xi_1,\xi_2) = \mu\{\pm\sin(2\alpha(r))\begin{pmatrix} \xi_1-1 & -\xi_2 \\ \xi_2 & \xi_1-1 \end{pmatrix}$$

$$-2\sin^2\alpha(r)\begin{pmatrix} \xi_2 & \xi_1-1 \\ 1-\xi_1 & \xi_2 \end{pmatrix}\}, \tag{4.1.9}$$

$$\alpha(r) = \frac{\sin A(r) + rA'(r)\cos A(r)}{\cos A(r) - rA'(r)\sin A(r)} - A(r), \tag{4.1.10}$$

$$\alpha(r) = \gamma A(r) + O(r^{\gamma+1}), r \to 0. \tag{4.1.11}$$

The problem (4.1.3)–(4.1.5) may be investigated using methods of asymptotic analysis. Let us assume that the displacement vectors $\boldsymbol{u}(r,\varphi), \boldsymbol{u}^\circ(r,\varphi)$ have the form

$$\boldsymbol{u}(r,\varphi) \sim r^\Lambda \boldsymbol{v}(\varphi);\ \ \boldsymbol{u}^\circ(r,\varphi) \sim r^\Lambda \boldsymbol{v}^\circ(t),\ \ t := A(r)^{-1}\varphi \in (-1,1), \tag{4.1.12}$$

where Λ is a constant and $\boldsymbol{v}$ is a smooth vector-valued function on the interval $(0, 2\pi)$.

From the first of the conditions (4.1.5) we can obtain the following relations:

$$\boldsymbol{v}(0) = \boldsymbol{v}^\circ(1), \boldsymbol{v}(2\pi) = \boldsymbol{v}^\circ(-1). \tag{4.1.13}$$

The substitution of the representation for $\boldsymbol{u}^\circ$ in (4.1.4) yields the following equality:

$$\mathbf{L}^\circ(r\frac{\partial}{\partial r}, \frac{\partial}{\partial \varphi})\boldsymbol{u}^\circ(r,\varphi) = \frac{r^\Lambda}{A^2}\mathbf{Q}^\circ \frac{d^2\boldsymbol{v}^\circ}{dt}(t) + O(r^{\Lambda+\gamma}), \tag{4.1.14}$$

where $\mathbf{Q}^\circ = \mathrm{diag}(\mu^\circ, 2\mu^\circ + \lambda^\circ)$, since

$$r\frac{\partial}{\partial r}(r^\Lambda \boldsymbol{v}^\circ(A(r)^{-1}\varphi)) = -r^{\Lambda+1}A'(r)A(r)^{-1}t\frac{d\boldsymbol{v}^\circ}{dt}(t) + \Lambda r^\Lambda \boldsymbol{v}^\circ(t) = O(r^\Lambda),$$

$$\frac{\partial}{\partial \varphi}(r^\Lambda \boldsymbol{v}^\circ(A(r)^{-1}\varphi)) = r^\Lambda A(r)^{-1}\frac{d\boldsymbol{v}^\circ}{dt}(t) = O(r^{\Lambda-\gamma}). \tag{4.1.15}$$

Hence, $\boldsymbol{v}^\circ$ is a linear function. Besides, from (4.1.7)–(4.1.12) we deduce that for $\varphi = \pm A(r)$,

$$r\boldsymbol{\sigma}^{(n)}(\boldsymbol{u}; r, \varphi) = r^\Lambda \mathbf{B}(\Lambda, \frac{\partial}{\partial \varphi})\boldsymbol{v}(\pi \mp \pi) + O(r^{\Lambda+\gamma}),$$

$$r\boldsymbol{\sigma}^{\circ,(n)}(\boldsymbol{u}; r, \varphi) = r^\Lambda A(r)^{-1}\mathbf{Q}^\circ \frac{d\boldsymbol{v}^\circ}{dt}(\pm 1) + O(r^\Lambda) = O(r^{\Lambda-\gamma}). \tag{4.1.16}$$

Thus, $d\boldsymbol{v}^\circ/dt = \mathbf{0}$, and, hence, $\boldsymbol{v}^\circ$ is a constant vector. Now, let us consider the second term of the asymptotic expansion of $\boldsymbol{u}^\circ$:

$$\boldsymbol{u}^\circ(r,\varphi) \sim r^\Lambda\{\boldsymbol{v}^\circ + A(r)\boldsymbol{w}^\circ(A(r)^{-1}\varphi)\}. \tag{4.1.17}$$

Taking into account the previous consideration, we see that $\boldsymbol{w}^\circ$ is a linear vector-valued function. Moreover,

$$r\boldsymbol{\sigma}^{\circ,(n)}(\boldsymbol{u}^\circ; r, \pm A(r)) = r^\Lambda\{\mathbf{B}^\circ(\Lambda, 0)\boldsymbol{v}^\circ$$

$$+ \mathbf{Q}^{\circ}\frac{d\boldsymbol{w}^{\circ}}{dt}(\pm 1)\} + O(r^{\Lambda+\gamma}). \tag{4.1.18}$$

Thus, using (4.1.5), (4.1.16), (4.1.18) and (4.1.13), the matching conditions for the angular part $\boldsymbol{v}$ can be derived:

$$\mathbf{B}(\Lambda, \frac{\partial}{\partial\varphi})\boldsymbol{v}(0) = \mathbf{B}(\Lambda, \frac{\partial}{\partial\varphi})\boldsymbol{v}(2\pi), \quad \boldsymbol{v}(0) = \boldsymbol{v}(2\pi). \tag{4.1.19}$$

Furthermore, the vector-valued function (4.1.12) satisfies the system (4.1.3) if and only if

$$\mathbf{L}(\Lambda, \frac{\partial}{\partial\varphi})\boldsymbol{v}(\varphi) = \mathbf{0}, \varphi \in (0, 2\pi). \tag{4.1.20}$$

Thus, the exponent Λ and the angular part $\boldsymbol{v}$ in (4.1.12) are an eigenvalue and the corresponding eigenvector of the problem (4.1.19), (4.1.20). Let us note that the eigenvalues in this case are integers, and for $\Lambda = k \neq 0$ there exist four eigenvectors

$$\boldsymbol{v}^{(k,1)} = (\cos[(1+k)\varphi], -\sin[(1+k)\varphi])^T,$$

$$\boldsymbol{v}^{(k,2)} = (\sin[(1+k)\varphi], \cos[(1+k)\varphi])^T,$$

$$\boldsymbol{v}^{(k,3)} = ((k-\varkappa)\cos[(1-k)\varphi], (k+\varkappa)\sin[(1-k)\varphi])^T, \tag{4.1.21}$$

$$\boldsymbol{v}^{(k,4)} = ((k-\varkappa)\sin[(1-k)\varphi], (k+\varkappa)\cos[(1-k)\varphi])^T,$$

where $\varkappa = (\lambda + 3\mu)(\lambda+\mu)^{-1}$. If $\Lambda = 0$, there are two eigenvectors and two associated vectors

$$\boldsymbol{v}^{(0,1)}(\varphi) = (\cos\varphi, -\sin\varphi)^T, \quad \boldsymbol{v}^{(1)}(\varphi) = (0, -\varkappa^{-1}\sin\varphi)^T,$$

$$\boldsymbol{v}^{(0,2)}(\varphi) = (\sin\varphi, \cos\varphi)^T, \quad \boldsymbol{v}^{(2)}(\varphi) = (0, \varkappa^{-1}\cos\varphi)^T. \tag{4.1.22}$$

The vectors (4.1.21), (4.1.22) correspond to solutions $\mathbf{U}^{(k,j)}$ and $\mathbf{U}^{(0,j)}, \mathbf{U}^{(j)}$ of the homogeneous equations (4.1.3):

$$\mathbf{U}^{(k,j)}(r,\varphi) = r^k \boldsymbol{v}^{(k,j)}(\varphi), (k = \pm 1, \pm 2, \ldots, \; j = 1,2,3,4); \tag{4.1.23}$$

$$\mathbf{U}^{(0,j)}(r,\varphi) = \boldsymbol{v}^{(0,j)}(\varphi),$$

$$\mathbf{U}^{(j)}(r,\varphi) = \boldsymbol{v}^{(0,j)}(\varphi)\ln r + \boldsymbol{v}^{(j)}(\varphi), (j = 1, 2). \tag{4.1.24}$$

Note that $\mathbf{U}^{(1)}, \mathbf{U}^{(2)}$ are columns of the Somigliana tensor. Vectors $\mathbf{U}^{(k,j)}, k > 0$, expressed in the Cartesian coordinate system, are homogeneous polynomials of degree k. If $k \leq 0$, these vectors are linear combinations of derivatives of order k of the Somigliana columns.

According to (4.1.13), (4.1.18) the asymptotic representations (4.1.12), (4.1.17) can be written as

$$\boldsymbol{u}(r,\varphi) \sim r^k \boldsymbol{v}^{(k,j)}(\varphi),$$

$$\boldsymbol{u}^\circ(r,\varphi) \sim r^k[1 + \varphi(\mathbf{Q}^\circ)^{-1}(\mathbf{B}(k, \frac{\partial}{\partial\varphi}) - \mathbf{B}^\circ(k,0))]\boldsymbol{v}^{(k,j)}(0). \tag{4.1.25}$$

The formula (4.1.25) contains only principal parts of the asymptotic expansions, which leave a discrepancy in the equations (4.1.4), (4.1.5). Thus, it is necessary to construct the next terms of asymptotic expansions. Let us describe this procedure. To simplify the presentation, we suppose that the function (4.1.2) is defined by the equality $A(r) = ar$. Then we consider the asymptotic series

$$\boldsymbol{u}(r,\varphi) \sim \sum_{p=0}^{+\infty} r^{k+p}\boldsymbol{w}^{(p)}(\varphi, \ln r),$$

$$\boldsymbol{u}^\circ(r,\varphi) \sim \sum_{p=0}^{+\infty} r^{k+p}\mathbf{Z}^{(p)}((ar)^{-1}\varphi, \ln r), \tag{4.1.26}$$

$$\boldsymbol{w}^{(0)}(\varphi) = \boldsymbol{v}^{(k,j)}(\varphi), \mathbf{Z}^{(0)} = \boldsymbol{w}^{(0)}(0),$$

$$\mathbf{Z}^{(1)}(t) = at(\mathbf{Q}^\circ)^{-1}(\mathbf{B}(k,\partial/\partial\varphi) - \mathbf{B}^\circ(k,0))\boldsymbol{v}^{(k,j)}(0) + \mathbf{C}_Z; \tag{4.1.27}$$

here $\mathbf{C}_Z$ is a constant vector. The equalities (4.1.27) are constructed in accordance with the representation (4.1.25).

Substitute (4.1.26) into the equations (4.1.3)–(4.1.5) and equate coefficients near like powers of r. The equation (4.1.3) gives

$$\mathbf{L}(k + p + \partial/\partial s, \partial/\partial\varphi)\boldsymbol{w}^{(p)}(\varphi, s) = \mathbf{0}, \varphi \in (0, 2\pi). \tag{4.1.28}$$

By (4.1.15), the operators $\mathbf{L}^\circ$ and $\mathbf{B}^\circ$ in the coordinates $(\rho,\varphi) = (r, (ar)^{-1}\varphi)$ have the following representation:

$$\mathbf{L}^\circ(r\partial/\partial r, \partial/\partial\varphi) = \mathbf{L}^\circ(\rho\partial/\partial\rho - t\partial/\partial t, (a\rho)^{-1}\partial/\partial t)$$

$$= \mathbf{Q}^\circ (a\rho)^{-2}\partial^2/\partial t^2 + (a\rho)^{-1}\mathbf{P}(\rho\partial/\partial\rho - t\partial/\partial t)\partial/\partial t + \mathbf{L}^\circ(\rho\partial/\partial\rho - t\partial/\partial t, 0),$$

$$\mathbf{B}^\circ(r\partial/\partial r, \partial/\partial\varphi) = (a\rho)^{-1}\mathbf{Q}^\circ\partial/\partial t + \mathbf{B}^\circ(\rho\partial/\partial\rho - t\partial/\partial t, 0), \quad (4.1.29)$$

where

$$\mathbf{P}(\eta) = \begin{pmatrix} 0 & (\lambda^\circ + \mu^\circ)\eta - (\lambda^\circ + 3\mu^\circ) \\ (\lambda^\circ + \mu^\circ)\eta + (\lambda^\circ + 3\mu^\circ) & 0 \end{pmatrix}.$$

Hence, from (4.1.4) we obtain

$$\mathbf{Q}^\circ \frac{\partial^2 \mathbf{Z}^{(p+1)}}{\partial t^2}(t,s) = -a\mathbf{P}(k+p-t\frac{\partial}{\partial t} + \frac{\partial}{\partial s})\frac{\partial \mathbf{Z}^{(p)}}{\partial t}(t,s)$$

$$- a^2\mathbf{L}^\circ(k+p-1-t\partial/\partial t + \partial/\partial s, 0)\mathbf{Z}^{(p-1)}(t,s), t \in (-1,1). \quad (4.1.30)$$

By expansion of the functions $\boldsymbol{w}^{(k)}$ in the Taylor series in a neighbourhood of points $\varphi = +0$ and $\varphi = 2\pi - 0$, using (4.1.5), we derive

$$\boldsymbol{w}^{(p)}(\pi \mp \pi, s) = \mathbf{Z}^{(p)}(\pm 1, s) - \mathbf{S}^\pm,$$

$$\mathbf{S}^\pm = \sum_{q=1}^{p} \frac{(\pm a)^q}{q!}\frac{\partial^q \boldsymbol{w}^{(p-q)}}{\partial \varphi^q}(\pi \pm \pi, s). \quad (4.1.31)$$

According to (4.1.9)–(4.1.11), the operators $\mathbf{D}^\pm$ and $\mathbf{D}^{\circ,\pm}$ from (4.1.7) admit the expansions

$$\mathbf{D}^\pm(r, r\frac{\partial}{\partial r}, \frac{\partial}{\partial\varphi}) \sim \sum_{q=1}^{\infty} r^q \mathbf{D}_{(q)}(r\frac{\partial}{\partial r}, \frac{\partial}{\partial\varphi}),$$

$$\mathbf{D}^{\circ,\pm}(r, r\frac{\partial}{\partial r}, \frac{\partial}{\partial\varphi}) \sim \sum_{q=0}^{\infty} r^q \mathbf{D}_{(q)}^{\circ,\pm}(t, \rho\frac{\partial}{\partial\rho}, \frac{\partial}{\partial t}), \quad \mathbf{D}_{(0)}^{\circ,\pm}(\eta, 0) = \mathbf{0}.$$

Hence, from the second equality in (4.1.5) we can derive

$$\mathbf{Q}^\circ \frac{\partial \mathbf{Z}^{(p+1)}}{\partial t}(\pm 1, s) = -a\mathbf{B}^\circ(\frac{\partial}{\partial t} + k + p + \frac{\partial}{\partial s}, 0)\mathbf{Z}^{(p)}(\pm 1, s)$$

$$+ a\mathbf{B}(k+p+\frac{\partial}{\partial s}, \frac{\partial}{\partial\varphi})\boldsymbol{w}^{(p)}(\pi \pm \pi)$$

$$+ a\{\sum_{q=1}^{p}\frac{(\pm a)^q}{q!}\mathbf{B}(k+p-q+\frac{\partial}{\partial s}, \frac{\partial}{\partial\varphi})\frac{\partial^q \boldsymbol{w}^{(p-q)}}{\partial\varphi^q}(\pi \pm \pi)$$

$$+\sum_{q=1}^{p}\sum_{h=0}^{p-q}\frac{(\pm a)^h}{h!}\mathbf{D}^{\pm}_{(q)}(t,k+p-q-h+\frac{\partial}{\partial s},\frac{\partial}{\partial\varphi})\frac{\partial^h \boldsymbol{w}^{(p-q-h)}}{\partial\varphi^h}(\pi\pm\pi)$$

$$-\sum_{q=0}^{p}\mathbf{D}^{\circ,\pm}_{(q)}(t,k+p-q+\frac{\partial}{\partial s},\frac{\partial}{\partial t})\mathbf{Z}^{(p-q)}(\pm 1,s)\}. \tag{4.1.32}$$

Let us describe the procedure used to determine the coefficients of the series (4.1.26). We assume that the vector-valued function $\mathbf{Z}^{(p)}$ is constructed up to a constant column $\mathbf{C}^{(p)}_Z$, and that the vector-valued functions $\boldsymbol{w}^{(0)}, \mathbf{Z}^{(0)}, \ldots, \boldsymbol{w}^{(p-1)}, \mathbf{Z}^{(p-1)}$ are known. Thus, equations for the functions $\boldsymbol{w}^{(p)}, \mathbf{C}^{(p)}_Z, \mathbf{Z}^{(p+1)}$ can be obtained. Denoting the right-hand side of (4.1.30) by $\mathbf{F}$ and the sums in the curly brackets in (4.1.32) by $\mathbf{T}^{\pm}$, we find that the problem (4.1.30),(4.1.32) has a solution provided the following condition holds:

$$\mathbf{B}(k+p+\frac{\partial}{\partial s},\frac{\partial}{\partial\varphi})\boldsymbol{w}^{(p)}(0,s)-\mathbf{B}(k+p+\frac{\partial}{\partial s},\frac{\partial}{\partial\varphi})\boldsymbol{w}^{(p)}(2\pi,s)$$

$$=\mathbf{T}^{-}(s)-\mathbf{T}^{+}(s)+\mathbf{B}^{\circ}(k+p-t\frac{\partial}{\partial t}+\frac{\partial}{\partial s},0)\mathbf{Z}^{(p)}(1,s)$$

$$-\mathbf{B}^{\circ}(k+p-t\frac{\partial}{\partial t}+\frac{\partial}{\partial s},0)\mathbf{Z}^{(p)}(-1,s)+\frac{1}{a}\int_{-1}^{1}\mathbf{F}(t,s)dt. \tag{4.1.33}$$

Note that the solution is determined up to a constant vector.

The equality (4.1.33) and the equation

$$\boldsymbol{w}^{(p)}(0,s)-\boldsymbol{w}^{(p)}(2\pi,s)=\mathbf{Z}^{(p)}(1,s)-\mathbf{Z}^{(p)}(-1,s)$$

$$-\mathbf{S}^{+}(s)+\mathbf{S}^{-}(s), \tag{4.1.34}$$

which is derived from (4.1.31), provide the boundary conditions for the system (4.1.28). As shown in [38], the problem (4.1.28), (4.1.33), (4.1.34) is solvable. Note that a procedure of calculation $\boldsymbol{w}^{(p)}$ is given in [38], [62], [89]. After determining $\mathbf{W}^{(p)}$, we can obtain $\mathbf{C}^{(p)}_Z$ from (4.1.31), and by solving the problem (4.1.30), (4.1.32) we can evaluate $\boldsymbol{w}^{(p+1)}$ up to an arbitrary vector term $\mathbf{C}^{(p+1)}_Z$.

The proof of the validity of the representations (4.1.26) uses the results of [60]; in the case of a second-order scalar operator, asymptotic representations of this type were obtained and investigated in [91].

Suppose that the function A in (4.1.1) is infinitely differentiable on $[0,1]$ and that $A(0)=0$. Then there are the following expansions for the solution of the problem concerning the deformation of a composite domain $\Omega \cup \omega$ by forces applied to the outer boundary of Ω :

$$\boldsymbol{u}(r,\varphi) = \mathbf{K}^{(0)} + \sum_{p=1}^{n} r^p \mathbf{W}^{(p)}(\varphi, \ln r) + O(r^{r+1-\delta}), \qquad (4.1.35)$$

$$\boldsymbol{u}^\circ(r,\varphi) = \sum_{p=1}^{n} r^p \mathbf{Z}^{(p)}(r^{-1}\varphi, \ln r) + O(r^{n+1-\delta}). \qquad (4.1.36)$$

Here δ is an arbitrary positive number, $\mathbf{K}^{(0)}$ is a constant vector, $\mathbf{W}^{(1)}(\varphi)$ is a linear combination of vector-valued functions $\boldsymbol{v}^{(1,j)}$ from (4.1.21) with coefficients $\mathbf{K}_j^1$, and $(Z_r^{(1)}, Z_\varphi^{(1)}) = (W_r^{(1)}(0), W_\varphi^{(1)}(0))$.

From (4.1.35), (4.1.36) we obtain the expansions for the components of the stress tensor. Considering only the principal term of the asymptotic expansions, we derive the following equalities

$$\sigma_{rr}(\boldsymbol{u};\boldsymbol{x}) = 2\mu\{K_1^{(1)}\cos 2\varphi + K_2^{(1)}\sin 2\varphi - 2K_3^{(1)}\} + O(r),$$

$$\sigma_{r\varphi}(\boldsymbol{u};\boldsymbol{x}) = 2\mu\{-K_1^{(1)}\sin 2\varphi + K_2^{(1)}\cos 2\varphi\} + O(r),$$

$$\sigma_{\varphi\varphi}(\boldsymbol{u};\boldsymbol{x}) = 2\mu\{-K_1^{(1)}\cos 2\varphi - K_2^{(1)}\sin 2\varphi - 2K_3^{(1)}\} + O(r); \qquad (4.1.37)$$

$$\sigma_{rr}^\circ(\boldsymbol{u};\boldsymbol{x}) = 2K_1^{(1)}(\mu^\circ + \lambda^\circ(\mu^\circ - \mu)(2\mu^\circ + \lambda^\circ)^{-1})$$

$$-K_3^{(1)} 4\mu(\lambda^\circ + 2\mu^\circ)^{-1}(\lambda^\circ + 2\mu^\circ(\mu^\circ + \lambda^\circ)(\mu + \lambda)^{-1}) + O(r),$$

$$\sigma_{r\varphi}^\circ(\boldsymbol{u};\boldsymbol{x}) = -2\mu K_2^{(1)} + O(r),$$

$$\sigma_{\varphi\varphi}^\circ(\boldsymbol{u};\boldsymbol{x}) = -2\mu(K_1^{(1)} + 2K_3^{(1)}) + O(r). \qquad (4.1.38)$$

The absence of the coefficient $K_4^{(1)}$ in these expressions is due to the fact that it premultiplies the term $(\varkappa + 1)(x_2, -x_1)$, which corresponds to a rotation.

4.1.2 Two-dimensional logarithmic narrow inclusions

Consider the same system of equations (4.1.3)–(4.1.5) in the case where the function A is replaced by

$$h(r) = |\ln r|^{-\gamma} h_0(|\ln r|^{-1}), \qquad (4.1.39)$$

$$h_0 \in \mathbf{C}^\infty[0,1), h_0(0) := a_0 > 0, \gamma > 0.$$

In this case, when $\gamma = 1$ and when the elastic materials of the inclusion and the elastic matrix have different shear moduli, there is a weak singularity of order $O(|\ln r|^\beta)$, $Re\beta > 0$. The exponent $\gamma \in (0,1)$ corresponds to a more complicated form of the stress singularity near the tip of the inclusion.

Let us consider an expansion of the displacement field $\boldsymbol{u}$ as a sum of terms of the form

$$\mathbf{U}(r,\varphi) = r^\Lambda \sum_{j=0}^{\infty} |\ln r|^{\beta-j} \mathbf{\Phi}^{(j)}(\varphi). \tag{4.1.40}$$

Here Λ, β are constants and $\mathbf{\Phi}^{(j)} \in \mathbf{C}^\infty[0,2\pi]$ are vector-valued functions, which are to be determined. Similar terms, corresponding to an elastic inclusion can be written as follows:

$$\mathbf{U}^\circ(r,\varphi) = r^\Lambda \sum_{j=0}^{\infty} |\ln r|^{\beta-j} \mathbf{\Psi}^{(j)}(h(r)^{-1}\varphi), \tag{4.1.41}$$

where $\mathbf{\Psi}^{(j)}$ are infinitely differentiable in $[-1,1]$ vector-valued functions. Let us introduce the notation $z = |\ln r|$. The following formulae allow us to calculate the Lamé transformations of the series (4.1.40), (4.1.41):

$$r\frac{\partial}{\partial r}\{r^\Lambda |\ln r|^\alpha \mathbf{\Psi}(h(r)^{-1}\varphi)\} = r^\Lambda z^\alpha \{(\Lambda - \alpha z^{-1})\mathbf{\Psi}(t)$$

$$-rh'(r)h(r)^{-1}t\frac{\partial}{\partial t}\mathbf{\Psi}(t)\}, \tag{4.1.42}$$

$$\frac{\partial}{\partial \varphi}\{r^\Lambda |\ln r|^\alpha \mathbf{\Psi}(h(r)^{-1}\varphi)\} = r^\Lambda z^\alpha h(r)^{-1}\frac{\partial}{\partial t}\mathbf{\Psi}(t),$$

$$r\frac{\partial}{\partial r}\{r^\Lambda |\ln r|^\alpha \mathbf{\Phi}(\varphi)\} = r^\Lambda z^\alpha (\Lambda - \alpha z^{-1})\mathbf{\Phi}(\varphi). \tag{4.1.43}$$

By (4.1.40), (4.1.43) we obtain

$$\mathbf{L}(r\frac{\partial}{\partial r}, \frac{\partial}{\partial \varphi})\mathbf{U}(r,\varphi) = r^\Lambda z^\beta \{\mathbf{L}(\Lambda, \frac{\partial}{\partial \varphi})\mathbf{\Phi}^{(0)}(\varphi)$$

$$+\frac{1}{z}\{\mathbf{L}(\Lambda,\frac{\partial}{\partial\varphi})\boldsymbol{\Phi}^{(1)}(\varphi)-\beta\frac{\partial}{\partial\Lambda}\mathbf{L}(\Lambda,\frac{\partial}{\partial\varphi})\boldsymbol{\Phi}^{(0)}(\varphi)\}+O(z^{-2})\}\quad(4.1.44)$$

The vector-valued function $\mathbf{U}^\circ$, corresponding to the displacements field in an elastic inclusion, satisfies the equality

$$\mathbf{L}^\circ(r\frac{\partial}{\partial r},\frac{\partial}{\partial\varphi})\mathbf{U}^\circ(r,\varphi)=r^\Lambda z^\beta h(z)^{-2}\{\mathbf{Q}^\circ\frac{d^2}{dt^2}\boldsymbol{\Psi}^{(0)}(t)+O(z^{-1})\}.\quad(4.1.45)$$

Considering the second interface condition in (4.1.5) in the form

$$r\boldsymbol{\sigma}^{(n\pm)}(\mathbf{U};r,\pm h(r)):=\{\mathbf{B}(r\frac{\partial}{\partial r},\frac{\partial}{\partial\varphi})\mp rh'(r)\mathbf{R}(r\frac{\partial}{\partial r},\frac{\partial}{\partial\varphi})\}\mathbf{U}(r,\pm h(r))$$

$$=\{\mathbf{B}^\circ(r\frac{\partial}{\partial r},\frac{\partial}{\partial\varphi})\mp rh'(r)\mathbf{R}^\circ(r\frac{\partial}{\partial r},\frac{\partial}{\partial\varphi})\}\mathbf{U}^\circ(r,\pm h(r))$$

$$:=r\boldsymbol{\sigma}^{\circ,(n\pm)}(\mathbf{U}^\circ;r,\pm h(r)),$$

with

$$\mathbf{R}(\xi_1,\xi_2)=\begin{pmatrix}(2\mu+\lambda)\xi_1+\lambda & \lambda\xi_2\\ \mu\xi_2 & \mu(\xi_1-1)\end{pmatrix},$$

we take into account the representations

$$r\boldsymbol{\sigma}^{(n\pm)}(\mathbf{U};r,\pm h(r))\sim r^\Lambda z^\beta\mathbf{B}(\Lambda,\frac{\partial}{\partial\varphi})\boldsymbol{\Phi}^{(0)}(\pi\mp\pi),$$

$$r\boldsymbol{\sigma}^{\circ,(n\pm)}(\mathbf{U}^\circ;r,\pm h(r))\sim r^\Lambda z^\beta h(r)^{-1}\mathbf{Q}^\circ\frac{d}{dt}\boldsymbol{\Psi}^{(0)}(\pm1).\quad(4.1.46)$$

The formulae (4.1.45), (4.1.46) yield $\frac{d^2}{dt^2}\boldsymbol{\Psi}^{(0)}(t)=\mathbf{0},t\in(-1,1)$; $\frac{d}{dt}\boldsymbol{\Psi}^{(0)}(\pm1)=\mathbf{0}$. Hence, $\boldsymbol{\Psi}^{(0)}=\mathbf{C}^{(0)}$ is a constant vector. By means of an argument similar to the one in the previous part, we derive the spectral problem (4.1.19), (4.1.20) for Λ and $\boldsymbol{\Phi}^{(0)}$. Taking into account that the eigenvalues of this problem are integers, and the solution of the original problem has the finite elastic energy, we shall consider the case of $\Lambda=1$.

Let us calculate the exponent β. Note that, if integer $\gamma>1$, our consideration will be similar to the previous part. The vector-valued function $\boldsymbol{\Phi}^{(0)}$ is determined by (4.1.21), with $k=1$. Vector-valued functions $\boldsymbol{\Psi}^{(1)},\dots,\boldsymbol{\Psi}^{(\gamma-1)}$ are constant vectors, and $\boldsymbol{\Psi}^{(\gamma)}(t)=\mathbf{A}t+\mathbf{B}$.

Thus, $\mathbf{\Phi}^{(1)}$ satisfies the continuity conditions (4.1.19), and, hence, $\beta = 0$ and $\mathbf{\Phi}^{(1)}$ is a linear combination of the vectors (4.1.21), with $k = 1$.

The case of $\gamma = 1$ is different from the previous one. Considering the next terms of the asymptotic expansions (4.1.45) and (4.1.46), we obtain that vector-valued function $\mathbf{\Psi}^{(1)}$ can be represented in the form

$$\Psi_r^{(1)}(t) = C_r^{(1)} + a_0 t \frac{\mu}{\mu^\circ} \frac{\partial}{\partial \varphi} \Phi_r^{(0)}(0),$$

$$\Psi_\varphi^{(1)}(t) = C_\varphi^{(1)} + \frac{a_0 t}{2\mu^\circ + \lambda^\circ} [2(\mu + \lambda - \mu^\circ - \lambda^\circ) \Phi_r^{(0)}(0)$$

$$+ (2\mu + \lambda) \frac{\partial}{\partial \varphi} \Phi_\varphi^{(0)}(0)]. \tag{4.1.47}$$

The equation (4.1.45) yields the following equality for the function $\mathbf{\Psi}^{(2)}$

$$a_0^{-1} \mathbf{Q}^\circ \frac{d^2}{dt^2} \mathbf{\Psi}^{(2)}(t) + \mathbf{S} \frac{d}{dt} \mathbf{\Psi}^{(1)}(t) = \mathbf{0}, t \in (0, 1), \tag{4.1.48}$$

where $\mathbf{S} = (S_{jk})_{j,k=1}^2, S_{11} = S_{22} = 0, S_{21} = -S_{12} = 2\mu^\circ$. Thus, the function $\mathbf{\Psi}^{(2)}$ is a polynomial of order two with respect to variable t. Using the equalities (4.1.47), (4.1.48), we derive a system of equations for the function $\mathbf{\Phi}^{(1)}$:

$$\mathbf{L}(1, \frac{\partial}{\partial \varphi}) \mathbf{\Phi}^{(1)}(\varphi) = \beta \frac{\partial}{\partial \Lambda} \mathbf{L}(1, \frac{\partial}{\partial \varphi}) \mathbf{\Phi}^{(0)}(\varphi), \varphi \in (0, 2\pi); \tag{4.1.49}$$

$$\mathbf{\Phi}^{(1)}(0) - \mathbf{\Phi}^{(1)}(2\pi) = \mathbf{\Psi}^{(1)}(1) - \mathbf{\Psi}^{(1)}(-1),$$

$$\mathbf{B}(1, \frac{\partial}{\partial \varphi}) \mathbf{\Phi}^{(1)}(0) - \mathbf{B}(1, \frac{\partial}{\partial \varphi}) \mathbf{\Phi}^{(1)}(2\pi)$$

$$= \mathbf{Q}^\circ (\frac{d}{dt} \mathbf{\Psi}^{(2)}(1) - \frac{d}{dt} \mathbf{\Psi}^{(2)}(-1))$$

$$+ 2(\mu^\circ + \lambda^\circ) \mathbf{J} (\mathbf{\Psi}^{(1)}(1) - \mathbf{\Psi}^{(1)}(-1)), \tag{4.1.50}$$

where $\mathbf{J} = (J_{jk})_{j,k=1}^2, J_{12} = 1, J_{11} = J_{22} = J_{21} = 0$. (Note that, if $\gamma > 1$, then the functions $\mathbf{\Psi}^{(1)}$ and $\frac{d}{dt} \mathbf{\Psi}^{(2)}$ are constant vectors, and the right-hand sides in (4.1.50) are equal to zero). We chose the factor β using the solvability condition for the system (4.1.49), (4.1.50):

$$\mathbf{N}c = \beta a_0 \mathbf{F}c. \tag{4.1.51}$$

Here $\mathbf{N}$ and $\mathbf{F}$ are matrices of order 4, and the list of elements, which are not equal to zero, has the following representation:

$$-F_{13} = F_{24} = -F_{31} = F_{42} = 8\pi\mu(\lambda + 2\mu)(\lambda + \mu)^{-1} := d^{-1},$$

$$N_{11} = 8(2\mu^\circ + \lambda^\circ)^{-1}(\mu - \mu^\circ)(\mu + \mu^\circ + \lambda^\circ),$$

$$N_{42} = -2N_{22} = -16(\mu^\circ)^{-1}(\mu - \mu^\circ)\mu,$$

$$N_{13} = 16[(\lambda + \mu)(2\mu^\circ + \lambda^\circ)]^{-1}(\mu - \mu^\circ)(\mu - \mu^\circ + \lambda - \lambda^\circ)\mu,$$

$$N_{31} = -16[(\lambda + \mu)(2\mu^\circ + \lambda^\circ)]^{-1}(\mu - \mu^\circ)(\lambda^\circ + \mu^\circ)(2\mu + \lambda),$$

$$N_{33} = 32[(\lambda + \mu)^2(2\mu^\circ + \lambda^\circ)]^{-1}(\mu - \mu^\circ + \lambda - \lambda^\circ)\mu\mu^\circ(2\mu + \lambda).$$

Finally, we obtain that the factor β and the vector $\mathbf{c}$, which determine the principal terms of the series (4.1.40) and (4.1.41), are the solutions of the eigenvalue problem (4.1.51). There are four of these solutions. Thus, we have described all the differences in the algorithms for the cases of $\gamma = 1$ and $\gamma > 1$.

Let us consider the question of the stress singularity. We shall denote the displacement fields with the asymptotic expansions (4.1.40), (4.1.41), with $\Lambda = 1$, by $\mathbf{U}^{(j)}, \mathbf{U}^{\circ,(j)}$. In accordance with the general results [57], the solution $\boldsymbol{u} \in W_2^1(\Omega)$ satisfies the following equality:

$$\boldsymbol{u}(\boldsymbol{x}) = \boldsymbol{b}^0 + \sum_{j=1}^{4} b_j \mathbf{U}^{(j)}(\boldsymbol{x}) + O(r^{2-\varepsilon}), r \to 0, \tag{4.1.52}$$

where $\boldsymbol{b}^0$ is a constant vector, $b_1, \ldots, b_4$ are constants, ε is positive. By (4.1.52) we have that, if $Re\beta < 0$, then there is no stress singularity in a neighbourhood of the inclusion tip. Thus, there is no stress singularity in the case of $\gamma > 1$.

Note that there is another behaviour of the stress components, when $\gamma = 1$. The roots of the polynomial $\beta \to \det(\mathbf{N} - \beta a_0 \mathbf{F})$ admit the following representation:

$$\beta_1 = 0, \quad \beta_2 = -2(\lambda + \mu)(\mu - \mu^\circ)[\pi a_0(\lambda + 2\mu^\circ)\mu^\circ]^{-1},$$

$$\beta_{3,4} = -(2da_0)^{-1}\{N_{13} + N_{31} \pm [(N_{13} - N_{31})^2 + 4N_{11}N_{33}]^{1/2}\}. \tag{4.1.53}$$

Suppose that $\beta_2 \le 0$ and that $Re\beta_q \le 0, q = 3, 4$. Then we have

$$\mu \ge \mu^\circ, \mu(\mu + \lambda)(\mu - \mu^\circ)(3\mu + \lambda)^{-1} \ge (\mu^\circ + \lambda^\circ)(\mu - \mu^\circ), \tag{4.1.54}$$

since $N_{13} + N_{31} \geq 0$.

Note that, if β_3, β_4 are complex values, then $(N_{13} - N_{31})^2 < -4N_{11}N_{33}$ and, hence, $(\mu - \mu^\circ)(\mu + \mu^\circ + \lambda^\circ) < -4\mu\mu^\circ(2\mu + \lambda)(\mu + \lambda - \mu^\circ - \lambda^\circ)(\mu + \lambda)^{-2}$. Thus, we have that $\mu + \lambda \leq \mu^\circ + \lambda^\circ$. Both roots β_3, β_4 are real nonpositive, provided $N_{13}N_{31} \geq N_{11}N_{33}$, $\mu + \lambda \leq \mu^\circ + \lambda^\circ$. These formulae satisfy (4.1.54) if and only if $\mu = \mu^\circ$. It is easy to check that in the case $\mu = \mu^\circ$ all roots are equal to zero. Thus, we have found that at least one of the numbers $\beta_2, \beta_3, \beta_4$ will have a positive real part, provided that the shear moduli of the inclusion and the elastic matrix are not equal. Hence, there is a stress singularity. Let us note that if the shear moduli are equal, then the stress components are bounded.

Finally, let us consider the case where γ is not an integer. If $\gamma \in (0, 1)$, then the following asymptotic representations hold:

$$\mathbf{U}^{(j)}(r, \varphi) \sim r\rho_j(r)\mathbf{\Phi}^{(j)}(\varphi),$$

$$\mathbf{U}^{\circ,(j)}(r, \varphi) \sim r\rho_j(r)\mathbf{\Phi}^{\circ,(j)}(h(r)^{-1}\varphi), \tag{4.1.55}$$

where $j = 1, 2, 3, 4$, $\rho_j = \exp(\beta_j(1 - \gamma)^{-1}|\ln r|^{1-\gamma})$ and β_j are the values (4.1.53). Vectors $\mathbf{U}^{(1)}, \mathbf{U}^{\circ,(1)}$ correspond to the rotation $(-x_2, x_1)$. Comparing the asymptotic expressions in this case with those obtained above, we see that the logarithmic terms $|\ln r|$ with noninteger indices depending on γ, appear in the asymptotic series. The weight factor is determined by considering the ordinary differential equation $r\frac{\partial}{\partial r}\rho = \beta h\rho$. In particular, $\rho_j = \text{const}|\ln r|^{\beta_j} + O(1)$ provided $h(r) = a_0|\ln r|^{-1}$, and $\rho_j = \text{const} + o(1)$ provided $h(r) = a_0|\ln r|^{-\gamma}, \gamma > 1$.

4.1.3 Asymptotics of elastic fields near narrow wedge-shaped inclusion

Let us assume now that in a neighbourhood of the point O the domain ω is specified by the inequalities $-\varepsilon < \varphi < \varepsilon$ instead of (4.1.1). Here ε is a small positive number. The asymptotic expansion of the solution of the problem (4.1.3)–(4.1.5) contains the terms $r^\Lambda \boldsymbol{v}(\varphi)$ in Ω and $r^\Lambda \boldsymbol{v}^\circ(\varphi)$ in ω. Here Λ is an eigenvalue, while $\boldsymbol{v}$ and $\boldsymbol{v}^\circ$ are eigenvectors of the spectral problem

$$\mathbf{L}(\Lambda, \frac{\partial}{\partial\varphi})\boldsymbol{v}(\varphi) = 0, \varphi \in (\varepsilon, 2\pi - \varepsilon),$$

$$\mathbf{L}^\circ(\Lambda, \frac{\partial}{\partial\varphi})\boldsymbol{v}^\circ(\varphi) = 0, \varphi \in (-\varepsilon, \varepsilon); \tag{4.1.56}$$

$$\boldsymbol{v}(\varepsilon) = \boldsymbol{v}^\circ(\varepsilon), \boldsymbol{v}(2\pi - \varepsilon) = \boldsymbol{v}^\circ(-\varepsilon); \tag{4.1.57}$$

$$\mathbf{B}(\Lambda, \frac{\partial}{\partial\varphi})\boldsymbol{v}(\varepsilon) = \mathbf{B}^\circ(\Lambda, \frac{\partial}{\partial\varphi})\boldsymbol{v}^\circ(\varepsilon),$$

$$\mathbf{B}(\Lambda, \frac{\partial}{\partial\varphi})\boldsymbol{v}(2\pi - \varepsilon) = \mathbf{B}^\circ(\Lambda, \frac{\partial}{\partial\varphi})\boldsymbol{v}^\circ(-\varepsilon). \tag{4.1.58}$$

Matrices $\mathbf{L}, \mathbf{L}^\circ$ and $\mathbf{B}, \mathbf{B}^\circ$ were specified above. Note that a similar problem has been studied in [19], where a transcendental equation for Λ is given.

Let us consider the behaviour of $\boldsymbol{v}, \boldsymbol{v}^\circ$ and Λ, as $\varepsilon \to 0$. Assume that the constants λ, μ and λ°, μ° of the two elastic media are of the same order. We shall find an asymptotic representation of the solution of the problem (4.1.56)–(4.1.58) in the following form:

$$\Lambda = \Lambda_0 + \varepsilon\Lambda_1 + \ldots; \quad \boldsymbol{v}(\varphi) = \boldsymbol{w}^{(0)}(\varphi) + \varepsilon\boldsymbol{w}^{(1)}(\varphi) + \ldots;$$

$$\boldsymbol{v}^\circ(\varphi) = \mathbf{Z}^{(0)} + \varepsilon\mathbf{Z}^{(1)}(\varepsilon^{-1}\varphi) + \varepsilon^2\mathbf{Z}^{(2)}(\varepsilon^{-1}\varphi) + \ldots. \tag{4.1.59}$$

Substituting the above expansions for Λ and $\boldsymbol{v}^\circ$ into the second system of equations in (4.1.56) we obtain

$$\mathbf{Q}^\circ\frac{d^2}{dt^2}\mathbf{Z}^{(1)}(t) = \mathbf{0}, t \in (-1, 1), \tag{4.1.60}$$

$$\mathbf{Q}^\circ\frac{d^2}{dt^2}\mathbf{Z}^{(2)}(t) = -\mathbf{P}(\Lambda_0)\frac{d}{dt}\mathbf{Z}^{(1)}(t) - \mathbf{L}^\circ(\Lambda_0, 0)\mathbf{Z}^{(0)}, t \in (-1, 1), \tag{4.1.61}$$

where matrices $\mathbf{Q}^\circ$ and $\mathbf{P}$ are defined in (4.1.29).

In a similar manner, substituting the expansions (4.1.59) for Λ and $\boldsymbol{v}$ into the first system of equations in (4.1.56), we find that

$$\mathbf{L}(\Lambda_0, \frac{\partial}{\partial\varphi})\boldsymbol{w}^{(0)}(\varphi) = \mathbf{0}, \varphi \in (0, 2\pi), \tag{4.1.62}$$

$$\mathbf{L}(\Lambda_0, \frac{\partial}{\partial\varphi})\boldsymbol{w}^{(1)}(\varphi) + \Lambda_1\mathbf{L}_1(\Lambda_0, \frac{\partial}{\partial\varphi})\boldsymbol{w}^{(0)}(\varphi) = \mathbf{0}, \varphi \in (0, 2\pi), \tag{4.1.63}$$

where $\mathbf{L}_1(\xi_1, \xi_2) = \frac{\partial}{\partial\xi_1}\mathbf{L}(\xi_1, \xi_2)$.

The interface conditions (4.1.57) yield the following equalities:

$$\boldsymbol{w}^{(0)}(0) = \boldsymbol{w}^{(0)}(2\pi) = \mathbf{Z}^{(0)}, \tag{4.1.64}$$

$$\boldsymbol{w}^{(1)}(0) + \frac{\partial}{\partial\varphi}\boldsymbol{w}^{(0)}(0) = \mathbf{Z}^{(1)}(1),$$

$$\boldsymbol{w}^{(1)}(2\pi) - \frac{\partial}{\partial\varphi}\boldsymbol{w}^{(0)}(2\pi) = \mathbf{Z}^{(1)}(-1). \tag{4.1.65}$$

Equating coefficients near like powers of ε in the left-hand and in the right-hand side of (4.1.58), we obtain the following relations for the terms of the expansions (4.1.59):

$$\mathbf{Q}^\circ \frac{d}{dt}\mathbf{Z}^{(1)}(\pm 1) = \mathbf{B}(\Lambda, \frac{\partial}{\partial\varphi})\boldsymbol{w}^{(0)}(\pi \pm \pi) - \mathbf{B}^\circ(\Lambda, 0)\mathbf{Z}^{(0)}, \tag{4.1.66}$$

$$\mathbf{Q}^\circ \frac{d}{dt}\mathbf{Z}^{(2)}(\pm 1) = \mathbf{B}(\Lambda_0, \frac{\partial}{\partial\varphi})\boldsymbol{w}^{(1)}(\pi \mp \pi) - \mathbf{B}^\circ(\Lambda_0, 0)\mathbf{Z}^{(1)}(\pm 1)$$

$$+ \Lambda_1 \mathbf{B}_1(\Lambda_0, \frac{\partial}{\partial\varphi})\boldsymbol{w}^{(0)}(\pi \mp \pi) \pm \mathbf{B}(\Lambda_0, \frac{\partial}{\partial\varphi})\frac{\partial}{\partial\varphi}\boldsymbol{w}^{(0)}(\pi \mp \pi)$$

$$- \Lambda_1 \mathbf{B}_1^\circ(\Lambda_0, 0)\mathbf{Z}^{(0)}. \tag{4.1.67}$$

Let us note that the system (4.1.60), (4.1.66) is solvable if

$$\mathbf{B}(\Lambda_0, \frac{\partial}{\partial\varphi})\boldsymbol{w}^{(0)}(0) = \mathbf{B}(\Lambda_0, \frac{\partial}{\partial\varphi})\boldsymbol{w}^{(0)}(2\pi), \tag{4.1.68}$$

and its solution is a linear function

$$\mathbf{Z}^{(1)}(t) = (\mathbf{Q}^\circ)^{-1}(\mathbf{B}(\Lambda_0, \frac{\partial}{\partial\varphi})\boldsymbol{w}^{(0)}(0) - \mathbf{B}^\circ(\Lambda_0, 0)\mathbf{Z}^{(0)})t + \mathbf{C}_w, \tag{4.1.69}$$

where $\mathbf{C}_w$ is an arbitrary constant vector.

Thus, Λ is an eigenvalue of the spectral problem (4.1.62), (4.1.64), (4.1.68), while the vector-valued function $\boldsymbol{w}^{(0)}$ is a linear combination of eigenvectors that correspond to $\Lambda_0 = k$ and are specified by (4.1.21), (4.1.22).

Let us write $\boldsymbol{w}^{(0)}$, corresponding to the first positive eigenvalue $\Lambda_0 = 1$, as a linear combination

$$\boldsymbol{w}^{(0)} = C_1\boldsymbol{v}^{(1,1)} + C_2\boldsymbol{v}^{(1,2)} + C_3\boldsymbol{v}^{(1,3)} + C_4\boldsymbol{v}^{(1,4)} \tag{4.1.70}$$

of the vectors (4.1.21). Equations (4.1.64), (4.1.65) and (4.1.68) yield

$$\boldsymbol{w}^{(1)}(0) - \boldsymbol{w}^{(1)}(2\pi) = V(\partial/\partial\varphi)\boldsymbol{w}^{(0)}(0),$$

$$V(\eta) = 2(\mathbf{Q}^\circ)^{-1}(\mathbf{B}(1, \eta) - \mathbf{B}^\circ(1, 0)) - 2\eta\mathbf{1}. \tag{4.1.71}$$

The problem (4.1.61), (4.1.67) for $\mathbf{Z}^{(2)}$ is solvable if and only if the following equalities are satisfied:

$$\mathbf{B}(1, \frac{\partial}{\partial\varphi})\boldsymbol{w}^{(1)}(0) - \mathbf{B}(1, \frac{\partial}{\partial\varphi})\boldsymbol{w}^{(1)}(2\pi) = 2T(\frac{\partial}{\partial\varphi})\boldsymbol{w}^{(0)}(0), \qquad (4.1.72)$$

$$T(\eta) = (\mathbf{B}^{\circ}(1,0) - \mathbf{P}(1))(\mathbf{Q}^{\circ})^{-1}(\mathbf{B}(1,\eta) - \mathbf{B}^{\circ}(1,0)) - \mathbf{B}(1,\eta)\eta.$$

Thus, we have obtained the interface conditions for $\boldsymbol{w}^{(1)}$.

Considering problem (4.1.63), (4.1.71), (4.1.72) and taking into account that its solution exists only when the right-hand side is orthogonal to the eigenvectors of the adjoint problem, we obtain in this way four equations, which can be written in the form of a system of algebraic equations for the vector $\mathbf{C} = (C_1, C_2, C_3, C_4)$ of coefficients of the linear combination (4.1.70):

$$(\mathbf{N} - \Lambda_1\mathbf{F})\mathbf{C} = \mathbf{0}, \qquad (4.1.73)$$

where $\mathbf{N}$ and $\mathbf{F}$ are the matrices from (4.1.51). Due to the fact that $\det\mathbf{F} \neq 0$, there exist four (including multiplicities) eigenvalues $\Lambda_1^{(q)}$ for system (4.1.73). These values make it possible to determine the second term in the asymptotic representation of eigenvalues of the problem (4.1.56)–(4.1.58):

$$\Lambda^{(q)} = 1 + \varepsilon\Lambda_1^{(q)} + o(\varepsilon), \varepsilon \to 0, \ (q = 1, 2, 3, 4). \qquad (4.1.74)$$

Let us note that the roots of a polynomial $\Lambda_1 \to \det(\mathbf{N} - \Lambda_1\mathbf{F})$ had been investigated in the previous part. In accordance with these results we see that there exists at least one element with a negative real part in the set $\{\Lambda_1^{(q)}\}_{q=1}^4$ of eigenvalues, provided the shear moduli of the inclusion and the elastic matrix are not equal. This case corresponds to the stress singularity at the vertex of an angular inclusion, and there is a well-known fact (see, for example, [19]) that the exponent of the singularity tends to zero, as $\varepsilon \to 0$.

Thus, we have described the influence of the shape of a narrow two-dimensional inclusion in a neighbourhood of an irregular point of the boundary on the behaviour of the stresses as boundary of the narrow inclusion changes its shape from power law behaviour via a logarithmic shape to an angular form.

4.1.4 The state of stress and strain in the vicinity of a three-dimensional cusp-shaped elastic inclusion

In this section we consider the stress-strain state of an elastic three-dimensional body in the vicinity of a cusp-shaped elastic inclusion. Let us note that an asymptotic expansion of the displacement field contains two types of terms: homogeneous vector polynomials satisfying the Lamé system in an elastic matrix, and boundary layer terms concentrated in the immediate vicinity of the boundary of the inclusion. We shall describe these boundary layer terms using auxiliary solutions of two-dimensional elasticity problems and of the problems of the anti-plane shear in a plane with an inclusion. The following asymptotic analysis shows that there is no stress singularity at the vertex of the elastic peak-shaped inclusion; however, the stress components are discontinuous functions.

Let us state the problem. Consider a doubly connected region G in $\mathbb{R}^3$ with boundary ∂G and suppose that this boundary is smooth everywhere except at the origin O, which is on the interior surface (see Fig. 4.2). Denote the region bounded by this interior surface by g, and assume that in a neighbourhood of O the set G is described by the relationships

$$h(x_3)^{-1}\boldsymbol{x}' \in \Upsilon; x_3 > 0, \tag{4.1.75}$$

where $\boldsymbol{x}' = (x_1, x_2)$; Υ is a plane domain with a smooth boundary $\partial\Upsilon$; h is a smooth function in $(0,1)$ which admits the representation

$$h(z) = az^{1+\gamma} + O(z^{1+\gamma+\delta}), \tag{4.1.76}$$

with positive numbers a, γ and δ.

Consider the three-dimensional system of equations for the displacement field $\boldsymbol{u}, \boldsymbol{u}^\circ$:

$$\mu\Delta\boldsymbol{u}(\boldsymbol{x}) + (\lambda+\mu)\nabla\nabla\cdot\boldsymbol{u}(\boldsymbol{x}) = \boldsymbol{0}; \boldsymbol{x} \in G; \tag{4.1.77}$$

$$\mu^\circ\Delta\boldsymbol{u}^\circ(\boldsymbol{x}) + (\lambda^\circ+\mu^\circ)\nabla\nabla\cdot\boldsymbol{u}^\circ(\boldsymbol{x}) = \boldsymbol{0}; \boldsymbol{x} \in g; \tag{4.1.78}$$

$$\boldsymbol{u}(\boldsymbol{x}) = \boldsymbol{u}^\circ(\boldsymbol{x}); \boldsymbol{\sigma}^{(n)}(\boldsymbol{u};\boldsymbol{x}) = \boldsymbol{\sigma}^{\circ,(n)}(\boldsymbol{u}^\circ;\boldsymbol{x}); \boldsymbol{x} \in \partial g\backslash O; \tag{4.1.79}$$

$$\boldsymbol{\sigma}^{(n)}(\boldsymbol{u};\boldsymbol{x}) = \boldsymbol{p}(\boldsymbol{x}); \boldsymbol{x} \in \partial G\backslash\partial g. \tag{4.1.80}$$

We are keeping here the previous notations assuming $\boldsymbol{p}$ to be an external self-balanced load.

Note that for small $\|\boldsymbol{x}\|$ the region g lies in any cone containing the Ox_3 axis. On a sphere of radius ρ it cuts out a set, which area is of order $O(\rho^{2(1+\gamma)})$, as $\rho \to 0$, and tends to zero faster than the area of the whole sphere. Thus, (see [57],[39]), the principal term of the asymptotic representation for the stress-strain state in a neighbourhood of the point O outside the cone corresponds to the elasticity problem for a homogeneous body without inclusion.

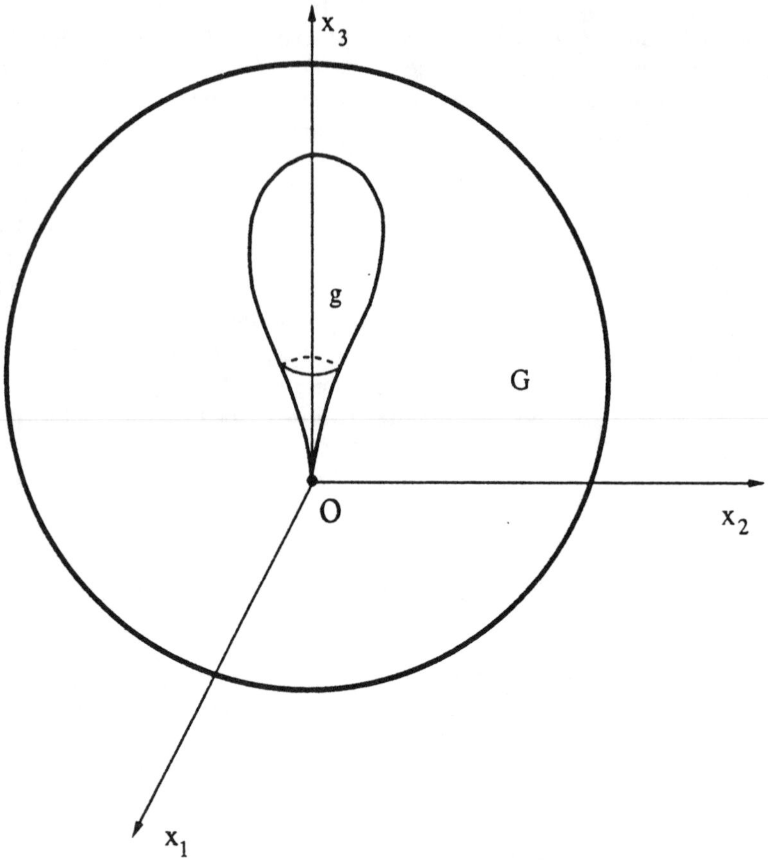

Fig. 4.2: A three-dimensional cusp-shaped elastic inclusion.

It can be shown by the Taylor expansion of the solution of the homogeneous Lamé system that in a neighbourhood of an internal point of the region the asymptotic form of the displacement field is a linear combination of vector-valued polynomial functions. The number of linearly independent homogeneous vector polynomials of degree m, satisfying the three-dimensional homogeneous Lamé system, is equal to

$3(2m+1)$. We denote these vector-valued functions by

$$\mathbf{V}^{(m,q)}(\boldsymbol{x}) = (V_1^{(m,q)}(\boldsymbol{x}), V_2^{(m,q)}(\boldsymbol{x}), V_3^{(m,q)}(\boldsymbol{x}))^T. \tag{4.1.81}$$

It is well-known that for the homogeneous (excluding a neighbourhood of the origin) Lamé system there exist solutions with singularities; for example, solutions corresponding to concentrated forces. These vector fields with singularities at an internal point of the region are expressed using columns $T^{(j)}, j = 1,2,3$ of the Somigliana tensor $T(\boldsymbol{x})$ which has the singularity $O(||\boldsymbol{x}||^{-1})$ as $||\boldsymbol{x}|| \to 0$. Moreover, the vectors

$$\sum_{j=1}^{3} V_j^{(m,q)}(\partial/\partial \boldsymbol{x}) T^{(j)}(\boldsymbol{x}), q = 1, \dots, 3(2m+1), \tag{4.1.82}$$

satisfy the homogeneous Lamé system away from the point O and are also linearly independent. In accordance with results of [38] the above-mentioned vector-valued functions (4.1.81), (4.1.82) describe all the power law solutions of the homogeneous Lamé system. Let us note that vector-valued functions $\mathbf{V}^{(p,q)}$ have finite elastic energy. First three solutions $\mathbf{V}^{(0,q)}$ correspond to rigid body translation. If $m = 1$, then $\mathbf{V}^{(1,q)}, q = 1, 2, \dots, 9$, may be represented by the vectors

$$x_j \boldsymbol{e}^{(j)},\ j = 1,2,3;\ \ x_1 \boldsymbol{e}^{(2)},\ \ x_2 \boldsymbol{e}^{(3)},\ \ x_3 \boldsymbol{e}^{(1)}; \tag{4.1.83}$$

$$x_2 \boldsymbol{e}^{(1)} - x_1 \boldsymbol{e}^{(2)},\ \ x_3 \boldsymbol{e}^{(2)} - x_2 \boldsymbol{e}^{(3)},\ \ x_1 \boldsymbol{e}^{(3)} - x_3 \boldsymbol{e}^{(1)}, \tag{4.1.84}$$

where $\boldsymbol{e}^{(j)}, j = 1,2,3$, are the unit Cartesian basis vectors in $\mathbb{R}^3$. The vectors (4.1.84) correspond to a rigid body rotation of the elastic domain. The remaining vectors (4.1.83), that satisfy the system (4.1.77), (4.1.78) and the first of the interface conditions, leave a discrepancy in the second condition (4.1.79).

To investigate the jumps in stresses on the boundary of the inclusion we shall introduce new variables in a neighbourhood of point O at $x_3 > 0$:

$$\boldsymbol{y} = (y_1, y_2) = h(x_3)^{-1}(x_1, x_2);\ \ z = x_3. \tag{4.1.85}$$

Thus, the region g is specified by the relationships $\boldsymbol{y} \in \Upsilon, z > 0$. The unit outer normal vector to ∂G is determined by the equality

$$\boldsymbol{n}(\boldsymbol{y}, z) = [1 + h'(z)^2 ||\boldsymbol{y}||^2]^{-1/2}(\nu_1, \nu_2, h'(z)||\boldsymbol{y}||)^T, \tag{4.1.86}$$

where $(\nu_1, \nu_2)^T$ is the unit internal normal vector to $\partial\Upsilon$.

Thus, we obtain the representation

$$\boldsymbol{\sigma}^{(n)}(\mathbf{V};\boldsymbol{x}) = \boldsymbol{\sigma}^{(1)}(\mathbf{V};\boldsymbol{x})\nu_1(\boldsymbol{y}) + \boldsymbol{\sigma}^{(2)}(\mathbf{V};\boldsymbol{x})\nu_2(\boldsymbol{y}) + O(z^\gamma).$$

The stress vectors $\boldsymbol{\sigma}^{(1)}_{(q)} := (\sigma^{(q)}_{11}, \sigma^{(q)}_{12})^T = \boldsymbol{\sigma}^{(1)}(\mathbf{V}^{(1,q)};\boldsymbol{x})$, $\boldsymbol{\sigma}^{(2)}_{(q)} := (\sigma^{(q)}_{12}, \sigma^{(q)}_{22})^T = \boldsymbol{\sigma}^{(2)}(\mathbf{V}^{(1,q)};\boldsymbol{x})$ have the components

$$\sigma^{(1)}_{11} = \sigma^{(2)}_{22} = 2\mu + \lambda;\ \ \sigma^{(1)}_{22} = \sigma^{(2)}_{11} = \sigma^{(3)}_{11} = \sigma^{(3)}_{22} = \lambda;$$

$$\sigma^{(4)}_{12} = \sigma^{(5)}_{13} = \sigma^{(6)}_{23} = \mu.$$

All other components are equal to zero. A similar procedure (with λ, μ replaced by λ°, μ°) determines the stresses at the boundary of the inclusion. Thus, the discrepancy of the vectors (4.1.83) in the second of the interface conditions (4.1.79) is described by the equality

$$\boldsymbol{\sigma}^{(n)}(\mathbf{V}^{(1,q)};\boldsymbol{x}) - \boldsymbol{\sigma}^{\circ,(n)}(\mathbf{V}^{(1,q)};\boldsymbol{x}) = (\boldsymbol{\sigma}^{(1)}_{(q)} - \boldsymbol{\sigma}^{\circ,(1)}_{(q)})\nu_1(\boldsymbol{y})$$

$$+ (\boldsymbol{\sigma}^{(2)}_{(q)} - \boldsymbol{\sigma}^{\circ,(2)}_{(q)})\nu_2(\boldsymbol{y}) + O(z^\gamma); \boldsymbol{y} \in \partial\Upsilon, z \to +0. \tag{4.1.87}$$

To compensate for the stress jumps (4.1.87) on the boundary of the inclusion we shall construct, using the coordinates (4.1.85), a boundary layer in the form

$$f(z)\boldsymbol{w}(\boldsymbol{y}),\ \ (h\boldsymbol{y}, z) \in G;\ \ f(z)\boldsymbol{w}^\circ(\boldsymbol{y}),\ \ (h\boldsymbol{y}, z) \in g, \tag{4.1.88}$$

where $f(z) = O(z^s)$, as $z \to +0$, $s \geq 0$. Taking into account the relationships

$$\frac{\partial}{\partial x_j} = \frac{1}{h(z)}\frac{\partial}{\partial y_j}, j = 1, 2;\ \ \frac{\partial}{\partial x_3} = \frac{\partial}{\partial z} - \sum_{j=1}^{2}\frac{h'(z)}{h(z)}y_j\frac{\partial}{\partial y_j}, \tag{4.1.89}$$

we obtain the equalities

$$\mathcal{L}^\circ(\frac{\partial}{\partial \boldsymbol{x}})\{f(z)\boldsymbol{w}^\circ(\boldsymbol{y})\} = \frac{f(z)}{h(z)^2}[\mathcal{L}^\circ(\frac{\partial}{\partial \boldsymbol{y}}, 0)\boldsymbol{w}^\circ(\boldsymbol{y}) + O(z^\gamma)], \boldsymbol{x} \in g,$$

$$\mathcal{L}(\frac{\partial}{\partial \boldsymbol{x}})\{f(z)\boldsymbol{w}(\boldsymbol{y})\} = \frac{f(z)}{h(z)^2}[\mathcal{L}(\frac{\partial}{\partial \boldsymbol{y}}, 0)\boldsymbol{w}(\boldsymbol{y}) + O(z^\gamma)], \boldsymbol{x} \in G, \tag{4.1.90}$$

where $\mathcal{L}, \mathcal{L}^\circ$ are matrix differential operators,

$$\mathcal{L}(\xi_1, \xi_2, 0) =$$

$$\begin{pmatrix} (2\mu+\lambda)\xi_1^2 + \mu\xi_2^2 & (\mu+\lambda)\xi_1\xi_2 & 0 \\ (\mu+\lambda)\xi_1\xi_2 & (2\mu+\lambda)\xi_2^2 + \mu\xi_1^2 & 0 \\ 0 & 0 & \mu(\xi_1^2+\xi_2^2) \end{pmatrix}, \qquad (4.1.91)$$

and the operator $\mathcal{L}^\circ$ is of the same form with λ, μ replaced by λ°, μ°.

Using equations (4.1.86), (4.1.89) for the interface condition in stresses we obtain

$$\sigma^{(n)}(f(z)\boldsymbol{w}(\boldsymbol{y}); \boldsymbol{x}) - \sigma^{\circ,(n)}(f(z)\boldsymbol{w}^\circ(\boldsymbol{y}); \boldsymbol{x})$$

$$= h(z)^{-1} f(z)[\mathcal{B}(\boldsymbol{y}, \frac{\partial}{\partial \boldsymbol{y}})\boldsymbol{w}(\boldsymbol{y}) - \mathcal{B}^\circ(\boldsymbol{y}, \frac{\partial}{\partial \boldsymbol{y}})\boldsymbol{w}^\circ(\boldsymbol{y})$$

$$+ O(z^\gamma)], \boldsymbol{x} \in \partial g, \qquad (4.1.92)$$

where

$$\mathcal{B}(\boldsymbol{y}, \xi_1, \xi_2) = \nu_1(\boldsymbol{y}) \begin{pmatrix} (2\mu+\lambda)\xi_1 & \lambda\xi_2 & 0 \\ \mu\xi_2 & \mu\xi_1 & 0 \\ 0 & 0 & \mu\xi_1 \end{pmatrix} \qquad (4.1.93)$$

$$+ \nu_2(\boldsymbol{y}) \begin{pmatrix} \mu\xi_2 & \mu\xi_1 & 0 \\ \lambda\xi_1 & (2\mu+\lambda)\xi_2 & 0 \\ 0 & 0 & \mu\xi_2 \end{pmatrix}.$$

Note that the above matrices (4.1.91) and (4.1.93) consist of operators corresponding to the plane problem of elasticity theory and to the problem of anti-plane shear.

Thus, $f(z) = h(z)$ and the vectors $\boldsymbol{w}(\boldsymbol{y}), \boldsymbol{w}^\circ(\boldsymbol{y})$ satisfy the boundary value problem

$$\mathcal{L}(\frac{\partial}{\partial \boldsymbol{y}}, 0)\boldsymbol{w}(\boldsymbol{y}) = \boldsymbol{0}, \boldsymbol{y} \in \mathbb{R}^2 \backslash \Upsilon;$$

$$\mathcal{L}^\circ(\frac{\partial}{\partial \boldsymbol{y}}, 0)\boldsymbol{w}^\circ(\boldsymbol{y}) = \boldsymbol{0}, \boldsymbol{y} \in \Upsilon; \qquad (4.1.94)$$

$$\boldsymbol{w}(\boldsymbol{y}) - \boldsymbol{w}^\circ(\boldsymbol{y}) = \boldsymbol{0}; \;\; \mathcal{B}(\boldsymbol{y}, \frac{\partial}{\partial \boldsymbol{y}})\boldsymbol{w}(\boldsymbol{y}) - \mathcal{B}^\circ(\boldsymbol{y}, \frac{\partial}{\partial \boldsymbol{y}})\boldsymbol{w}^\circ(\boldsymbol{y})$$

$$= -(\sigma^{(1,q)} - \sigma^{\circ,(1,q)})\nu_1(\boldsymbol{y}) - (\sigma^{(2,q)} - \sigma^{\circ,(2,q)})\nu_2(\boldsymbol{y}), \boldsymbol{y} \in \Upsilon. \quad (4.1.95)$$

Therefore, we can see that the boundary layer corresponding to vectors (4.1.83) in a neighbourhood of a cusp-shaped inclusion has the form

$$h(x_3)\mathbf{W}^{(q)}(h(x_3)^{-1}\boldsymbol{x}'), \boldsymbol{x} \in G; \;\; h(x_3)\mathbf{W}^{\circ,(q)}(h(x_3)^{-1}\boldsymbol{x}'), \boldsymbol{x} \in g, \tag{4.1.96}$$

where the vector-valued functions

$$\mathbf{W}^{(j)}(\boldsymbol{y}) = (w_1^{(j)}(\boldsymbol{y}), w_2^{(j)}(\boldsymbol{y}), 0)^T; \mathbf{W}^{(j+4)}(\boldsymbol{y}) = (0, 0, w_3^{(j)}(\boldsymbol{y}))^T, j = 1, 2;$$

$$\mathbf{W}^{(3)}(\boldsymbol{y}) = (w_1^{(0)}(\boldsymbol{y}), w_2^{(0)}(\boldsymbol{y}), 0)^T;$$

$$\mathbf{W}^{(4)}(\boldsymbol{y}) = 2^{-1/2}(w_1^{(3)}(\boldsymbol{y}), w_2^{(3)}(\boldsymbol{y}), 0)^T \tag{4.1.97}$$

are expressed in terms of solutions of the following auxiliary problems:

$$\mu\Delta \boldsymbol{w}^{(j)}(\boldsymbol{y}) + (\lambda + \mu)\nabla\nabla \cdot \boldsymbol{w}^{(j)}(\boldsymbol{y}) = \boldsymbol{0}, \boldsymbol{y} \in \mathbb{R}^2 \backslash \Upsilon;$$

$$\mu^\circ \Delta \boldsymbol{w}^{\circ,(j)}(\boldsymbol{y}) + (\lambda^\circ + \mu^\circ)\nabla\nabla \cdot \boldsymbol{w}^{\circ,(j)}(\boldsymbol{y}) = \boldsymbol{0}, \boldsymbol{y} \in \Upsilon; \tag{4.1.98}$$

$$\boldsymbol{w}^{(j)}(\boldsymbol{y}) - \boldsymbol{w}^{\circ,(j)}(\boldsymbol{y}) = \boldsymbol{0},$$

$$\boldsymbol{\sigma}^{(n)}(\boldsymbol{w}^{(j)}; \boldsymbol{y}) - \boldsymbol{\sigma}^{\circ,(n)}(\boldsymbol{w}^{\circ,(j)}; \boldsymbol{y}) = -\boldsymbol{p}^{(j)}(\boldsymbol{y}), \boldsymbol{y} \in \partial\Upsilon, \tag{4.1.99}$$

$$\boldsymbol{w}^{(j)} = (w_1^{(j)}, w_2^{(j)})^T, j = 1, 2, 3;$$

and

$$\mu\Delta w_3^{(j)}(\boldsymbol{y}) = 0, \boldsymbol{y} \in \mathbb{R}^2 \backslash \Upsilon; \mu^\circ \Delta w_3^{\circ,(j)}(\boldsymbol{y}) = 0, \boldsymbol{y} \in \Upsilon; \tag{4.1.100}$$

$$w_3^{(j)}(\boldsymbol{y}) - w_3^{\circ,(j)}(\boldsymbol{y}) = 0;$$

$$\mu \frac{\partial w_3^{(j)}}{\partial \nu}(\boldsymbol{y}) - \mu^\circ \frac{\partial w_3^{\circ,(j)}}{\partial \nu}(\boldsymbol{y}) = (\mu^\circ - \mu)\nu_j, \boldsymbol{y} \in \partial\Upsilon, \;\; j = 1, 2. \tag{4.1.101}$$

Here

$$\boldsymbol{p}^{(1)} = (\nu_1(2\mu + \lambda - 2\mu^\circ - \lambda^\circ), \nu_2(\lambda - \lambda^\circ))^T;$$

$$\boldsymbol{p}^{(2)} = (\nu_1(\lambda - \lambda^\circ), \nu_2(2\mu + \lambda - 2\mu^\circ - \lambda^\circ))^T; \tag{4.1.102}$$

$$\boldsymbol{p}^{(3)} = 2^{1/2}(\nu_2(\mu - \mu^\circ), \nu_1(\mu - \mu^\circ))^T.$$

Since

$$\int_{\partial\Upsilon} \boldsymbol{p}^{(j)}(\boldsymbol{y})ds = \boldsymbol{0}, \;\; \int_{\partial\Upsilon} [p_2^{(j)}(\boldsymbol{y})y_1 - p_1^{(j)}(\boldsymbol{y})y_2]ds = 0, \;\; j = 1, 2, 3,$$

and the mean value at $\partial\Upsilon$ in the right-hand side of the second interface condition (4.1.101) is equal to zero, the following asymptotic representations hold:

$$\boldsymbol{w}^{(j)}(\boldsymbol{y}) = \sum_{k=1}^{3} q_{jk} Y^j(\frac{\partial}{\partial \boldsymbol{y}}) S(\boldsymbol{y}) + O(\|\boldsymbol{y}\|^{-2}), \|\boldsymbol{y}\| \to \infty, \quad (4.1.103)$$

$$w_3^{(j)}(\boldsymbol{y}) = \frac{1}{2\pi\mu} \sum_{k=1}^{2} d_{jk} \|\boldsymbol{y}\|^{-2} y_k + O(\|\boldsymbol{y}\|^{-2}), \|\boldsymbol{y}\| \to \infty. \quad (4.1.104)$$

Here $S(\boldsymbol{y})$ is the Somigliana tensor related to the fundamental solution of the two–dimensional Lamé system.

The form $d(\boldsymbol{l}) = \sum_{j,k=1}^{2} d_{jk} l_j l_k$ is defined in [107] as the dipole form, corresponding to the virtual mass $D(\boldsymbol{l}) = \sum_{j,k=1}^{2} D_{jk} l_j l_k$ in the direction $\boldsymbol{l}$, where

$$D_{jk} = \mu \int_{\mathbb{R}^2 \setminus \Upsilon} \nabla w_3^{(j)}(\boldsymbol{y}) \cdot \nabla w_3^{(k)}(\boldsymbol{y}) d\boldsymbol{y} + \mu^\circ \int_{\Upsilon} \nabla w_3^{\circ,(j)}(\boldsymbol{y}) \cdot \nabla w_3^{\circ,(k)}(\boldsymbol{y}) d\boldsymbol{y}.$$

Using the Greenhill analogy in problems of anti-plane shear and flow of the ideal liquid [1], D may be interpreted as the energy of an additional flow, connected with the body Υ, introduced into initially uniform flow with direction $\boldsymbol{l}$. The Green formula yields the following relationship between the matrices $\mathbf{D} = (D_{jk})_{j,k=1}^2$ and $\boldsymbol{d} = (d_{jk})_{j,k=1}^2$:

$$\mathbf{D} = \boldsymbol{d} - \text{mes}\Upsilon(\mu - \mu^\circ)\mathbf{1},$$

where $\mathbf{1}$ is the unit matrix.

Note that the coefficients q_{jk} in (4.1.103) are the elements of the Pólya–Szegö matrix, which was investigated in [3], [131].

In the special case where $\mu + \lambda = \mu^\circ + \lambda^\circ$, we have that $\boldsymbol{p}^{(1)} = -\boldsymbol{p}^{(2)}$ and, hence, the solutions $\boldsymbol{w}^{(1)} = -\boldsymbol{w}^{(2)}$, and they are not linearly independent. There exists an additional solution $(w_1^{(0)}, w_2^{(0)})^T$ of the problem (4.1.98), (4.1.99) with $\boldsymbol{p}^{(0)}(\boldsymbol{y}) = (\lambda - \lambda^\circ)(\nu_1(\boldsymbol{y}), \nu_2(\boldsymbol{y}))^T$ that is linearly independent of the vector $\boldsymbol{w}^{(1)} = -\boldsymbol{w}^{(2)}$. If $\lambda + \mu \neq \lambda^\circ + \mu^\circ$ it is expressed in terms of $\boldsymbol{w}^{(1)}, \boldsymbol{w}^{(2)}$,

$$\boldsymbol{w}^{(0)}(\boldsymbol{y}) = (\lambda - \lambda^\circ)[2(\lambda + \mu - \lambda^\circ - \mu^\circ)]^{-1}[\boldsymbol{w}^{(1)}(\boldsymbol{y}) + \boldsymbol{w}^{(2)}(\boldsymbol{y})].$$

Finally, let us consider the following vector-valued functions:

$$\mathbf{X}^{(q)}(\boldsymbol{x}) = \mathbf{V}^{(1,q)}(\boldsymbol{x}) + \chi(x_3^{-1}\|\boldsymbol{x}'\|)h(x_3)\mathbf{W}^{(q)}(h(x_3)^{-1}\boldsymbol{x}'); \quad (4.1.105)$$

$$\mathbf{X}^{\circ,(q)}(\boldsymbol{x}) = \mathbf{V}^{(1,q)}(\boldsymbol{x}) + h(x_3)\mathbf{W}^{\circ,(q)}(h(x_3)^{-1}\boldsymbol{x}'). \quad (4.1.106)$$

Here $\chi \in \mathbf{C}^\infty(\mathbb{R}^1)$ is a cut–off function such that $\chi(\rho) = 1$ if $\rho < 1/2$ and $\chi(\rho) = 0$ if $\rho > 1$; the second term in the right-hand side of (4.1.106) is defined only for $x_3 > 0$.

It is important to note that vector-valued functions $\mathbf{X}^{(q)}$ leave a discrepancy in the second interface condition (4.1.78), which is of order $O(\|\boldsymbol{x}\|^\gamma)$; the corresponding discrepancy in the Lamé system (4.1.76), (4.1.77) is $O(\|\boldsymbol{x}\|^{2\gamma+1})$ outside of K and $O(\|\boldsymbol{x}\|^{-1}(1+x_3^{-1-\gamma}\|\boldsymbol{x}'\|)^{-2})$ in K.

Finally, the asymptotic representation for the solutions of the problem (4.1.76)–(4.1.78) in the vicinity of the vertex of the elastic inclusion g has the form

$$\boldsymbol{u}(\boldsymbol{x}) = \mathbf{C}^{(1)} + \mathbf{C}^{(2)} \times \boldsymbol{x} + \sum_{q=1}^{6} c_q \mathbf{X}^{(q)}(\boldsymbol{x}) + O(\|\boldsymbol{x}\|^\beta), \boldsymbol{x} \in G, \quad (4.1.107)$$

$$\boldsymbol{u}^\circ(\boldsymbol{x}) = \mathbf{C}^{(1)} + \mathbf{C}^{(2)} \times \boldsymbol{x} + \sum_{q=1}^{6} c_q \mathbf{X}^{\circ,(q)}(\boldsymbol{x}) + O(\|\boldsymbol{x}\|^\beta), \boldsymbol{x} \in g,$$

where $\beta = \min\{2, 2\gamma+1\}; \mathbf{C}^{(1)}, \mathbf{C}^{(2)}$ are constant vectors; $c_1, \ldots, c_6$ are scalar constants; these constants depend on the data of the whole problem.

Therefore, the corresponding components of the stress tensor are described by the equalities

$$\sigma_{jk} = \sum_{q=1}^{6} [\sigma_{jk}(\mathbf{V}^{(1,q)}; \boldsymbol{x}) + \chi(x_3^{-1}\|\boldsymbol{x}'\|)\boldsymbol{\Sigma}_{jk}^{(q)}(\boldsymbol{y})]$$

$$+ O(\|\boldsymbol{x}\|^\gamma), \boldsymbol{x} \in G. \quad (4.1.108)$$

Here $\boldsymbol{\Sigma}_{jk}^{(q)}$ are the terms of the boundary layer type,

$$\boldsymbol{\Sigma}_{kk}^{(q)}(\boldsymbol{y}) = (2\mu+\lambda)\frac{\partial W_k^{(q)}}{\partial y_k}(\boldsymbol{y}) + \lambda\frac{\partial W_{3-k}^{(q)}}{\partial y_{3-k}}(\boldsymbol{y});$$

$$\Sigma_{k3}^{(q)}(y) = \mu \frac{\partial W_3^{(q)}}{\partial y_k}(y); k = 1, 2; \Sigma_{33}^{(q)}(y) = \lambda[\frac{\partial W_1^{(q)}}{\partial y_1}(y) + \frac{\partial W_2^{(q)}}{\partial y_2}(y)];$$

$$\Sigma_{12}^{(q)}(y) = \mu[\frac{\partial W_1^{(q)}}{\partial y_2}(y) + \frac{\partial W_2^{(q)}}{\partial y_1}(y)]. \tag{4.1.109}$$

Corresponding representations for the stress components in g have the same form with λ, μ replaced by λ°, μ°.

It should be mentioned that all above asymptotic considerations are valid even in the case of $\lambda^\circ = \mu^\circ = 0$, which corresponds to a three-dimensional peak-shaped cavity. The case of an absolutely rigid inclusion is not included in these considerations.

Taking into account the above asymptotic representations for the stress tensor components we can show that in a neighbourhood of the tip of a three-dimensional cusp-shaped inclusion these components are bounded. Moreover, in the exterior of a conical neighbourhood of the Ox_3 axis the following asymptotic relation holds:

$$\sigma_{jk}(u; x) = \sum_{q=1}^{6} c_q \sigma_{jk}(\mathbf{V}^{(1,q)}; x) + O(\|x\|^\gamma).$$

Thus, the components of the stress tensor σ are continuous in this set.

On the other hand, in a cusp-shaped neighbourhood of the inclusion and in the inclusion itself, the components σ_{jk} are discontinuous at the tip of the cusp. The value $\lim_{\|x\|\to 0} \sigma_{jk}$ depends on the direction, tangent to the boundary of the inclusion g. Thus, for $x' = yh(x_3)$, where y is a fixed point from $\mathbb{R}^2 \backslash \Upsilon$, the value $\lim_{x_3 \to 0} \sigma_{jk}(u; x)$ is equal to

$$\sum_{q=1}^{6} c_q[\sigma_{jk}(\mathbf{V}^{(1,q)}; x) + \Sigma_{jk}^{(q)}(y)]. \tag{4.1.110}$$

Similar equality is used for the calculation of the limits for the stress components σ_{jk} in g, as $x_3 \to 0$, and $x' = yh(x_3), y \in \Upsilon$.

Therefore, the above results show that the stresses in a neighbourhood of a three-dimensional pointed elastic inclusion are bounded. However, the stresses lose continuity at the vertex of the inclusion on approaching that point along a trajectory which is tangent to the axis of the inclusion.

4.1.5 The stress-strain state near the tip of an absolutely rigid three-dimensional inclusion inserted into an elastic body

In this part we shall prove that in contrast to the case of a three-dimensional elastic inclusion or cavity, the components of the stress tensor in an elastic body with an absolutely rigid inclusion have a singularity at the vertex O of the cusp.

Now we consider the same doubly connected domain G as in the previous section and assume that on the surface $\Gamma = \partial g$, containing the point O, the elastic body is in the contact, without slip or separation, with the fixed absolutely rigid inclusion.

The displacement vector $\boldsymbol{u}$ satisfies the Lamé system (4.1.77), the loading condition (4.1.80) and the following boundary condition on the surface of the inclusion:

$$\boldsymbol{u}(\boldsymbol{x}) = \boldsymbol{\Psi}(\boldsymbol{x}), \boldsymbol{x} \in \Gamma. \tag{4.1.111}$$

We assume that $\boldsymbol{\Psi} = \mathbf{0}$. We shall study the asymptotic behaviour of the displacement field in the vicinity of the point O. As it was mentioned above, the intersection of the domain G with the surface of a sphere of a small radius δ has diameter $\varepsilon = O(\delta^{1+\gamma})$, and the corresponding area goes to zero faster than the total surface area of the sphere, as $\delta \to 0$. It was also mentioned that the study of the boundary-layer-type solutions requires the determination of solutions of auxiliary two-dimensional problems for an elastic plane with small defects. The papers [33], [87] deal with asymptotic representations of solutions to problems in regions with small defects, as $\varepsilon \to +0$. In particular, it was found that a dependence on $\ln \varepsilon$ appears in the corresponding asymptotic expansions for two-dimensional problems. Taking into account these results and introducing the spherical variables (ρ, θ, φ), with $\rho = ||\boldsymbol{x}||, \theta \in [0, \pi], \varphi \in [0, 2\pi)$, we assume that

$$\boldsymbol{u}(\boldsymbol{x}) \sim |\ln \rho|^{\beta} \boldsymbol{C}, \tag{4.1.112}$$

where $\boldsymbol{C}$ is the constant vector of rigid body displacements, $\beta \in \mathbb{R}^1$.

The vector-valued function (4.1.112) leaves a discrepancy in the boundary condition (4.1.111). Introducing the coordinates (4.1.85) we compensate for this discrepancy by a solution of the boundary layer type. This vector field can be represented in the form $|\ln \rho|^{\beta} \boldsymbol{w}(\boldsymbol{y})$, where the vector-valued function $\boldsymbol{w}$ satisfies the equations

$$\mathcal{L}(\frac{\partial}{\partial \boldsymbol{y}}) \boldsymbol{w}(\boldsymbol{y}) = \mathbf{0}, \boldsymbol{y} \in \mathbb{R}^2 \backslash \Upsilon; \;\; \boldsymbol{w}(\boldsymbol{y}) = -\boldsymbol{C}, \boldsymbol{y} \in \partial \Upsilon. \tag{4.1.113}$$

Following [54], [89], we introduce auxiliary solutions $\mathbf{W}^{(j)}, j = 1, 2, 3,$ of the homogeneous problem (4.1.113) with a logarithmic behaviour at infinity. The third component of the vector $\mathbf{W}^{(3)} = (0, 0, W_3^{(3)})^T$ is a function which is harmonic in $\mathbb{R}^2 \backslash \bar{\Upsilon}$ and is equal to zero on $\partial\Upsilon$. It admits the asymptotic representation

$$W_3^{(3)}(\boldsymbol{y}) = -\alpha^{-1} \ln \|\boldsymbol{y}\| + A_3 + O(\|\boldsymbol{y}\|^{-1}), \|\boldsymbol{y}\| \to \infty, \tag{4.1.114}$$

with constant A_3 and $\alpha = 2\pi\mu(1 + \varkappa)\varkappa^{-1}$.

First two vectors $\mathbf{W}^{(j)} = (W_1^{(j)}, W_2^{(j)}, 0)^T, j = 1, 2,$ satisfy the homogeneous two-dimensional Lamé system of equations in $\mathbb{R}^2 \backslash \Upsilon$, vanish on $\partial\Upsilon$, and admit the following asymptotic representation:

$$W_k^{(j)}(\boldsymbol{y}) = S_{jk}(\boldsymbol{y}) + A_{jk} + O(\|\boldsymbol{y}\|^{-1}), \|\boldsymbol{y}\| \to \infty, \tag{4.1.115}$$

where $S_{jk}(\boldsymbol{y}) = -\delta_{jk}\alpha^{-1} \ln \|\boldsymbol{y}\| + (\alpha\varkappa)^{-1} y_j y_k \|\boldsymbol{y}\|^{-2}$ are the components of the two-dimensional Somigliana tensor. In accordance with the terminology of [45], the value $\exp(\alpha A_3)$ is called the logarithmic capacity or the external conformal radius of the manifold Υ, and the 2×2 matrix with the elements A_{jk} is the Wiener elastic capacity matrix [7].

Consider the following vector-valued function:

$$\mathbf{V}(\boldsymbol{x}) = -\boldsymbol{C} + \alpha(\ln h(z) - \ln \rho)^{-1} \sum_{k=1}^{3} C_k \mathbf{W}^{(k)}(\boldsymbol{y}). \tag{4.1.116}$$

Taking into account the fact that in a conical neighbourhood $K_d = \{\boldsymbol{x} : \theta < d < \pi/2\}$ of the Ox_3 axis

$$\ln x_3 = \ln \rho + O(1), \rho \to 0, \tag{4.1.117}$$

and using (4.1.76), (4.1.114) and (4.1.115), we obtain the following formula which is valid in $K_d \backslash K_{d/2}$:

$$\mathbf{V}(\boldsymbol{x}) = -\boldsymbol{C} + (\ln h(z) - \ln \rho)^{-1} \sum_{k=1}^{3} c_k \{(\ln h(z) - \ln \rho)\boldsymbol{e}^{(k)}$$

$$+ \alpha\boldsymbol{\Phi}^{(k)}(\rho^{-1}\boldsymbol{x}') + O(\rho^{\gamma})\} = O(|\ln \rho|^{-1}), \rho \to 0, \tag{4.1.118}$$

where $\boldsymbol{e}^{(k)}$ are the unit basis vectors in $\mathbb{R}^3$, and $\boldsymbol{\Phi}^{(j)} = (\Phi_1^{(j)}, \Phi_2^{(j)}, 0)^T,$ $j = 1, 2; \boldsymbol{\Phi}^{(3)} = (0, 0, \Phi_3^{(3)})^T$ are the principal parts of the asymptotic representations (4.1.114), (4.1.115).

The vector-valued function (4.1.116), which vanishes far from the axis Ox_3, may be used to represent a boundary layer. Thus, the asymptotic representation of the vector $\boldsymbol{u}$ has the form

$$\boldsymbol{u}(\boldsymbol{x}) \sim |\ln\rho|^{\beta}(\boldsymbol{C} + \mathbf{V}(\boldsymbol{x})\chi(\rho^{-1}\boldsymbol{x})), \qquad (4.1.119)$$

where χ is a smooth cut–off function which is equal to 1 on $K_{d/2}$ and 0 outside K_d.

Now, to determine the exponent β of the singularity we consider the second term of the asymptotic expansion of the displacement field. Representing the Lamé operator in spherical coordinates in the form $\rho^{-2}\mathcal{P}(\theta, \varphi, \rho\partial/\partial\rho, \partial/\partial\theta, \partial/\partial\varphi)$, we introduce the operator-function $\mathcal{P}(\Lambda) : \Lambda \to \mathcal{P}(\theta, \varphi, \Lambda, \partial/\partial\theta, \partial/\partial\varphi)$. If $\rho \to 0$, then

$$\mathcal{P}(|\ln\rho|^{\beta}\boldsymbol{C}) = -\beta|\ln\rho|^{\beta-1}\mathcal{P}'(0)\boldsymbol{C} + O(|\ln\rho|^{\beta-2}). \qquad (4.1.120)$$

The principal term of the boundary layer leaves a discrepancy in (4.1.77). Taking into account that $h(x_3) = (\gamma+1)\ln\rho + O(1)$ in K_d, from (4.1.118) we deduce the relationship

$$\mathbf{V}(\boldsymbol{x}) = \alpha(\gamma\ln\rho)^{-1}\sum_{k=1}^{3} C_k\boldsymbol{\Phi}^{(k)}(\rho^{-1}\boldsymbol{x}')$$

$$+ O(|\ln\rho|^{-2}), \quad \boldsymbol{x} \in K_d\backslash K_{d/2}. \qquad (4.1.121)$$

Note that in the vicinity of the point $\mathcal{N}$ the operator $\mathcal{P}(0)$ admits the representation

$$\rho^{-2}\mathcal{P}(0) = \mathcal{P}_0(\frac{\partial}{\partial\boldsymbol{x}'}) + \mathcal{M}(\boldsymbol{x}', \frac{\partial}{\partial\boldsymbol{x}'}), \qquad (4.1.122)$$

where $\mathcal{M}$ is a second-order differential operator with smooth coefficients, and $\mathcal{P}_0$ is the principal homogeneous part of the operator (4.1.122).

The equations (4.1.121), (4.1.122) yield the asymptotic equality

$$\rho^{-2}\mathcal{P}(|\ln\rho|^{\beta}\mathbf{V}(\boldsymbol{x})) = -\alpha\gamma^{-1}|\ln\rho|^{\beta-1}$$

$$\times\{\mathcal{M}\chi + [\mathcal{P}_0, \chi]\}\sum_{k=1}^{3} C_k\boldsymbol{\Phi}^{(k)}(\rho^{-1}\boldsymbol{x}') + O(|\ln\rho|^{\beta-2}). \qquad (4.1.123)$$

Here, $[\mathcal{A}, \mathcal{B}] = \mathcal{A}\mathcal{B} - \mathcal{B}\mathcal{A}$ is the commutator of the operators $\mathcal{A}$ and $\mathcal{B}$. Let us denote the term containing $|\ln\rho|^{\beta-1}$ in the right-hand side of (4.1.123) by $-\alpha\gamma^{-1}|\ln\rho|^{\beta-1}\mathcal{F}$.

Taking into account the equalities (4.1.120) and (4.1.123), we find the second term of the asymptotic expansion of the displacement field $\boldsymbol{u}$ in the form $|\ln\rho|^{\beta-1}\boldsymbol{v}(\rho^{-1}\boldsymbol{x})$, where the vector-valued function $\boldsymbol{v}$ on the unit sphere $\mathbb{S} = \{\boldsymbol{x} \in \mathbb{R}^3 : \|\boldsymbol{x}\| = 1\}$ satisfies the equation

$$\mathcal{P}(0)\boldsymbol{v} = -\beta\mathcal{P}'(0)\mathcal{C} + \alpha\gamma^{-1}\mathcal{F}. \tag{4.1.124}$$

Proposition 5. *1. The system $\mathcal{P}(0)\boldsymbol{v} = \boldsymbol{f}$ on $\mathbb{S}$ is solvable if and only if*

$$\int_{\mathbb{S}} \boldsymbol{f} \cdot T^{(0,j)} ds = 0, j = 1, 2, 3. \tag{4.1.125}$$

Here, $T^{(0,j)}$ are the traces on $\mathbb{S}$ of the columns $T^{(j)}$ of the three-dimensional Somigliana tensor, $T_k^{(j)} = 1/2(\alpha\|\boldsymbol{x}\|)^{-1}\{\delta_{jk} + \varkappa^{-1}x_j x_k\|\boldsymbol{x}\|^{-2}\}$. The solution $\boldsymbol{v}$ is determined to within an arbitrary constant vector.

2. The following equalities hold:

$$\int_{\mathbb{S}} T^{(0,j)} \cdot \mathcal{P}'(0)\mathcal{C} ds = C_j; \quad \int_{\mathbb{S}} T^{(0,j)} \cdot \mathbf{F}^{(k)} ds = -\alpha_k T_k^{(j)}(\mathcal{N}), \tag{4.1.126}$$

where

$$\alpha_1 = \alpha_2 = 1, \alpha_3 = \varkappa(1+\varkappa)^{-1}, \mathbf{F}^{(k)} = (\mathcal{M}\chi + [\mathcal{P}_0, \chi])\boldsymbol{\Phi}^{(k)}.$$

3. The exponent β of the singularity is equal to $-(2\gamma)^{-1}$.

Proof. The self-adjointness of the Lamé system and definition of the operator-function $\mathcal{P}(\Lambda)$ yield the equality $\mathcal{P}(0)^* = \mathcal{P}(-1)$. Moreover, the homogeneous solutions of the Lamé system in $\mathbb{R}^3 \backslash O$ of orders 0 and 1 are represented only by constant vectors and by linear combinations of the Somigliana tensor T. Thus, the equality (4.1.125) is proved.

Consider the second part of the statement. Let $\zeta \in \mathbf{C}_0^\infty(-1,1)$ be a function which is equal to 1 near the origin. Thus, for a ball $\mathcal{D}_r = \{\boldsymbol{x} \in \mathbb{R}^3 : \|\boldsymbol{x}\| < r\}$ the following equalities are valid:

$$\int_{\mathcal{D}_1} T^{(j)} \cdot \mathcal{L}(\zeta\mathcal{C}) d\boldsymbol{x} = \int_{\mathcal{D}_1} \zeta\mathcal{C} \cdot \mathcal{L}T^{(j)} d\boldsymbol{x}$$

$$= -\int_{\mathcal{D}_1} \zeta\mathcal{C} \cdot \boldsymbol{e}^{(j)}\delta(\boldsymbol{x}) d\boldsymbol{x} = -C_j, \tag{4.1.127}$$

where δ is the Dirac delta function.

We take into account the fact that the left-hand side of the above relation may be written in the form:

$$\int_{\mathcal{D}_1} T^{(j)} \cdot \mathcal{L}(\zeta \mathcal{C}) d\boldsymbol{x} = \lim_{r\to 0} \int_{\mathcal{D}_1 \setminus \mathcal{D}_r} T^{(j)} \cdot \mathcal{L}(\zeta \mathcal{C}) d\boldsymbol{x}$$

$$= \lim_{r\to 0} \int_r^1 \int_{\mathbb{S}} T^{(j)} \cdot \mathcal{P}(\rho \frac{\partial}{\partial \rho})(\zeta \mathcal{C}) d\rho ds \qquad (4.1.128)$$

$$= - \lim_{r\to 0} \int_{\mathbb{S}} \{\rho T^{(0,j)} (\frac{1}{2} \mathcal{P}''(0) \rho \frac{\partial}{\partial \rho} + \mathcal{P}'(0)) \mathcal{C} - \frac{1}{2} \mathcal{C} \frac{\partial}{\partial \rho} \rho T^{(0,j)} \}_{\rho = r} ds$$

$$= - \int_{\mathbb{S}} T^{(0,j)} \cdot \mathcal{P}'(0) \mathcal{C} ds.$$

In these transformations we used the following equalities: $\mathcal{P}^*(0) T^{(0,j)} = \mathbf{0}$ for $\rho > 0$; $\rho T^{(j)} = T^{(0,j)}$; $\mathcal{P}(\Lambda) = \mathcal{P}(0) + \mathcal{P}'(0)\Lambda + 1/2 \mathcal{P}''(0)\Lambda^2$. Thus, relations (4.1.127), (4.1.128) yield the first formula of (4.1.126).

Using the definition of the vectors $\mathbf{\Phi}^{(k)}$, we deduce from (4.1.122) that $\mathcal{P}(0)(\chi \mathbf{\Phi}^{(k)}) = \mathbf{F}^{(k)} - \alpha_k \delta(\theta) e^{(k)}$. Hence, we obtain the equalities

$$\int_{\mathbb{S}} T^{(0,j)} \cdot \mathbf{F}^{(k)} ds = \int_{\mathbb{S}} T^{(0,j)} \cdot \mathcal{P}(0)(\chi \mathbf{\Phi}^{(k)}) ds - \{\alpha_k T_k^{(0,j)}\}_{\theta=0}$$

$$= \int_{\mathbb{S}} \chi \mathbf{\Phi}^{(k)} \cdot \mathcal{P}(-1) T^{(0,j)} ds - \alpha_k T_k^{(j)}(\mathcal{N}) = -\alpha_k T_k^{(j)}(\mathcal{N}),$$

which prove the second formula (4.1.126).

Let us remark that the solvability conditions (4.1.125) for the system (4.1.124) yield

$$- \beta C_j - \alpha \gamma^{-1} \sum_{k=1}^{3} \alpha_k T_k^{(j)}(\mathcal{N}) C_k = 0, j = 1, 2, 3. \qquad (4.1.129)$$

Substituting the representations for $T_k^{(j)}(\mathcal{N})$ into (4.1.129), we transform them to the form $\beta C_j + 1/2 \gamma^{-1} C_j = 0$. Thus, $\beta = -(2\gamma)^{-1}$. □

Next, let us consider the behaviour of a stress-strain state in the vicinity of the tip of the inclusion. Using the previous considerations, we deduce that the displacement field $\boldsymbol{u}$, which has finite elastic energy, has the representation

$$\boldsymbol{u}(\boldsymbol{x}) = |\ln\rho|^{\beta}\{\boldsymbol{C} - \chi(\rho^{-1}\boldsymbol{x})(\boldsymbol{C} + \alpha\gamma^{-1}|\ln\rho|^{-1}\sum_{k=1}^{3} C_k \mathbf{W}^{(k)}(h^{-1}\boldsymbol{x}'))\}$$

$$+ O((|\theta| + |\ln\rho|^{-1})|\ln\rho|^{\beta-1}), \rho \to 0. \qquad (4.1.130)$$

It is important to note that there is a factor $|\ln\rho|^{\beta}$ with a negative exponent β, which depends on the geometry of the peak-shaped inclusion. On the other hand, elasticity problems in regions with angular or conic inclusions or notches have solutions with singularities of the form $\rho^{\Lambda}\Phi(\varphi, \theta, \ln\rho)$, where Λ is a complex number and Φ is a polynomial of the function $\ln\rho$ with coefficients depending smoothly on the angle variables.

Consider the following two regions in the vicinity of the singular point O : the exterior of the conical neighbourhood K_d and the peak-shaped set $\Pi_A = \{\boldsymbol{x} : x_3 > 0, \|\boldsymbol{x}'\| < Ax_3^{1+\gamma}\}$, containing the ray Ox_3. Outside K_d the equation (4.1.130) yields the asymptotic equality

$$\boldsymbol{u}(\boldsymbol{x}) = |\ln\rho|^{\beta}\boldsymbol{C} + O(|\ln\rho|^{\beta-1}), \rho \to 0, \boldsymbol{x} \in G\backslash K_d, \qquad (4.1.131)$$

while inside Π_A the following relation is valid:

$$\boldsymbol{u}(\boldsymbol{x}) = \alpha\gamma^{-1}|\ln\rho|^{\beta-1}\sum_{k=1}^{3} C_k \mathbf{W}^{(k)}(h(x_3)^{-1}\boldsymbol{x}') + O(|\ln\rho|^{\beta-2}), \qquad (4.1.132)$$

$$\text{as } \rho \to 0, \ \boldsymbol{x} \in \Pi_A \cap G.$$

Thus, the asymptotic behaviour of the components of the stress tensor is determined from the representation

$$\sigma_{jk}(\boldsymbol{u}; \boldsymbol{x}) = \frac{4\pi\mu(\lambda + 2\mu)}{(\lambda + 3\mu)\alpha\gamma\rho^{1+\gamma}}|\ln\rho|^{\beta-1}\sum_{m=1}^{3} C_m \Sigma_{jk}^{(m)}(h(x_3)^{-1}\boldsymbol{x}') \qquad (4.1.133)$$

$$+ O(|\ln\rho|^{\beta-2}(\rho^{1+\gamma} + \|\boldsymbol{x}'\|)^{-1}), \rho \to 0, \boldsymbol{x} \in \Pi_A \cap G,$$

where $(\Sigma_{jk}^{(m)}(\boldsymbol{y}))_{j,k=1}^{2}$ is a two-dimensional stress tensor specified by

$$\Sigma_{3j}^{(m)}(\boldsymbol{y}) = \mu\frac{\partial W_3^{(m)}}{\partial y_j}(\boldsymbol{y}), j = 1, 2;$$

$$\Sigma_{33}^{(m)}(\boldsymbol{y}) = \lambda(\frac{\partial W_1^{(m)}}{\partial y_1}(\boldsymbol{y}) + \frac{\partial W_2^{(m)}}{\partial y_2}(\boldsymbol{y})). \tag{4.1.134}$$

To find asymptotic representations for the components of the stress tensor, as $\rho \to 0$, outside the set K_d we take into account the second term of the expansion of the displacement field. Let us note that the vector-valued function $\boldsymbol{v} + \chi\alpha\gamma^{-1}(C_1\boldsymbol{\Phi}^{(1)} + C_2\boldsymbol{\Phi}^{(2)} + C_3\boldsymbol{\Phi}^{(3)})$ coincides with the linear combination

$$\boldsymbol{\Xi} = \mathbf{B} + \alpha\gamma^{-1}\sum_{m=1}^{3} C_m\alpha_m\mathbf{G}^{(m)}, \tag{4.1.135}$$

where $\mathbf{B}$ is a constant vector, $\mathbf{G}^{(m)}$ are the generalized Green's functions with singularities at the point $\mathcal{N}$, which satisfy the systems of equations $\mathcal{P}(0)\mathbf{G}^{(m)} + (\delta(\theta) - \mathcal{P}'(0))\boldsymbol{e}^{(m)} = \mathbf{0}$ on $\mathbb{S}$. Thus, (4.1.131) may be written as

$$\boldsymbol{u}(\boldsymbol{x}) = |\ln\rho|^{\beta}(\boldsymbol{C} + |\ln\rho|^{-1}\boldsymbol{\Xi} + O(|\ln\rho|^{-2})), \rho \to 0, \boldsymbol{x} \in G\backslash K_d. \tag{4.1.136}$$

Differentiating (4.1.136), we obtain the final asymptotic representation for the stress tensor components

$$\sigma_{jk}(\boldsymbol{u};\boldsymbol{x}) = \gamma^{-1}|\ln\rho|^{\beta-1}\{-\frac{1}{2}\rho^{-2}(\mu(C_jx_k + C_kx_j) + \delta_{jk}\lambda\boldsymbol{C}\cdot\boldsymbol{x})$$

$$+ \alpha\rho^{-1}\sum_{m=1}^{3} C_m\alpha_m\tau_{jk}^{(m)}\} + O(\rho^{-1}|\ln\rho|^{\beta-2}),\ \rho \to 0,\ \boldsymbol{x} \in G\backslash K_d, \tag{4.1.137}$$

where $\tau_{jk}^{(m)}$ are the angular parts of the components of the stress tensor corresponding to the fields $\mathbf{G}^{(m)}$.

Comparison of relations (4.1.133) and (4.1.137) shows that the highest stress concentration is in a thin layer near the cusp-shaped boundary of the inclusion in the vicinity of the vertex. If the inclusion gets thinner there, the increase on the exponent γ leads to the increase of the order of the stress singularity $O(|\ln\rho|^{\beta-1}\rho^{-1-\gamma})$. Note that the diameter of the zone of stress concentration is of order $O(\rho^{\gamma})$. Due to this fact, the field $\boldsymbol{u}$ has finite elastic energy. It should be observed that the coefficients C_j in the above expansions are similar to the stress intensity factors and depend on the whole data of the problem.

The equality (4.1.133) shows that the character of fracture near the vertex of a cusp-shaped inclusion is determined by the solutions $\mathbf{W}^{(j)}$ of the two-dimensional problem in the exterior of the cross-section Υ of the domain g. Note that these vector-valued functions also arise in problems related to such characteristics of the domain Υ as logarithmic capacity and the Wiener's elastic capacity ([45], [7]).

In the present section we have studied a set of elasticity problems related with a cusp-shaped boundary of an imperfection in an elastic body. The first three parts deal with two-dimensional cases of sharp elastic inclusions. We have proved that there is no stress singularity at the tip of the cusp-shaped elastic inclusion, which boundary has a geometry satisfying a power law. It should be taken into account that, in contrast with the above-mentioned problem, the case of a cusp-shaped cavity in an elastic body is characterized by a singularity of order $O(r^{-1/2})$ in stress components. The elasticity problems about a narrow wedge-shaped inclusion and a cusp-shaped one with a logarithmic geometry of a boundary have been also investigated. This consideration allows us to describe the influence of the shape of a narrow two-dimensional elastic inclusion on the behaviour of stresses in a neighbourhood of an irregular point of the boundary.

In the fourth part we have considered three-dimensional cusp-shaped elastic inclusion and have analyzed asymptotic behaviour of a stress-strain state in the vicinity of the vertex of the inclusion. As in case of three-dimensional cusp-shaped cavity, in the present situation there is no stress singularity at the irregular point of a boundary. Nevertheless, using a solution of a boundary-layer-type problem, we have proved that stress components lose continuity at the vertex of the inclusion on approaching this point along a curve which is tangent to the axis of the inclusion.

In the final part, which deals with an absolutely rigid three-dimensional inclusion inserted into an elastic body, we have analyzed the stress singularity exponent behaviour at the irregular point of a cusp-shaped boundary. By means of asymptotic analysis and boundary layer investigation we have shown that the highest stress concentration of order $O(|\ln\rho|^{\beta-1}\rho^{-1-\gamma})$ is in a thin layer near the cusp-shaped boundary of the inclusion in the vicinity of the vertex.

4.2 On the stress concentration near soft and rigid cusp-shaped inclusions

In the present section we study the stress–strain state of an elastic domain containing a cusp-shaped inclusion with the Young modulus which is relatively large or small in comparison with the Young modulus of elastic matrix. We construct the boundary layer to describe the formation of the square root singularity at the limit case of a cusp-shaped cavity or an absolutely rigid inclusion.

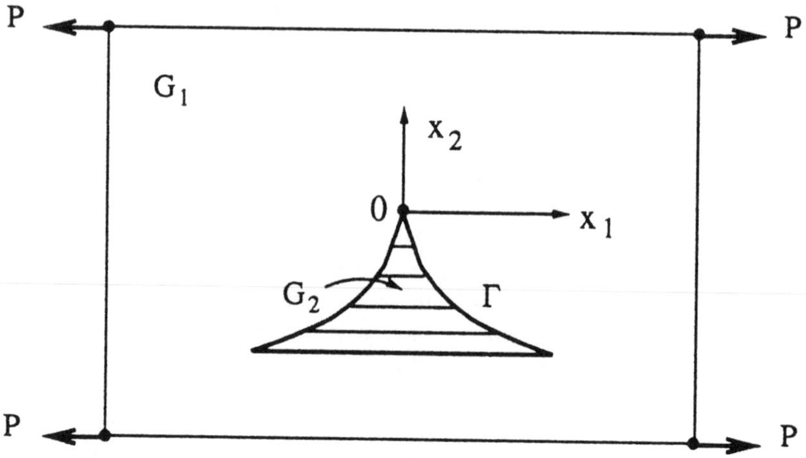

Fig. 4.3: The sample for numerical experiment; $P = 0.5p \operatorname{diam} G_1$.

4.2.1 Motivation and numerical experiment

In the first section of the present chapter it was proved that at the tip of a two-dimensional power cusp-shaped elastic inclusion components of the stress tensor are bounded and do not have the square root singularity. The finite element calculations were performed for an elastic domain specified in Fig. 4.3. The plots of normalized stresses σ_{11}/p, σ_{22}/p, calculated on the axis of symmetry in the elastic matrix, are presented in Fig. 4.4a and Fig. 4.4b.

The Poisson ratio was chosen to be $\nu = 0.1$ for the elastic matrix and $\nu^\circ = 0.2$ for the cusp-shaped inclusion. The ratio of the Young moduli E°/E is equal to 0.05, 0.25, 4 and 20 for the curves 1, 2, 3 and 4, respectively. The results of calculations show that for the case of elastic moduli of the same order of magnitude the increment in the stress components is small near the singular point on the interface boundary. This increment increases with the increasing of $\max\{E^\circ/E,\ E/E^\circ\}$.

It is known that in the limit $E^\circ/E \to 0$ (or $E^\circ/E \to \infty$) the solution of the problem on the cusp-shaped inclusion converges weakly (in the energy Hilbert space) to the solution corresponding to a cavity (or an absolutely rigid inclusion). In both cases the stress components have the square root singularity. It is important to understand the formation of this singularity. In order to solve the problem we have to perform the analysis of the boundary layer which occurs in the cases of relatively soft or rigid cusp-shaped inclusions.

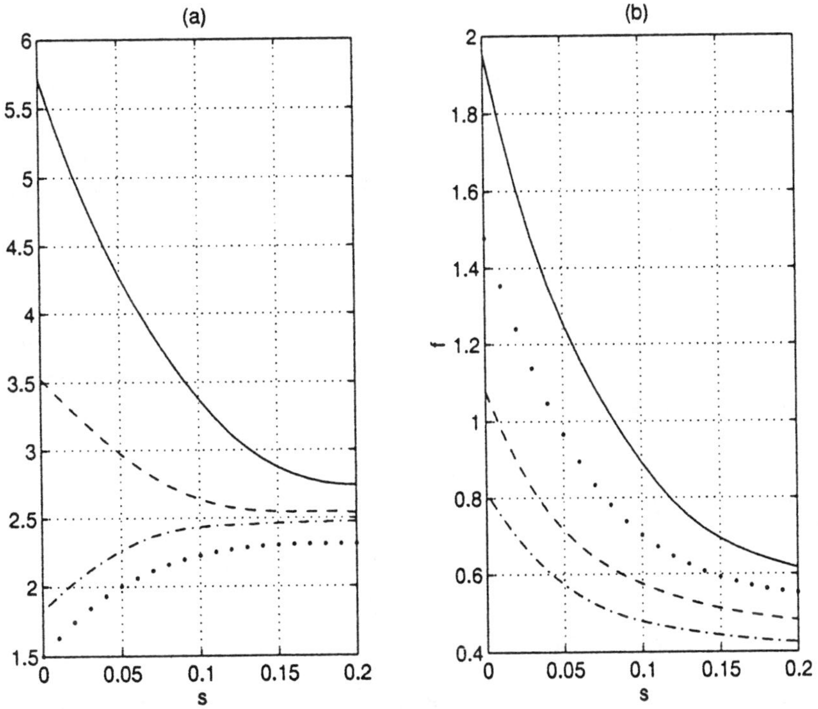

Fig. 4.4: Normalized stress in the vicinity of the cusp:
1–solid line, 2–dashed line, 3–dash-dotted line,
4–dotted line;
$s = x_2/\text{diam}\, G_2$, (a) $f = \sigma_{11}/p$, (b) $f = \sigma_{22}/p$.

4.2.2 Governing equations

Let G_1 denote a doubly connected domain, and assume that ∂G_1 is smooth everywhere except for the origin O which is located on the inner contour Γ.

Let G_2 denote a simply connected region, surrounded by Γ, such that in a neighbourhood of the origin it is specified by the following inequalities:

$$-H(x_1) < x_2 < H(x_1), \quad x_1 > 0, \tag{4.2.1}$$

where $H \in \mathbb{C}^\infty[0,1]$, $H(0) = H'(0) = 0$. In polar coordinates with the centre at O one has

$$-A(r) < \varphi < A(r), \tag{4.2.2}$$

where A is assumed to be smooth on $[0,1]$, and $A(0) = 0$.

Suppose that the shape of the interface boundary is subjected to the power law, which means that

$$H(x_1) = a x_1^{1+\gamma} + O(x_1^{2+\gamma}), \quad A(r) = a r^\gamma + O(r^{1+\gamma}), \tag{4.2.3}$$

where $a > 0$ and γ is a natural number.

Consider the plane-strain problem in a compound domain $G_1 \cup G_2$ and assume that the Lamé constants λ°, μ° of the cusp-shaped inclusion are specified by

$$\mu^\circ = \varepsilon m, \quad \lambda^\circ = \varepsilon l, \tag{4.2.4}$$

where the quantities l, m have the same order of magnitude as λ, μ, ε is a positive nondimensional parameter. Clearly, as $\varepsilon \to \infty$, we have an absolutely rigid inclusion. As $\varepsilon \to 0$, we deal with a cusp-shaped cavity.

The displacement fields $\boldsymbol{u}^\circ$ and $\boldsymbol{u}$ in the inclusion and elastic matrix are subjected to the homogeneous Lamé system of equations (the body force density is assumed to be zero):

$$\mu \Delta \boldsymbol{u}(\boldsymbol{x}) + (\lambda + \mu) \text{grad div } \boldsymbol{u}(\boldsymbol{x}) = \boldsymbol{0}, \quad \boldsymbol{x} \in G_1, \tag{4.2.5}$$

$$\mu^\circ \Delta \boldsymbol{u}^\circ(\boldsymbol{x}) + (\lambda^\circ + \mu^\circ) \text{grad div } \boldsymbol{u}^\circ(\boldsymbol{x}) = \boldsymbol{0}, \quad \boldsymbol{x} \in G_2. \tag{4.2.6}$$

The ideal contact conditions hold on the interface boundary:

$$\boldsymbol{u}(\boldsymbol{x}) = \boldsymbol{u}^\circ(\boldsymbol{x}), \quad \boldsymbol{\sigma}^{(n)}(\boldsymbol{u}; \boldsymbol{x}) = \boldsymbol{\sigma}^{\circ,(n)}(\boldsymbol{u}^\circ; \boldsymbol{x}), \quad \boldsymbol{x} \in \Gamma. \tag{4.2.7}$$

On the outer boundary $\partial G_1 \backslash \Gamma$ we prescribe self-balanced tractions

$$\boldsymbol{\sigma}^{(n)}(\boldsymbol{u}; \boldsymbol{x}) = \boldsymbol{p}(\boldsymbol{x}), \quad \boldsymbol{x} \in \partial G_1 \backslash \Gamma, \tag{4.2.8}$$

$$\int_{\partial G_1 \backslash \Gamma} \boldsymbol{p}(\boldsymbol{x}) ds = \boldsymbol{0}, \quad \int_{\partial G_1 \backslash \Gamma} \boldsymbol{x} \times \boldsymbol{p}(\boldsymbol{x}) ds = \boldsymbol{0}. \tag{4.2.9}$$

The objective is to analyze the asymptotic approximation of the displacement field in the cases when the parameter ε is small or large.

4.2.3 The limit boundary value problems

Introduce two small parameters δ and α such that

$$\delta = \varepsilon, \quad \alpha = \varepsilon^{1/\gamma}, \quad \text{as } \varepsilon \to 0,$$

and

$$\delta = \varepsilon^{-1}, \quad \alpha = \varepsilon^{-1/\gamma}, \quad \text{as } \varepsilon \to \infty.$$

The solution of (4.2.5)–(4.2.8) is sought in the form of the asymptotic series

$$\boldsymbol{u}(\delta; \boldsymbol{x}) \sim \sum_{j=0}^{\infty} \delta^j \boldsymbol{v}^{(j)}(\boldsymbol{x}), \quad \boldsymbol{u}^{\circ}(\delta; \boldsymbol{x}) \sim \sum_{j=0}^{\infty} \delta^j \boldsymbol{w}^{(j)}(\boldsymbol{x}). \tag{4.2.10}$$

Substitute (4.2.10) into (4.2.5)–(4.2.8) and equate coefficients near like powers of δ. Consequently, we obtain the recurrent sequence of problems with respect to the vectors $\boldsymbol{v}^{(j)}$ and $\boldsymbol{w}^{(j)}$. In particular, for the case of a soft inclusion we have

$$\boldsymbol{L}(\partial/\partial \boldsymbol{x}) \boldsymbol{v}^{(j)}(\boldsymbol{x}) = \boldsymbol{0}, \ \boldsymbol{x} \in G_1; \tag{4.2.11}$$

$$\boldsymbol{\sigma}^{(n)}(\boldsymbol{v}^{(j)}; \boldsymbol{x}) = \delta_{0j} \boldsymbol{p}(\boldsymbol{x}), \quad \boldsymbol{x} \in \partial G_1 \backslash \Gamma, \tag{4.2.12}$$

$$\boldsymbol{\sigma}^{(n)}(\boldsymbol{v}^{(j)}; \boldsymbol{x}) = \boldsymbol{\Sigma}^{(n)}(\boldsymbol{w}^{(j-1)}; \boldsymbol{x}), \quad \boldsymbol{x} \in \Gamma, \tag{4.2.13}$$

$$\boldsymbol{L}^{\circ}(\partial/\partial \boldsymbol{x}) \boldsymbol{w}^{(j)}(\boldsymbol{x}) = \boldsymbol{0}, \quad \boldsymbol{x} \in G_2, \tag{4.2.14}$$

$$\boldsymbol{w}^{(j)}(\boldsymbol{x}) = \boldsymbol{v}^{(j)}(\boldsymbol{x}), \quad \boldsymbol{x} \in \Gamma. \tag{4.2.15}$$

For the case of a rigid inclusion the interface condition (4.2.13) should be replaced by

$$\boldsymbol{\Sigma}^{(n)}(\boldsymbol{w}^{(j)}; \boldsymbol{x}) = \boldsymbol{\sigma}^{(n)}(\boldsymbol{v}^{(j-1)}; \boldsymbol{x}), \quad \boldsymbol{x} \in \Gamma, \tag{4.2.16}$$

where δ_{0j} is the Kronecker delta, $\boldsymbol{L}, \boldsymbol{L}^{\circ}$ are the matrix differential operators corresponding to the Lamé system, $\boldsymbol{\Sigma}^{(n)} := \varepsilon^{-1} \boldsymbol{\sigma}^{\circ,(n)}$, and all quantities with negative indices are assumed to be zero.

4.2.4 Asymptotics near the cusp

The leading order term $\boldsymbol{v}^{(0)}$ from (4.2.10) satisfies the homogeneous Lamé system in G_1 and the traction boundary condition (4.2.8) on $\partial G_1 \backslash \Gamma$. On the contour Γ zero tractions ($\varepsilon \to 0$) or zero displacements ($\varepsilon \to \infty$) are prescribed. In both cases the vector $\boldsymbol{v}^{(0)}$ admits the following asymptotic representation:

$$\boldsymbol{v}^{(0)}(\boldsymbol{x}) = r^{1/2}(K_1 \boldsymbol{\Phi}^{(1)}(\varphi) + K_2 \boldsymbol{\Phi}^{(2)}(\varphi)) + O(r), \qquad (4.2.17)$$

where K_j, $j = 1, 2$, are the stress-intensity factors, and $\boldsymbol{\Phi}^{(k)}$, $k = 1, 2$, are smooth vector functions of the angular variable, particular form of $\boldsymbol{\Phi}^{(k)}$ depends on the type of boundary conditions on Γ.

Consider the displacement field within the inclusion. First, analyze the case of a soft inclusion. The vector-valued function $\boldsymbol{w}^{(0)}$ satisfies the boundary value problem (4.2.14), (4.2.15). Introduce the coordinates (x_1, t), where

$$t = H(x_1)^{-1} x_2, \quad t \in [-1, 1].$$

The following differential relations hold:

$$d/dx_1 = \partial/\partial x_1 - H'(x_1) H^{-1}(x_1) t \partial/\partial t;$$

$$d/dx_2 = H^{-1}(x_1) \partial/\partial t. \qquad (4.2.18)$$

Thus,

$$\boldsymbol{L}^\circ(\partial/\partial \boldsymbol{x}) \boldsymbol{w}^{(0)}(\boldsymbol{x}) = H(x_1)^{-2} \boldsymbol{Q}^\circ (\partial^2 \boldsymbol{w}^{(0)}/\partial t^2)(x_1, t)$$

$$+ O(x_1^{-2-\gamma}), \qquad (4.2.19)$$

where the matrix $\boldsymbol{Q}^\circ$ is specified by

$$\boldsymbol{Q}^\circ = \text{diag}\,(\mu^\circ, 2\mu^\circ + \lambda^\circ).$$

Consequently,

$$\frac{\partial^2 \boldsymbol{w}^{(0)}}{\partial t^2} = \boldsymbol{0},$$

and the components of $\boldsymbol{w}^{(0)}$ depend linearly on t. The coefficients can be determined from the boundary conditions, and, as a result, we have

$$\boldsymbol{w}^{(0)}(\boldsymbol{x}) = 1/2 r^{1/2} \sum_{j=1}^{\infty} K_j \{ t(\boldsymbol{\Phi}^{(j)}(0) - \boldsymbol{\Phi}^{(j)}(2\pi)) + \boldsymbol{\Phi}^{(j)}(0)$$

$$+ \mathbf{\Phi}^{(j)}(2\pi)\} + O(r). \tag{4.2.20}$$

Now, consider the inclusion which is relatively rigid in comparison with the elastic matrix. As $j = 0$, the right-hand sides in (4.2.14), (4.2.16) are equal to zero, and, therefore $\boldsymbol{w}^{(0)}$ is rigid-body displacement (it includes a rigid-body translation and rotation). To provide uniqueness, we assume that the displacement field is equal to zero at the origin, and therefore,

$$\boldsymbol{w}^{(0)} = \mathbf{0}.$$

The boundary condition for the term $\boldsymbol{w}^{(1)}$ is nonhomogeneous, and in the vicinity of the cusp

$$\boldsymbol{\sigma}^{(n)}(\boldsymbol{v}^{(0)}; \boldsymbol{x}) = r^{-1/2}(K_1\mathbf{\Psi}^{(1)}(\varphi) + K_2\mathbf{\Psi}^{(2)}(\varphi)) + O(1), \quad r \to 0. \tag{4.2.21}$$

The asymptotic algorithm for the displacement field within G_2 is similar to the derivation of the beam equation on the basis of the asymptotic analysis of the elasticity equations in a thin domain (see, for example [22], [42], [49], [90], [127]). Let us present the final asymptotic form (it can be verified by direct substitution into (4.2.14), (4.2.16)). The vector $\boldsymbol{w}^{(1)}$ admits the asymptotic representation

$$\boldsymbol{w}^{(1)}(x_1, t) = x_1^{-3\gamma+1/2}(\boldsymbol{c} + O(x_1^{1/2})), \quad \boldsymbol{c} = (c_1, c_2),$$

$$c_1 = 0, \; c_2 = 2(K_1\Psi_2^{(1)}(0) + K_2\Psi_2^{(2)}(0))[3a^3D(9\gamma^2 - 1/4)]^{-1},$$

$$D = m(l+m)[3(l+2m)]^{-1}. \tag{4.2.22}$$

Consider the vector $\boldsymbol{v}^{(1)}$ which satisfies the boundary value problem (4.2.11)–(4.2.13) or (4.2.11), (4.2.12), (4.2.15). In the first case formulae (4.2.18), (4.2.20) yield that the right-hand side of the boundary condition (4.2.13), $j = 1$, has the order $O(r^{-\gamma-1/2})$. Therefore (it follows, for example, from the results [38]), for the boundary value problem (4.2.11)–(4.2.13)

$$\boldsymbol{v}^{(1)}(\boldsymbol{x}) = O(r^{1/2-\gamma}), \quad \text{as } r \to 0.$$

If $\gamma > 1/2$, then this field produces infinite elastic energy.

It follows from (4.2.22) that the solution of (4.2.11), (4.2.12), (4.2.15) has the order $O(r^{-3\gamma+1/2})$ in a neighbourhood of the cusp. Therefore, in the case of the relatively rigid inclusion, the elastic energy is infinite for $\gamma > 1/6$.

4.2.5 Scaled coordinates near the tip of the cusp

The interface boundary Γ is not smooth, and the boundary layer is needed for the asymptotic description of the displacement field. We restrict the analysis by considering the leading order terms of the asymptotic expansions.

To describe the boundary layer, introduce the scaled coordinates

$$\boldsymbol{\xi} = \alpha^{-\beta}\boldsymbol{x}, \quad \beta > 0. \tag{4.2.23}$$

As $\alpha \to 0$, the region G_1 transforms into

$$K = \mathbb{R}^2 \backslash \{\boldsymbol{\xi} : \xi_1 > 0,\ \xi_2 = 0\}, \tag{4.2.24}$$

and the domain G_2 is given by the inequalities

$$\xi_1 > 0, \quad |\xi_2| < \alpha^{-\beta} H(\alpha^\beta \xi_1). \tag{4.2.25}$$

For the points of G_2 introduce the variable

$$\tau = \frac{\alpha^\beta \xi_2}{H(\alpha^\beta \xi_1)}.$$

The exponent β is chosen in such a way that the leading parts of $\boldsymbol{u}, \boldsymbol{u}^\circ$, represented in coordinates (4.2.23), have the same order of magnitude.

First, suppose that the inclusion is soft. Then, due to (4.2.17), (4.2.20),

$$\boldsymbol{u}(\delta, \boldsymbol{x}) \sim \boldsymbol{v}^{(0)}(\boldsymbol{x}), \boldsymbol{v}^{(0)}(\boldsymbol{x}) = O(r^{1/2}) = \alpha^{\beta/2} O(\rho^{1/2}), \tag{4.2.26}$$

$$\boldsymbol{u}^\circ(\delta, \boldsymbol{x}) \sim \boldsymbol{w}^{(0)}(\boldsymbol{x}), \boldsymbol{w}^{(0)}(\boldsymbol{x}) = O(r^{1/2}) = \alpha^{\beta/2} O(\rho^{1/2}); \tag{4.2.27}$$

here $\rho = ||\boldsymbol{\xi}||$. Thus, for the case of a soft inclusion $\beta = 1$.

If the inclusion is rigid, then $\boldsymbol{w}^{(0)} = \boldsymbol{0}$, and the formula (4.2.27) should be replaced by

$$\boldsymbol{u}^\circ(\delta, \boldsymbol{x}) \sim \delta \boldsymbol{w}^{(1)}(\boldsymbol{x}), \tag{4.2.28}$$

$$\delta \boldsymbol{w}^{(1)}(\boldsymbol{x}) = \delta O(r^{1/2 - 3\gamma}) = \alpha^{-\beta(3\gamma - 1/2) + \gamma} O(\rho^{1/2 - 3\gamma}).$$

The comparison of (4.2.26) and (4.2.28) yields

$$\frac{\beta}{2} = \gamma - \beta(3\gamma - 1/2),$$

and, therefore, $\beta = 1/3$.

4.2.6 Boundary layer in a neighbourhood of the soft inclusion

The leading part of the boundary layer is sought in the form

$$\alpha^{1/2}\boldsymbol{V}(\boldsymbol{\xi}) \quad \text{and} \quad \alpha^{1/2}\boldsymbol{W}(\xi_1,\tau)$$

in the elastic matrix and inclusion, respectively. Consider the interface boundary conditions (4.2.7), represented in coordinates (4.2.23), with $\beta = 1$.

As $\alpha \to 0$, the first condition yields

$$\boldsymbol{V}(\xi_1, \pm 0) = \boldsymbol{W}(\xi_1, \pm 1). \tag{4.2.29}$$

Assume that $||\boldsymbol{\xi}|| = O(1)$. Then

$$\boldsymbol{n}(\alpha\boldsymbol{\xi}) = (0, \pm 1)^T + O(\delta), \quad \boldsymbol{\sigma}^{(n)}(\boldsymbol{V}; \alpha, \boldsymbol{\xi}) = \pm\boldsymbol{\sigma}^{(2)}(\boldsymbol{V}; \xi_1, \pm 0) + O(\delta).$$

Due to (4.2.18), (4.2.23), the second condition on the interface boundary reduces to

$$\boldsymbol{\sigma}^{(2)}(\boldsymbol{V}; \xi_1, \pm 0) = \boldsymbol{Q} a^{-1}\xi_1^{-1-\gamma}(\partial \boldsymbol{W}/\partial\tau)(\xi_1, \pm 1), \tag{4.2.30}$$

where $\boldsymbol{Q} = \operatorname{diag}\{m, 2m + l\}$.

We obtain from the Lamé system (4.2.5), $i = 2$, that the vector function $\tau \to \boldsymbol{W}(\xi_1, \tau)$ is linear,

$$\boldsymbol{W}(\xi_1, \tau) = \boldsymbol{A}_1(\xi_1)\tau + \boldsymbol{A}_0(\xi_1). \tag{4.2.31}$$

The components of $\boldsymbol{A}_j$, $j = 0, 1$, are defined from the system (4.2.29), (4.2.30),

$$\boldsymbol{A}_j(\xi_1) = 1/2[\boldsymbol{V}(\xi_1, +0) + (-1)^j \boldsymbol{V}(\xi_1, -0)], \quad (j = 0, 1). \tag{4.2.32}$$

The system is solvable if and only if

$$\boldsymbol{\sigma}^{(2)}(\boldsymbol{V}; \xi_1, \pm 0) - 1/2 a^{-1}\xi_1^{-1-\gamma}\boldsymbol{Q}(\boldsymbol{V}(\xi_1, +0) - \boldsymbol{V}(\xi_1, -0)) = \boldsymbol{0}. \tag{4.2.33}$$

Thus, the leading order term of the boundary layer in the elastic matrix satisfies the Lamé system

$$\mu\Delta \boldsymbol{V}(\boldsymbol{\xi}) + (\lambda + \mu)\text{grad div } \boldsymbol{V}(\boldsymbol{\xi}) = \boldsymbol{0}, \quad \boldsymbol{\xi} \in \mathbb{R}^2 \backslash L, \tag{4.2.34}$$

and the condition (4.2.33) on $L = \{\boldsymbol{\xi} : \xi_1 > 0, \ \xi_2 = 0\}$.

We shall analyze the solution of (4.2.33), (4.2.34) at infinity and in a neighbourhood of the origin.

Since $\gamma > 0$, the second term in (4.2.33) is small in comparison with the first one, as $||\boldsymbol{\xi}|| \to \infty$. Therefore, at infinity the leading order approximation of (4.2.33) has the form

$$\boldsymbol{\sigma}^{(2)}(\boldsymbol{V}; \xi_1, 0) = \boldsymbol{0}.$$

Thus, the solution of (4.2.33), (4.2.34) admits the following asymptotic representation at infinity,

$$\boldsymbol{V}(\boldsymbol{\xi}) = ||\boldsymbol{\xi}||^{1/2}(K_1 \boldsymbol{\Phi}^{(1)}(\varphi) + K_2 \boldsymbol{\Phi}^{(2)}(\varphi))$$

$$+ O(||\boldsymbol{\xi}||^{-1/2}), \quad ||\boldsymbol{\xi}|| \to \infty, \tag{4.2.35}$$

and corresponds to the problem on an infinite elastic plane with a semi-infinite cut with zero tractions.

The comparison of (4.2.35) and (4.2.17) gives that $\boldsymbol{v}^{(0)}(\boldsymbol{x})$ and $\alpha^{1/2}\boldsymbol{V}(\alpha^{-1}\boldsymbol{x})$ are equal in the region $||\boldsymbol{x}|| = O(\alpha^{1/2})$, and, therefore, the matching condition of the boundary layer and the leading term of the outer expansion holds.

At the limit $||\boldsymbol{\xi}|| \to 0$ the relations (4.2.33) are asymptotically equivalent to

$$\boldsymbol{\sigma}^{(2)}(\boldsymbol{V}; \xi_1, +0) = \boldsymbol{\sigma}^{(2)}(\boldsymbol{V}; \xi_1, -0); \quad \boldsymbol{V}(\xi_1, +0) = \boldsymbol{V}(\xi_1, -0). \tag{4.2.36}$$

These conditions correspond to the ideal contact on the interface boundary, and, therefore (also, see [5])

$$\boldsymbol{V}(\boldsymbol{\xi}) = \boldsymbol{l}(\boldsymbol{\xi}) + O(||\boldsymbol{\xi}||^{\min\{2, 1+\gamma\}}), \quad \text{as } ||\boldsymbol{\xi}|| \to 0,$$

where $\boldsymbol{l}$ is a linear vector-valued function. Thus, the boundary layer does not have a singularity in stress components at the tip of the cusp.

4.2.7 Special solutions of the homogeneous Neumann problem in an infinite strip of variable thickness

As $x_1 \to 0$, the set (4.2.1) should be interpreted as a "thin region." The asymptotic representation of solutions of the elasticity problems with the traction boundary conditions on the lateral surface requires auxiliary solutions of the plane-strain problem in an infinite strip. Often these vector functions are called the Jordan chains (see, for example, [22], [42], [49], [90], [127]). The analysis of the displacement field in a thin beam requires two groups of auxiliary vector functions (two Jordan chains). The first group includes four elements and corresponds to the bending of a thin beam; the second group has two elements and describes the longitudinal deformations. Consequently, we have to deal with two types of the boundary layer for the plane-strain problem on a cusp-shaped inclusion which is relatively rigid in comparison with the elastic matrix. The first boundary layer removes the singularity related to the displacement in the Ox_2 direction, and it exists in the region of the diameter $O(\alpha^{1/3})$. The second boundary layer is related to the longitudinal displacements and the diameter of the boundary layer region has the order $O(\alpha)$.

The first Jordan's chain is specified by

$$\boldsymbol{J}^{(0,1)} = (0, W_2), \quad \boldsymbol{J}^{(1,1)} = (-\tau\partial_1 W_2, 0),$$

$$\boldsymbol{J}^{(2,1)} = (0, 1/2(l+2m)^{-1} l\tau^2 \partial_1^2 W_2), \tag{4.2.37}$$

$$\boldsymbol{J}^{(3,1)} = (1/2(l+2m)^{-1}(\partial_1((l+2/3m)\tau^3 - h^2))\partial_1 W_2,\ 0),$$

and the second group is

$$\boldsymbol{J}^{(0,2)} = (W_1, 0), \ \boldsymbol{J}^{(1,2)} = (0, -l(l+2m)^{-1}\tau\partial_1 W_1). \tag{4.2.38}$$

Here W_j, $j = 1, 2$, are the functions of ξ_1; $\partial_1 = \partial/\partial_{\xi_1}$. The Jordan chains for a thin beam of a constant thickness were constructed in [42], [90].

4.2.8 Boundary layer for the bending mode

In accordance with (4.2.26), (4.2.28) the leading order terms of the boundary layer in the elastic matrix and inclusion will be sought in the form

$$\alpha^{1/6}\boldsymbol{V}^{(1)}(\boldsymbol{\xi}) \ \text{ and } \ \alpha^{1/6}\boldsymbol{W}^{(1)}(\delta, \xi_1, \tau).$$

The following representation holds:

$$\boldsymbol{W}^{(1)}(\delta,\xi_1,\tau)=\sum_{j=0}^{3}\delta^{j/3}\boldsymbol{W}^{(j,1)}(\xi_1,\tau)+\delta^{4/3}\boldsymbol{W}^{(1),*}(\xi_1,\tau). \qquad (4.2.39)$$

The solvability condition for the boundary value problem with respect to $\boldsymbol{W}^{(1),*}$ yields the analog of the beam equation. The vectors $\boldsymbol{W}^{(j,1)}$ can be represented in the form

$$\boldsymbol{W}^{(j,1)}(\xi_1,\tau)=\sum_{k=0}^{j}\boldsymbol{J}^{(j-k,1)}(W_2^{(1)}(\xi_1,0)). \qquad (4.2.40)$$

Here $\boldsymbol{J}^{(p,1)}(W_2^{(1)})$ are vector-valued functions specified by (4.2.37), with $h(\xi_1)=a\xi_1^{1+\gamma}$.

As before, the first contact condition (4.2.7) gives

$$\boldsymbol{V}^{(1)}(\xi_1,\pm 0)=\boldsymbol{W}^{(0,1)}(\xi_1). \qquad (4.2.41)$$

The representation (4.2.39) can be substituted into the Lamé system (4.2.5), $i=2$, and the analysis of the term of order $\alpha^{-2/3}\delta^{-1/3}$ (it must be zero) gives the ordinary differential equation with respect to $\boldsymbol{W}^{(1),*}$. The second contact condition (4.2.7) provides the boundary conditions for $\boldsymbol{W}^{(1),*}$. Thus, we obtain the Neumann boundary value problem which is solvable if and only if the following equality holds:

$$\frac{m(l+m)}{3(l+2m)}a^3\frac{\partial^2}{\partial\xi_1^2}\xi_1^{3+3\gamma}\frac{\partial^2 W_2^{(0,1)}}{\partial\xi_1^2}(\xi_1)+\sigma_{22}(\boldsymbol{V}^{(1)};\xi_1,+0) \qquad (4.2.42)$$

$$-\ \sigma_{22}(\boldsymbol{V}^{(1)};\xi_1,-0)=0.$$

Introduce the notation

$$D=\frac{m(l+m)}{3(l+2m)}.$$

Relations (4.2.41), (4.2.42) yield the conditions for the vector function $\boldsymbol{V}^{(1)}$ on L

$$\frac{m(l+m)}{6(l+2m)}a^3\frac{\partial^2}{\partial\xi_1^2}\xi_1^{3+3\gamma}\frac{\partial^2}{\partial\xi_1^2}(V_2^{(1)}(\xi_1,+0)+V_2^{(1)}(\xi_1,-0))$$

$$=\sigma_{22}(\boldsymbol{V}^{(1)};\xi_1,+0)-\sigma_{22}(\boldsymbol{V}^{(1)};\xi_1,-0),$$

$$V_2^{(1)}(\xi_1,+0) = V_2^{(1)}(\xi_1,-0), \quad V_1^{(1)}(\xi_1,\pm 0) = 0. \qquad (4.2.43)$$

Thus, the first boundary layer is specified by solutions of the problem (4.2.34), (4.2.43) in $\mathbb{R}^2 \backslash L$ and characterized by the asymptotic representation (4.2.35), where the angular parts $\mathbf{\Phi}^{(j)}$ correspond to the absolutely rigid contour L with zero displacements (also, see formula (4.1) in [103]).

In accordance with (4.2.41) the displacement field within the inclusion is given by

$$\frac{1}{2}(\mathbf{V}^{(1)}(\xi_1,+0) + \mathbf{V}^{(1)}(\xi_1,-0)).$$

The leading order terms at the faces of L have the form

$$\mathbf{V}^{(1)}(\xi_1,\pm 0) \sim |\xi_1|^{1/2-3\gamma} \frac{8}{3Da^3(36\gamma^2-1)} \{K_1\Psi_2^{(1)}(0) + K_2\Psi_2^{(2)}(0)\}\mathbf{e}^{(2)},$$

and, therefore,

$$\mathbf{W}^{(0,1)}(\xi_1) = |\xi_1|^{1/2-3\gamma} \frac{8}{3Da^3(36\gamma^2-1)} \{K_1\Psi_2^{(1)}(0)$$

$$+ K_2\Psi_2^{(2)}(0)\}\mathbf{e}^{(2)} + O(|\xi_1|^{-1/2-3\gamma}), \quad |\xi_1| \to +\infty. \qquad (4.2.44)$$

The comparison of (4.2.35), (4.2.44) with (4.2.17), (4.2.22) shows the matching conditions for the outer and inner expansions are satisfied.

In the vicinity of the origin the asymptotic expansion of $\mathbf{V}^{(1)}$ is different. As $||\boldsymbol{\xi}|| \to 0$, the leading order approximations of the boundary conditions (4.2.43) are

$$V_1^{(1)}(\xi_1,\pm 0) = 0; \ V_2^{(1)}(\xi_1,+0) = V_2^{(1)}(\xi_1,-0),$$

$$\sigma_{22}(\mathbf{V}^{(1)};\xi_1,+0) = \sigma_{22}(\mathbf{V}^{(1)};\xi_1,-0). \qquad (4.2.45)$$

These relations show that we have the continuity of the normal components of tractions and displacements across the thin inclusion. However, the longitudinal component of displacement is zero on the boundary. This type of the boundary conditions was discussed in [5].

Near the origin the field $\mathbf{V}^{(1)}$ is given by

$$\mathbf{V}^{(1)}(\boldsymbol{\xi}) = ||\boldsymbol{\xi}||^{1/2} K_1 \mathbf{\Phi}^{(1)}(\varphi) + O(||\boldsymbol{\xi}||), \quad \text{as } ||\boldsymbol{\xi}|| \to 0, \qquad (4.2.46)$$

and the first term in (4.2.46) solves the problem (4.2.34), (4.2.45) everywhere in $\mathbb{R}^2 \backslash L$.

4.2.9 Second boundary layer in the vicinity of the relatively rigid cusp-shaped inclusion

One can see that the boundary layer, constructed in the previous subsection does not compensate a singularity at the tip of the inclusion. The explanation is simple. There is another boundary layer (let us call it "the inner boundary layer") which should be represented in the coordinates (4.2.23), $\beta = 1$. To avoid any confusion, we denote these coordinates by $\boldsymbol{\xi}^*$. The asymptotic algorithm follows the analysis described in the previous sections. Thus, we shall present the answer.

The leading order terms of the inner boundary layer within the inclusion and elastic matrix are

$$\alpha^{1/2}\boldsymbol{W}^{(2)}(\delta, \xi_1^*, \tau^*) \text{ and } \alpha^{1/2}\boldsymbol{V}^{(2)}(\boldsymbol{\xi}^*),$$

where

$$\boldsymbol{W}^{(2)}(\delta, \xi_1^*, \tau^*) = \boldsymbol{W}^{(0,2)}(\xi_1^*) + \delta \boldsymbol{W}^{(1,2)}(\xi_1^*, \tau^*) + \delta^2 \boldsymbol{W}^{(2),*}(\xi_1^*, \tau^*).$$

The notations are similar to (4.2.39), (4.2.40) with the index 1 replaced by 2. The first contact conditions (4.2.7) and the solvability condition of the Neumann boundary value problem on $[-1, 1]$ with respect to $\boldsymbol{W}^{(2),*}$ give the following equalities:

$$\boldsymbol{V}^{(2)}(\xi_1, +0) = \boldsymbol{V}^{(2)}(\xi_1, -0); \ \sigma_{22}(\boldsymbol{V}^{(2)}; \xi_1, +0) = \sigma_{22}(\boldsymbol{V}^{(2)}; \xi_1, -0),$$

$$-4\frac{l(l+m)}{l+2m} a \frac{\partial}{\partial \xi_1} \xi_1^{1+\gamma} \frac{\partial}{\partial \xi_1} (V_1^{(2)}(\xi_1, +0) + V_1^{(2)}(\xi_1, -0))$$

$$= \sigma_{21}(\boldsymbol{V}^{(2)}; \xi_1, +0) - \sigma_{21}(\boldsymbol{V}^{(2)}; \xi_1, -0). \tag{4.2.47}$$

To leading order approximation, boundary conditions (4.2.47) written at infinity have the form (4.2.45), and, therefore, there exists a solution of (4.2.34), (4.2.47) which admits the asymptotic form

$$\boldsymbol{V}^{(2)}(\boldsymbol{\xi}^*) = ||\boldsymbol{\xi}^*||^{1/2} K_1 \boldsymbol{\Phi}^{(1)}(\varphi) + O(||\boldsymbol{\xi}^*||^{-1/2} \ln ||\boldsymbol{\xi}^*||),$$

$$\text{as } ||\boldsymbol{\xi}^*|| \to \infty. \tag{4.2.48}$$

Due to (4.2.46), (4.2.48), the first and the second boundary layers can be matched in the region $\{\boldsymbol{x} : \|\boldsymbol{x}\| = O(\alpha^{2/3})\}$. Thus, in a neighbourhood of the point $\boldsymbol{\xi}^* = \mathbf{0}$ the equalities (4.2.47) are asymptotically equivalent to (4.2.36), and the stress tensor components, corresponding to $\boldsymbol{V}^{(2)}, \boldsymbol{W}^{(2)}$, do not have singularities at the origin.

4.2.10 Uniform asymptotic approximation

Introduce a smooth cut-off function χ such that $\chi(z) = 0$ for $z > 1$, and $\chi(z) = 1$ for $z < 1/2$.

For the case of a soft inclusion the leading order approximations of the displacement fields in the elastic matrix and the cusp-shaped inclusion are given by

$$\boldsymbol{u}(\varepsilon, \boldsymbol{x}) \sim \boldsymbol{v}^{(0)}(\boldsymbol{x})(1 - \chi(\alpha^{-1/2}r)) + \alpha^{1/2}\chi(\alpha^{-1/2}r)\boldsymbol{V}(\alpha^{-1}\boldsymbol{x}),$$

$$\boldsymbol{u}^\circ(\varepsilon, \boldsymbol{x}) \sim \boldsymbol{w}^{(0)}(\boldsymbol{x})(1 - \chi(\alpha^{-1/2}r)) + \alpha^{1/2}\chi(\alpha^{-1/2}r) \tag{4.2.49}$$

$$\times \; (\boldsymbol{A}_1(\alpha^{-1}\boldsymbol{x})H^{-1}(x_1)x_2 + \boldsymbol{A}_0(\alpha^{-1}\boldsymbol{x})),$$

where $\boldsymbol{v}^{(0)}, \boldsymbol{w}^{(0)}$ and $\boldsymbol{V}$ are solutions of (4.2.11)–(4.2.15) and (4.2.34), (4.2.33); the quantities $\boldsymbol{A}_j$, $j = 0, 1$, are specified by (4.2.32).

For the case of a rigid inclusion the displacement field in the elastic matrix is

$$\boldsymbol{u}(\varepsilon, \boldsymbol{x}) \sim \boldsymbol{v}^{(0)}(\boldsymbol{x})(1 - \chi(\alpha^{-1/6}r)) + \alpha^{1/6}\chi(\alpha^{-1/6}r) \tag{4.2.50}$$

$$\times \; (1 - \chi(\alpha^{-1/2}r))\boldsymbol{V}^{(1)}(\alpha^{-1/3}\boldsymbol{x}) + \alpha^{1/2}\chi(\alpha^{-1/2}r)\boldsymbol{V}^{(2)}(\alpha^{-1}\boldsymbol{x}),$$

where the fields $\boldsymbol{v}^{(0)}$ and $\boldsymbol{V}^{(1)}, \boldsymbol{V}^{(2)}$ satisfy (4.2.11), (4.2.12), (4.2.15), (4.2.34), (4.2.43) and (4.2.34), (4.2.47).

Let us discuss the asymptotic formulae (4.2.49)–(4.2.50). Suppose that the ratio of the Young moduli E/E° is small or large. Then in some region, outside a neighbourhood of the radius $O(\alpha^\beta)$ with the centre at the origin, the stress components increase like $O(r^{-1/2})$ for small r. For a rigid inclusion the parameters α and β are defined by the equalities

$$\alpha = E(E^\circ)^{-1}, \quad \beta = 1/3;$$

for a soft inclusion these relations are

$$\alpha = E^\circ E^{-1}, \quad \beta = 1$$

(it follows from formulae (4.2.49)–(4.2.50) and (4.2.17)).

Within the region of the diameter $O(\alpha^\beta)$ the shear stress components reach the level $O(E\alpha^{-\beta/2})$ and keep this order of magnitude up to the tip of the inclusion. If the inclusion is soft, then this asymptotic behaviour holds for diagonal components of the stress tensor as well. For a rigid inclusion the diagonal components continue their growth (with the exponent $-1/2$) up to the region

$$\{\boldsymbol{x} : ||\boldsymbol{x}|| < \text{const } \alpha\},$$

where their order of magnitude is $O(E/\sqrt{\alpha})$.

In any case the stresses are bounded, and the stress concentration in a neighbourhood of the origin is characterized by the boundary layer.

From Fig. 4.4 one can see qualitatively that the boundary layer for soft and rigid cusp-shaped inclusions has different width; it has the order $O(\alpha)$ and $O(\alpha^{1/3})$ respectively.

Consequently, in the case of rigid inclusions the square root asymptotics occur for smaller values of α than in the case of soft inclusions.

4.3 Exercises

1. Consider an anti-plane shear of a domain with a cusp-shaped elastic inclusion, specified in Section 4.1.1. The third component of the displacement vector satisfies the equations

$$\Delta u_3(\boldsymbol{x}) = 0, \ \boldsymbol{x} \in \Omega, \quad \Delta u_3^\circ(\boldsymbol{x}) = 0, \ \boldsymbol{x} \in \omega,$$

and the interface boundary conditions

$$u_3(\boldsymbol{x}) = u_3^\circ(\boldsymbol{x}), \ \mu \frac{\partial u_3}{\partial n}(\boldsymbol{x}) = \mu^\circ \frac{\partial u_3^\circ}{\partial n}(\boldsymbol{x}), \ \boldsymbol{x} \in \Gamma,$$

where μ, μ° are the shear moduli of the matrix and elastic inclusion, and the cusp-shaped boundary Γ is given by (4.1.2). A self-balanced load is applied on the external contour

$$\mu \frac{\partial u_3}{\partial n} = p(\boldsymbol{x}), \ \boldsymbol{x} \in \partial\Omega \setminus \Gamma,$$

$$\int_{\Gamma} p(\boldsymbol{x}) ds = 0.$$

Determine the asymptotic behaviour of the displacement field in a neighbourhood of the vertex of the cusp. This problem is easier than the one discussed in Section 4.1.1.

2. Consider a plane-strain state of the cusp–shaped elastic region ω with the boundary Γ, defined by (4.1.2). The two–dimensional displacement vector satisfies the homogeneous Lamé system and the traction boundary conditions

$$\boldsymbol{\sigma}^{(n)}(\boldsymbol{u}; \boldsymbol{x}) = \mathbf{p}(\boldsymbol{x}), \ \ \boldsymbol{x} \in \Gamma,$$

where the principal force and moment vectors of the load $\mathbf{p}$ are zero. Describe the leading order terms of the longitudinal and transversal components of the displacement vector in a neighbourhood of the cusp. In order to do this, one can apply the REDUCE program from the Appendix and the asymptotic algorithm presented in Section 3.3.2 for the derivation of equations of thin beams.

5

Integral Characteristics in Elasticity Problems for Nonhomogeneous Bodies

In the present chapter we study the integral characteristics of finite inclusions or cavities in elastic media. The main attention is paid to the Pólya-Szegö matrix that characterizes a perturbation of a linear displacement field when a finite defect is introduced in an elastic continuum. The components of this matrix depend on the shape of the defect and elastic moduli of the materials, and they are independent of the external load. In Section 5.1 we show how to construct the Pólya-Szegö matrix and discuss its application to the evaluation of the energy integral. Section 5.2 shows another example, where we look at the asymptotics of the eigenvalues for a problem on the vibration of an elastic body with a small defect. The interaction of a crack with a small cavity (or elastic inclusion), including the case of periodic dilute composite, is considered in Section 5.3.

5.1 Integral characteristics of elastic inclusions and cavities in two-dimensional theory of elasticity

Here we describe the integral characteristics of elastic inclusions and cavities. The Pólya-Szegö matrices for a finite elliptical cavity and for cracks are constructed in the case of the two-dimensional Lamé operator. These are named by analogy with their equivalents in potential theory, developed by Pólya and Szegö [107]; their main properties were investigated in [69], [131], [130].

The matrix of Pólya and Szegö, being an essential integral characteristic of a defect, appears in various asymptotic expansions; examples include the variation of the natural frequency of oscillations [56] of a body due to a defect, the potential strain energy increment of an elastic body with a defect [131], the effective elastic modulus of a composite material with widely separated inclusions [130].

In the present section, the Pólya-Szegö matrix is constructed explicitly for certain cases of solution of the two-dimensional Lamé system of equations governing plane-strain deformation of an isotropic elastic body.

5.1.1 Governing equations

Let us consider a simply connected compact set $\bar{G} \subset \mathbb{R}^2$ with a piecewise-smooth boundary and use the notation $\Omega := \mathbb{R}^2 \backslash \bar{G}$. Assume that the displacement fields $\boldsymbol{u}, \boldsymbol{u}^\circ$ satisfy the system of equations

$$\mu \Delta \boldsymbol{u}(\boldsymbol{x}) + (\lambda + \mu)\nabla\nabla \cdot \boldsymbol{u}(\boldsymbol{x}) = \boldsymbol{0}, \boldsymbol{x} \in \Omega, \tag{5.1.1}$$

$$\mu^\circ \Delta \boldsymbol{u}^\circ(\boldsymbol{x}) + (\lambda^\circ + \mu^\circ)\nabla\nabla \cdot \boldsymbol{u}^\circ(\boldsymbol{x}) = \boldsymbol{0}, \boldsymbol{x} \in G, \tag{5.1.2}$$

$$\boldsymbol{\sigma}^{(n)}(\boldsymbol{u}; \boldsymbol{x}) - \boldsymbol{\sigma}^{\circ,(n)}(\boldsymbol{u}^\circ; \boldsymbol{x}) = \boldsymbol{\Psi}(\boldsymbol{x}),$$

$$\boldsymbol{u}(\boldsymbol{x}) - \boldsymbol{u}^\circ(\boldsymbol{x}) = \boldsymbol{\Phi}(\boldsymbol{x}), \boldsymbol{x} \in \partial G, \tag{5.1.3}$$

$$\boldsymbol{u}(\boldsymbol{x}) \to \boldsymbol{u}^\infty(\boldsymbol{x}), ||\boldsymbol{x}|| \to \infty, \tag{5.1.4}$$

where $\boldsymbol{\sigma}^{(n)}(\boldsymbol{u}; \boldsymbol{x}) = n_1(\boldsymbol{x})((2\mu + \lambda)\frac{\partial u_1}{\partial x_1}(\boldsymbol{x}) + \lambda \frac{\partial u_2}{\partial x_2}(\boldsymbol{x}), \mu(\frac{\partial u_1}{\partial x_2}(\boldsymbol{x}) + \frac{\partial u_2}{\partial x_1}(\boldsymbol{x})))^T + n_2(\boldsymbol{x})(\mu(\frac{\partial u_1}{\partial x_2}(\boldsymbol{x}) + \frac{\partial u_2}{\partial x_1}(\boldsymbol{x})), (2\mu + \lambda)\frac{\partial u_2}{\partial x_2}(\boldsymbol{x}) + \lambda \frac{\partial u_1}{\partial x_1}(\boldsymbol{x}))^T$; $\boldsymbol{n}$ is the unit inner normal to ∂G. Thus, the domain G is occupied by an elastic inclusion with the Lamé elastic constants λ°, μ°, while the remaining part Ω of the elastic plane has the Lamé constants λ, μ. Note that, if $\boldsymbol{\Phi} = \boldsymbol{0}, \mu^\circ = \lambda^\circ = 0$, the equations (5.1.3) have the sense of conditions on the boundary of the cavity G.

The results of [38], [104] lead to the following statement.

Proposition 6. *The solution* $\boldsymbol{u}, \boldsymbol{u}^\circ$ *of the system of equations (5.1.1)–(5.1.3), which satisfies the condition (5.1.4) in the form* $\boldsymbol{u}(\boldsymbol{x}) = O(|\ln(\|\boldsymbol{x}\|)|), \|\boldsymbol{x}\| \to \infty$, *admits the following representation:*

$$\boldsymbol{u}(\boldsymbol{x}) = -\mathbf{F}\cdot T(\boldsymbol{x}) + \frac{1}{2}\mathbf{M}\cdot\nabla\times T(\boldsymbol{x}) + \sum_{j=1}^{3} c_i \mathbf{V}^{(j)}(\frac{\partial}{\partial \boldsymbol{x}})T(\boldsymbol{x})$$

$$+ O(\|\boldsymbol{x}\|^{-2}), \|\boldsymbol{x}\| \to \infty. \tag{5.1.5}$$

Here T *is the Somigliana tensor and* $\mathbf{V}^{(j)}(\frac{\partial}{\partial \boldsymbol{x}}), j = 1, 2, 3$, *are differential operators corresponding to the vector-valued functions*

$$\mathbf{V}^{(i)}(\boldsymbol{x}) = x_i \boldsymbol{e}^{(i)},\ i = 1, 2;\ \mathbf{V}^{(3)}(\boldsymbol{x}) = 2^{-1/2}(x_2 \boldsymbol{e}^{(1)} + x_1 \boldsymbol{e}^{(2)}). \tag{5.1.6}$$

By the results of [130] the coefficients in (5.1.5) satisfy the following equations:

$$\alpha_j c_j = \int_{\partial G} (\boldsymbol{\Phi}(\boldsymbol{x})\cdot\boldsymbol{\sigma}^{(n)}(\mathbf{U}^{(j)}; \boldsymbol{x}) - \boldsymbol{\Psi}(\boldsymbol{x})\cdot\mathbf{U}^{(j)}(\boldsymbol{x}))ds, \tag{5.1.7}$$

where $\alpha_1 = \alpha_2 = 1, \alpha_3 = 2^{1/2}$;

$$F_k = -\int_{\partial G} \boldsymbol{\Phi}(\boldsymbol{x})\cdot\boldsymbol{\sigma}^{(n)}(x_k \boldsymbol{e}^{(k)}; \boldsymbol{x})ds;$$

$$M_k = -\int_{\partial G} \boldsymbol{\Phi}(\boldsymbol{x})\cdot\boldsymbol{\sigma}^{(n)}(\boldsymbol{e}^{(k)}\times\boldsymbol{x}; \boldsymbol{x})ds, k = 1, 2. \tag{5.1.8}$$

Here $\boldsymbol{e}^{(k)}, k = 1, 2$, are unit basis vectors; $\mathbf{U}^{(j)}, j = 1, 2$, are the vector fields which satisfy the homogeneous equations (5.1.1)–(5.1.3) and have the following asymptotic behaviour at infinity:

$$\mathbf{U}^{(i)}(\boldsymbol{x}) = \mathbf{V}^{(j)}(\boldsymbol{x}) + \sum_{k=1}^{3} m_{jk}(G)\sum_{p=1}^{2} V_p^{(k)}(\frac{\partial}{\partial \boldsymbol{x}})T^{(p)}(\boldsymbol{x})$$

$$+ O(\|\boldsymbol{x}\|^{-2}), \tag{5.1.9}$$

where $T^{(p)}, p = 1, 2$, are the columns of the Somigliana tensor. The main properties of the symmetric matrix $(m_{jk})_{j,k=1}^{3}$ were studied in [69], [131], [130].

5.1.2 The Pólya-Szegö matrix of an elliptical cavity

Let us consider the complex Kolosov-Muskhelishvili potentials $\phi(\zeta), \psi(\zeta)$ [85], corresponding to the problem (5.1.1)–(5.1.4) for the special solutions $\mathbf{U}^{(j)}$, which were defined in Section 5.1.1, in the case of an elliptical cavity G. It should be noted that in this case it is necessary to consider (5.1.1) outside the domain G, with the condition (5.1.3) in the form

$$\boldsymbol{\sigma}^{(n)}(\boldsymbol{u}; \boldsymbol{x}) = \mathbf{0}, \boldsymbol{x} \in \partial G. \tag{5.1.10}$$

We use the assumption that the conformal mapping of the exterior of the unit disk $\{\zeta : |\zeta| > 1\} = \mathbb{R}^2 \backslash \bar{D}$ onto the domain Ω,

$$z = \omega(\zeta) = c_1\zeta + c_0 + c_{-1}\zeta^{-1}, \tag{5.1.11}$$

is known. We have then the relations

$$\phi(\zeta) = a_1\zeta + a_0 + a_{-1}\zeta^{-1} + \ldots ; \psi(\zeta) = b_1\zeta + b_0 + b_{-1}\zeta^{-1} + \ldots, \tag{5.1.12}$$

where a_i, b_i are complex coefficients.

If the complex potentials are known, then the solution $\boldsymbol{u}$ is defined by the relation [85]

$$2\mu(u_1 + iu_2) = \varkappa\phi(\zeta) - \frac{\omega(\zeta)}{\overline{\omega'(\zeta)}}\overline{\phi'(\zeta)} - \overline{\psi(\zeta)}, \tag{5.1.13}$$

with $\varkappa = (\lambda + 3\mu)(\lambda + \mu)^{-1}$.

Without loss of generality it is possible to assume $c_0 = a_0 = b_0 = 0$. The constants a_1, b_1 are determined from the conditions at infinity. More precisely, the special solutions $\alpha_j \mathbf{U}^{(j)}, j = 1, 2, 3$, correspond to the coefficients $a_1^{(j)}, b_1^{(j)}$, and the following equations hold:

$$\frac{a_1^{(i)}}{c_1} = \frac{1}{2}(\lambda + \mu), \frac{b_1^{(j)}}{c_1} = (-1)^j\mu, j = 1, 2, a_1^{(3)} = 0, \frac{b_1^{(3)}}{c_1} = 2i\mu. \tag{5.1.14}$$

The relations (5.1.13)–(5.1.14) lead to

$$2\mu(u_1 + iu_2) = \varkappa(a_1\zeta + a_{-1}\zeta^{-1}) - \zeta\frac{c_1}{\bar{c_1}}\bar{a_1} - \frac{\bar{a_1}c_{-1}}{\bar{c_1}}\zeta^{-1}$$

$$+ \frac{\zeta}{\bar{\zeta}^2}\frac{c_1}{\bar{c_1}}(\bar{a}_{-1} - \frac{\bar{c}_{-1}}{\bar{c_1}}a_1) - \bar{b}_1\bar{\zeta} - \bar{b}_{-1}\bar{\zeta}^{-1} + O(|\zeta|^{-2}), |\zeta| \to \infty. \tag{5.1.15}$$

It is necessary to compute the coefficients of the terms in (5.1.15), which decrease as $O(|\zeta|^{-1})$, as $|\zeta| \to \infty$. In these terms only the coefficients a_1, a_{-1}, b_1, b_{-1} of the expansion (5.1.12) appear. The unknown coefficients a_{-1}, b_{-1} are determined from the equation

$$\phi(\sigma) + \frac{\omega(\sigma)}{\omega'(\sigma)}\overline{\phi'(\sigma)} + \overline{\psi(\sigma)} = 0, |\sigma| = 1, \qquad (5.1.16)$$

which is valid for the complex potentials, corresponding to the solution $\mathbf{U}^{(j)}$ of the homogeneous problem (5.1.1), (5.1.10), (5.1.9). Let us multiply the equality (5.1.16) by the factor $\sigma[2\pi i(\sigma - \zeta)]^{-1}$. Then we integrate this expression around the unit circle $\gamma = \{\sigma : |\sigma| = 1\}$, choosing the positive direction of γ, so that the region D lies on the left. Here we can use the well-known statement [85], concerning the calculation of the Cauchy-type integrals.

Proposition 7. 1°. *Let $f(\zeta)$ be holomorphic in D and continuous in $\bar{D}$ with the possible exclusion of the points $\zeta_1, \zeta_2, \ldots, \zeta_n \in D$, where it may have poles with the principal parts $G_1(\zeta), G_2(\zeta), \ldots, G_n(\zeta)$. Then*

$$\frac{1}{2\pi i}\int_\gamma \frac{f(\sigma)d\sigma}{\sigma - \zeta} = \begin{cases} f(\zeta) - G_1(\zeta) - G_2(\zeta) - \ldots - G_n(\zeta) & ,\zeta \in D, \\ -G_1(\zeta) - G_2(\zeta) - \ldots - G_n(\zeta) & ,\zeta \in \mathbb{R}^2 \backslash \bar{D}. \end{cases}$$

2°. *Let $f(\zeta)$ be holomorphic in $\mathbb{R}^2 \backslash \bar{D}$ and continuous in $\mathbb{R}^2 \backslash D$, possibly except the points $\zeta_1, \zeta_2, \ldots, \zeta_n \in \mathbb{R}^2 \backslash \bar{D}$ and the point $\zeta = \infty$, where it may have poles with the principal parts $G_1(\zeta), G_2(\zeta), \ldots, G_n(\zeta), G_\infty(\zeta)$. Then*

$$\frac{1}{2\pi i}\int_\gamma \frac{f(\sigma)d\sigma}{\sigma - \zeta} = \begin{cases} G_1(\zeta) + \cdots + G_n(\zeta) + G_\infty(\zeta) & ,\zeta \in D, \\ -f(\zeta) + G_1(\zeta) + \ldots + G_n(\zeta) + G_\infty(\zeta) & ,\zeta \in \mathbb{R}^2 \backslash \bar{D}. \end{cases}$$

Note that $\sigma\phi(\sigma)$ is the boundary value of a function which is holomorphic in $\mathbb{R}^2 \backslash \bar{D}$ except the point $\zeta = \infty$, where it has the pole with the principal part $a_1\zeta^2 + a_{-1}$. Taking into account the relation $\overline{\psi(\sigma)} = \bar{\psi}(1/\sigma)$, where $|\sigma| = 1$, we shall obtain that $\sigma\overline{\psi(\sigma)}$ is the boundary value of a function which is holomorphic in D. Also, retaining an arbitrary finite number of terms in the representation (5.1.11), we can consider the function $\frac{\sigma\omega(\sigma)}{\omega'(\sigma)}\overline{\phi'(\sigma)}$ as the boundary value of a function which is holomorphic in D except the pole $\zeta = 0$, where we define the principal part, using the expansion of $[\bar{\omega}'(1/\zeta)]^{-1}$ in the form of Taylor's series in the neighbourhood of $\zeta = 0$. Then, with the application of

Proposition 7 we obtain the representations

$$a_{-1} = -c_{-1}\bar{A} - \bar{c}_1\bar{B},$$

$$b_{-1} = -(A+\bar{A})c_1 - A\frac{|c_{-1}|^2}{|c_1|^2}(c_1+\bar{c_1}) - \frac{c_{-1}Bc_1}{\bar{c}_1}, \tag{5.1.17}$$

with $A = a_1/\bar{c}_1, B = b_1/c_1$. Also, we take into account the relation

$$\zeta = \frac{z}{c_1} - \frac{c_{-1}}{z} + O(|z|^{-3}), z \to \infty.$$

To construct the Pólya-Szegö matrix we need the representation of the Green tensor $\mathbf{T}$ in $\mathbb{R}^2$ for the Lamé system

$$\mathbf{T}(\boldsymbol{x}) = q\begin{pmatrix} -2\varkappa \ln R + \frac{2x_1^2}{R^2} & \frac{2x_1x_2}{R^2} \\ \frac{2x_1x_2}{R^2} & -2\varkappa \ln R + \frac{2x_2^2}{R^2} \end{pmatrix}, \tag{5.1.18}$$

where $R = (x_1^2+x_2^2)^{1/2}, q = (\lambda+\mu)[8\pi\mu(\lambda+2\mu)]^{-1}$.

Using (5.1.18) we obtain the relations

$$q^{-1}(V_1+iV_2) = 2i(-\varkappa\frac{\bar{z}}{R^2} + \frac{z^3}{R^4}), \tag{5.1.19}$$

with $(V_1, V_2)^T = \frac{\partial T^{(1)}}{\partial x_2} + \frac{\partial T^{(2)}}{\partial x_1}$;

$$q^{-1}(V_1+iV_2) = (D_2-D_1)\frac{z^3}{R^4} + (1-\varkappa)(D_1+D_2)\frac{z}{R^2}$$

$$+ \varkappa(D_2-D_1)\frac{\bar{z}}{R^2}, \tag{5.1.20}$$

with $(V_1, V_2)^T = D_1\frac{\partial T^{(1)}}{\partial x_1} + D_2\frac{\partial T^{(2)}}{\partial x_2}$.

Considering the asymptotic behaviour at infinity of the vector-valued functions $\mathbf{U}^{(j)}$, described by (5.1.9), we can see that the vanishing terms in (5.1.15) have the representation (5.1.20), when $j = 1, 2$, and the representation (5.1.19), when $j = 3$. These equations allow us to define the coefficients in the sums (5.1.9) and to construct the Pólya-Szegö matrix $\boldsymbol{m} = (m_{jk})_{j,k=1}^3$

$$\boldsymbol{m} = 8\pi(\lambda+2\mu)\begin{pmatrix} 2h_r - \Sigma & \Xi & h_i \\ \Xi & -2h_r - \Sigma & h_i \\ h_i & h_i & -\frac{2\mu}{\lambda+\mu}|c_1|^2 \end{pmatrix}, \tag{5.1.21}$$

where

$$h_r = Re(c_1 c_{-1}), \quad h_i = Im(c_1 c_{-1}),$$

$$\Sigma = \frac{\lambda+\mu}{2\mu}(|c_1|^2 + |c_{-1}|^2) + \frac{\mu}{\lambda+\mu}|c_1|^2,$$

$$\Xi = \frac{\mu}{2(\lambda+\mu)}(|c_1|^2 - |c_{-1}|^2) - \frac{\lambda(\lambda+2\mu)}{2\mu(\lambda+\mu)}(|c_1|^2 + |c_{-1}|^2).$$

Note that the equation $Im(c_1 c_{-1}) = 0$ defines the canonical system of coordinates, where the matrix $\boldsymbol{m}$ has a block–diagonal structure. Let us note also that it is possible to construct the rank-four tensor using components m_{jk} of the Pólya-Szegö matrix.

5.1.3 Illustrative examples

1. First, consider the particular case of elliptic hole G oriented parallel to the coordinate axes. The conformal mapping of the set $\{\zeta : |\zeta| > 1\}$ to the outside of the domain $\mathbb{R}^2 \backslash \bar{G}$ is

$$z = \omega(\zeta) = C\{\zeta + \frac{m}{\zeta}\}; m, C \in \mathbb{R}^1.$$

Thus, the Pólya-Szegö matrix may be represented as

$$\boldsymbol{m} = \frac{C^2}{4\mu q(1-\varkappa)}\mathrm{diag}\{(\lambda+\mu)\begin{pmatrix} m(m-f)+1 & m(m+f)+1 \\ m(m+f)+1 & m(m+f)+1 \end{pmatrix}$$

$$+\mu\begin{pmatrix} f-2m & -f-2m \\ -f-2m & f+2m \end{pmatrix}, 2\mu f\}, \tag{5.1.22}$$

where $f = 2\mu(\lambda+\mu)^{-1}$.

Hence, substituting the value 1 or -1 instead of m in (5.1.22), we can obtain the equalities for the Pólya-Szegö matrices, corresponding to cracks oriented parallel to the coordinates' axes. The case of a circular cavity corresponds to $m = 0$.

2. Let us note that the Pólya-Szegö matrix of a crack M, given by the relation $M = \{(x_1, x_2) : x_2 = 0, |x_1| < a\}$, may be obtained by direct consideration of the exact solution [114] of the problem on the loading of the plane with the cut. Obviously, this representation corresponds to the formula (5.1.22), when $m = 1$, and has the form

$$\boldsymbol{m} = -\frac{a^2\pi(\lambda+2\mu)}{2\mu(\lambda+\mu)}\begin{pmatrix} \lambda^2 & (2\mu+\lambda)\lambda & 0 \\ (2\mu+\lambda)\lambda & (2\mu+\lambda)^2 & 0 \\ 0 & 0 & 2\mu^2 \end{pmatrix}.$$

In [130] it was shown that if a defect with the Pólya-Szegö matrix $\boldsymbol{m}$ appears in an infinite elastic plane, then the potential strain energy increment $\Delta\mathcal{E}$ is equal to $\frac{1}{2}\boldsymbol{l}^\infty \cdot \boldsymbol{m}\boldsymbol{l}^\infty$. Here the vector $\boldsymbol{l}^\infty$ has the components $l_j^\infty = \lim_{\|\boldsymbol{x}\|\to\infty} \mathbf{V}^{(j)}(\partial/\partial\boldsymbol{x})\boldsymbol{u}(\boldsymbol{x}), j = 1,2,3$. Hence the potential strain energy increment for the elastic plane with the crack M satisfies the equality

$$\Delta\mathcal{E} = -\frac{a^2\pi(\lambda+2\mu)}{4\mu(\lambda+\mu)}((\sigma_{12}^\infty)^2 + (\sigma_{22}^\infty)^2), \tag{5.1.23}$$

where σ_{ij} are the components of the stress tensor. The equality (5.1.23) allows us to define the orientation of a crack which corresponds to an extremal value of $\Delta\mathcal{E}$. Assume, for example, that the Cartesian system of coordinates is chosen such that the axes Ox_1 and Ox_2 coincide with the principal axes of the stress tensor $\boldsymbol{\sigma}^\infty$, and the angle of the slope of the crack with respect to the axis Ox_1 equals α. Then (5.1.23) assumes the form

$$\Delta\mathcal{E} = -\frac{a^2\pi(\lambda+2\mu)}{4\mu(\lambda+\mu)}((\sigma_{11}^\infty \sin\alpha)^2 + (\sigma_{22}^\infty \cos\alpha)^2). \tag{5.1.24}$$

Therefore, if $|\sigma_{11}^\infty| > |\sigma_{22}^\infty|$, the right–hand side of (5.1.24) attains a maximal value at $\alpha = 0$ and it attains a minimal value at $\alpha = \pi/2$.

5.2 Vibration of elastic solids with small holes

In this section we deal with a two-dimensional problem on vibration of an elastic domain with a small cavity and with the vibration problem of a plate weakened by a small cavity. The asymptotic representation of the fundamental frequency of oscillation is presented.

5.2.1 An eigenvalue problem in a two-dimensional domain with a small hole

Let Ω and ω be two-dimensional domains with smooth boundaries. Also, assume that these domains include the origin O. Introduce the sets

$$\omega_\varepsilon = \{\boldsymbol{x} \in \mathbb{R}^2 : \varepsilon^{-1}\boldsymbol{x} \in \omega\}; \quad \Omega_\varepsilon = \Omega \backslash \overline{\omega}_\varepsilon,$$

which depend on a small positive parameter ε. The material, occupying Ω_ε, is characterized by the Lamé constants λ, μ and the density ρ.

The eigenvalue problem is formulated as follows:

$$\boldsymbol{L}(\frac{\partial}{\partial \boldsymbol{x}})\boldsymbol{u}(\varepsilon, \boldsymbol{x}) + \Lambda(\varepsilon)\boldsymbol{u}(\varepsilon, \boldsymbol{x}) = \boldsymbol{0}, \quad \boldsymbol{x} \in \Omega_\varepsilon, \tag{5.2.1}$$

$$\boldsymbol{\sigma}^{(n)}(\boldsymbol{u}; \boldsymbol{x}) = \boldsymbol{0}, \quad \boldsymbol{x} \in \partial\Omega_\varepsilon. \tag{5.2.2}$$

Here $\boldsymbol{L}$ is a two-dimensional Lamé operator with components $L_{jk} = \mu\delta_{jk}\Delta + (\lambda+\mu)\partial^2/\partial x_j \partial x_k$; $\sigma_j^{(n)} = \sigma_{jk} n_k$, $j, k = 1, 2$; $\boldsymbol{n}$ is the vector of the outward normal to Ω_ε; δ_{jk} is the Kronecker delta; $\Lambda(\varepsilon) = \rho\vartheta(\varepsilon)^2$, $\vartheta(\varepsilon)$ is the vibration frequency.

Following [56], we seek asymptotic expansions of the solution of problem (5.2.1), (5.2.2) in the form

$$\Lambda(\varepsilon) = \sum_{j=0}^{\infty} \varepsilon^j \Lambda_j, \quad \boldsymbol{u}(\varepsilon, \boldsymbol{x}) = \sum_{j=0}^{\infty} \varepsilon^j \left(\boldsymbol{U}^{(j)}(\boldsymbol{x}) + \boldsymbol{w}^{(j)}\left(\frac{\boldsymbol{x}}{\varepsilon}\right)\right), \tag{5.2.3}$$

where Λ_0 is a simple eigenvalue, while $\boldsymbol{U}^{(0)}(\boldsymbol{x})$ is the corresponding eigenfunction of the problem of vibration of the domain Ω without a hole, $||\boldsymbol{U}^{(0)}(\boldsymbol{x}),\ L_2(\Omega)|| = 1$. Note that $\boldsymbol{U}^{(0)}(\boldsymbol{x})$ leaves a discrepancy of order $O(\varepsilon)$ in the traction boundary condition on $\partial\omega_\varepsilon$. Consequently,

$$\boldsymbol{w}^{(0)}(\boldsymbol{\xi}) = \boldsymbol{0}, \quad \boldsymbol{\xi} = \varepsilon^{-1}\boldsymbol{x} \in \mathbb{R}^2 \backslash \omega, \quad \boldsymbol{U}^{(1)}(\boldsymbol{x}) = \boldsymbol{0}, \quad \boldsymbol{x} \in \Omega, \quad \Lambda_1 = 0.$$

The discrepancy should be compensated by means of the boundary layer term $\boldsymbol{w}^{(1)}$ which decays at infinity and satisfies the following boundary value problem in an infinite domain:

$$\boldsymbol{L}(\frac{\partial}{\partial \boldsymbol{\xi}})\boldsymbol{w}^{(1)}(\boldsymbol{\xi}) = \boldsymbol{0}, \quad \boldsymbol{\xi} \in \mathbb{R}^2 \backslash \overline{\omega};$$

$$\boldsymbol{\sigma}^{(n)}(\boldsymbol{w}^{(1)}; \boldsymbol{\xi}) = -\boldsymbol{\sigma}^{(n)}(\boldsymbol{x}^T \nabla \boldsymbol{U}^{(0)}; \boldsymbol{\xi}), \quad \boldsymbol{\xi} \in \partial\omega.$$

It is known (see, for example [38], [62]) that the following asymptotic formula holds:

$$\boldsymbol{w}^{(1)}(\boldsymbol{\xi}) = \sum_{q=1}^{3} C_q \sum_{k=1}^{3} m_{qk} \mathbf{V}^{(1,k)}\left(\frac{\partial}{\partial \boldsymbol{\xi}}\right) T(\boldsymbol{\xi}) + O(||\boldsymbol{\xi}||^{-2})$$

$$= \mathbf{D}(\boldsymbol{\xi}) + O(||\boldsymbol{\xi}||^{-2}). \tag{5.2.4}$$

Here $T(\boldsymbol{\xi})$ is a two-dimensional Somigliana tensor, $T_{jk}(\boldsymbol{\xi}) = -a\delta_{jk}\ln||\boldsymbol{\xi}|| + ab||\boldsymbol{\xi}||^{-2}\xi_j\xi_k, j,\ k = 1,2;\ a = (\lambda+3\mu)[4\pi\mu(\lambda+2\mu)]^{-1}$, $b = (\lambda+\mu)(\lambda+3\mu)^{-1}$; $\mathbf{V}^{(1,k)}(\partial/\partial\boldsymbol{\xi})$ are differential operators of the form $(\partial/\partial\xi_1, 0)^T$, $(0, \partial/\partial\xi_2)^T$, $2^{-1/2}(\partial/\partial\xi_2, \partial/\partial\xi_1)^T$, and C_q are coefficients from the expansion

$$\boldsymbol{U}^{(0)}(\boldsymbol{x}) = \boldsymbol{U}^{(0)}(\mathbf{0}) + b_U(x_2, -x_1)^T + C_1\mathbf{V}^{(1,1)}(\boldsymbol{x}) + C_2\mathbf{V}^{(1,2)}(\boldsymbol{x})$$

$$+ C_3\mathbf{V}^{(1,3)}(\boldsymbol{x}) + \frac{1}{2}\mathfrak{L}(\boldsymbol{U}^{(0)};\boldsymbol{x}) + \dots,$$

$$\mathfrak{L}(\boldsymbol{U}^{(0)};\boldsymbol{x}) = \left(\sum_{i,j=1}^{2} x_i x_j \frac{\partial^2 \boldsymbol{U}_k^{(0)}}{\partial x_i \partial x_j}(\mathbf{0})\right)_{k=1}^{2}.$$

The coefficients m_{qk} form a negative definite symmetric matrix $\boldsymbol{m} = (m_{qk})_{q,k=1}^{3}$ such that

$$\begin{aligned}
&(m_{qk})_{q,k=1}^{2} = -(N_{qk})_{q,k=1}^{2} - (2\mu\mathbf{I} + \lambda\mathbf{E})\text{mes}\omega;\\
&(m_{13}, m_{23}) = -(N_{13}, N_{23}), m_{33} = -N_{33} - 2\mu\text{mes}\omega;\\
&N_{qk} = \sum_{i,j=1}^{2}\int_{\mathbb{R}^2\setminus\omega} \varepsilon_{ij}(\mathbf{V}^{(q)};\boldsymbol{x})\sigma_{ij}(\mathbf{V}^{(k)};\boldsymbol{x})d\boldsymbol{x},
\end{aligned} \tag{5.2.5}$$

where $\mathbf{E}$ is a 2×2 matrix all of whose elements are equal to $1; \mathbf{V}^{(q)}, q = 1,2,3$, is the solution of the boundary value problem

$$\boldsymbol{L}(\frac{\partial}{\partial\boldsymbol{\xi}})\mathbf{V}^{(q)}(\boldsymbol{\xi}) = \mathbf{0}, \quad \boldsymbol{\xi}\in\mathbb{R}^2\setminus\overline{\omega};$$

$$\boldsymbol{\sigma}^{(n)}(\mathbf{V}^{(q)};\boldsymbol{\xi}) = -\boldsymbol{\sigma}^{(n)}(\mathbf{V}^{(1,q)};\boldsymbol{\xi}),\ \boldsymbol{\xi}\in\partial\omega.$$

For the function $\boldsymbol{w}^{(2)}(\boldsymbol{\xi})$ we obtain

$$\boldsymbol{L}(\frac{\partial}{\partial\boldsymbol{\xi}})\boldsymbol{w}^{(2)}(\boldsymbol{\xi}) = \mathbf{0}, \quad \boldsymbol{\xi}\in\mathbb{R}^2\setminus\overline{\omega};$$

$$\boldsymbol{\sigma}^{(n)}(\boldsymbol{w}^{(2)};\boldsymbol{\xi}) = -\frac{1}{2}\boldsymbol{\sigma}^{(n)}(\mathfrak{L}(\boldsymbol{U}^{(0)};\boldsymbol{x});\boldsymbol{\xi}), \quad \boldsymbol{\xi}\in\partial\omega.$$

As $||\boldsymbol{\xi}||\to\infty$, we have the asymptotic expansion

$$\boldsymbol{w}^{(2)}(\boldsymbol{\xi}) = T\mathfrak{C} + \mathbf{C}_0 + O(||\boldsymbol{\xi}||^{-1}),$$

where $\mathfrak{C}, \mathbf{C}_0$ are constant vectors. The results [62] yield that $\mathfrak{C}$ is given by

$$\begin{aligned}\mathfrak{C} &= [\mu\Delta \boldsymbol{U}^{(0)}(\mathbf{0}) + (\lambda+\mu)\nabla\nabla\cdot \boldsymbol{U}^{(0)}(\mathbf{0})]\text{mes}\omega \\ &= -\Lambda_0 \boldsymbol{U}^{(0)}(\mathbf{0})\text{mes}\omega.\end{aligned}$$

The quantity Λ_2 can be obtained from the solvability condition of the boundary value problem for $\boldsymbol{U}^{(2)}$,

$$\boldsymbol{L}(\frac{\partial}{\partial \boldsymbol{x}})\boldsymbol{U}^{(2)}(\boldsymbol{x}) + \Lambda_0 \boldsymbol{U}^{(2)}(\boldsymbol{x}) + \Lambda_2 \boldsymbol{U}^{(0)}(\boldsymbol{x})$$

$$+ \Lambda_0(\mathbf{D}(\boldsymbol{x}) + T(\boldsymbol{x})\mathfrak{C} + a(\ln\varepsilon)\mathfrak{C}) = \mathbf{0}, \ \boldsymbol{x} \in \Omega,$$

$$\boldsymbol{\sigma}^{(n)}(\boldsymbol{U}^{(2)}; \boldsymbol{x}) = -\boldsymbol{\sigma}^{(n)}(\mathbf{D}(\boldsymbol{x}) + T(\boldsymbol{x})\mathfrak{C}; \boldsymbol{x}), \ \boldsymbol{x} \in \partial\Omega.$$

Evaluating the integrals, we obtain

$$\Lambda_2 = \mathbf{C}^T \boldsymbol{m}\mathbf{C} + \Lambda_0 ||\boldsymbol{U}^{(0)}(\mathbf{0})||^2 \text{mes}\omega, \tag{5.2.6}$$

where $\mathbf{C} = (C_q)_{q=1}^3$. Thus, $\Lambda(\varepsilon) = \Lambda_0 + \varepsilon^2\Lambda_2 + O(\varepsilon^2)$. Note that the sign of Λ_2 depends on the relationship between $\mathbf{C}^T\boldsymbol{m}\mathbf{C}$ and $\Lambda_0||\boldsymbol{U}^{(0)}(\mathbf{0})||^2$. If the hole is close to the boundary, then the modulus of the vector $\mathbf{C}$ is small (due to the boundary condition, it is equal to zero on $\partial\Omega$). In this case $\Lambda_2 > 0$. Thus, when a small hole appears near the boundary of the domain, the fundamental frequency of oscillation increases. Assume that $\boldsymbol{U}^{(0)}(\mathbf{0}) = \mathbf{0}$ (there exists a stationary point in the domain Ω without a hole). Then, due to (5.2.6)

$$\Lambda_2 = \mathbf{C}^T\boldsymbol{m}\mathbf{C} < 0.$$

Therefore, in this case the fundamental frequency of oscillation decreases.

5.2.2 Vibration of an elastic plate with a small hole

Consider the following eigenvalue problem in the domain Ω_ε :

$$\begin{aligned}&\boldsymbol{L}(\tfrac{\partial}{\partial \boldsymbol{x}})\boldsymbol{u}(\varepsilon,\boldsymbol{x}) + \Lambda(\varepsilon)\boldsymbol{u}(\varepsilon,\boldsymbol{x}) = \mathbf{0}, \quad \boldsymbol{x} \in \Omega_\varepsilon, \\ &\boldsymbol{u}(\varepsilon,\boldsymbol{x}) = \mathbf{0}, \ \boldsymbol{x} \in \partial\omega_\varepsilon; \ \boldsymbol{\sigma}^{(n)}(\boldsymbol{u};\boldsymbol{x}) = \mathbf{0}, \ \boldsymbol{x} \in \partial\Omega.\end{aligned} \tag{5.2.7}$$

The first limiting problem

$$\boldsymbol{L}(\frac{\partial}{\partial \boldsymbol{x}})\boldsymbol{v}(\boldsymbol{x}) + \Lambda\boldsymbol{v}(\boldsymbol{x}) = \mathbf{0}, \quad \boldsymbol{x} \in \Omega; \quad \boldsymbol{\sigma}^{(n)}(\boldsymbol{v};\boldsymbol{x}) = \mathbf{0}, \quad \boldsymbol{x} \in \partial\Omega, \tag{5.2.8}$$

has triple eigenvalue $\Lambda = 0$, which corresponds to rigid-body displacements $\boldsymbol{v}^{(1)}(\boldsymbol{x}) = (1,0)^T$, $\boldsymbol{v}^{(2)}(\boldsymbol{x}) = (0,1)^T$, $\boldsymbol{v}^{(3)}(\boldsymbol{x}) = 2^{-1/2}(x_2, -x_1)^T$.

The objective is to determine the asymptotic approximations of the eigenvalues $\Lambda^{(k)}(\varepsilon)$, $k = 1,2,3$, of problem (5.2.7) which are close to $\Lambda = 0$.

Vector-valued functions $\boldsymbol{w}^{(0,k)}$, $k = 1,2,3$, of the boundary layer type satisfy the boundary value problem

$$\boldsymbol{L}(\frac{\partial}{\partial \boldsymbol{\xi}})\boldsymbol{w}^{(0,k)}(\boldsymbol{\xi}) = \boldsymbol{0}, \ \boldsymbol{\xi} \in \mathbb{R}^2 \setminus \overline{\omega}; \ \boldsymbol{w}^{(0,k)}(\boldsymbol{\xi}) = -\boldsymbol{v}^{(k)}(\boldsymbol{0}), \ \boldsymbol{\xi} \in \partial\omega. \quad (5.2.9)$$

A solution of (5.2.9) does not decay at infinity. Thus, the zero-order approximation to $\boldsymbol{u}^{(k)}(\varepsilon, \boldsymbol{x})$ cannot be provided by $\boldsymbol{v}^{(k)}(\boldsymbol{x})$, and we need to change the algorithm and the structure of the asymptotic expansion.

Note that, for $k = 3$, problem (5.2.9) has a trivial solution, and the asymptotic approximation can be constructed in a way which is similar to the previous section.

Consider $k = 1,2$. There exist two solutions of homogeneous problem (5.2.9), for which, as $||\boldsymbol{\xi}|| \to \infty$, we have the expansions $\boldsymbol{\Phi}^{(j)}(\boldsymbol{\xi}) = T^{(j)}(\boldsymbol{\xi}) + \mathfrak{C}^{(j)} + O(||\boldsymbol{\xi}||^{-1})$, $j = 1,2$, where $T^{(j)}(\boldsymbol{\xi})$ are the columns of a two-dimensional Somigliana tensor, while $\mathfrak{C}^{(j)}$ are constant vectors.

It is assumed that diam ω_ε/diam $\Omega_\varepsilon \ll 1$, and at a distance of order $O(1)$ from ω_ε the components of the vector $\boldsymbol{u}^{(k)}(\varepsilon, \boldsymbol{x})$ should be close to components of $\boldsymbol{v}^{(k)}$. Thus, we take the solution of (5.2.9) in the form

$$\boldsymbol{w}^{(0,k)}(\boldsymbol{\xi}, z) = a^{-1} z \boldsymbol{\Phi}^{(k)}(\boldsymbol{\xi}),$$

where $z = (\ln \varepsilon)^{-1}$.

Next, we seek the approximation of order $O(\varepsilon^{1-\delta})$ to $\boldsymbol{u}^{(k)}(\varepsilon, \boldsymbol{x})$ and $\Lambda^{(k)}(\varepsilon)$ in the form

$$\begin{aligned} \boldsymbol{u}^{(k)}(\varepsilon, \boldsymbol{x}) &\sim \boldsymbol{u}^{(0,k)}(\varepsilon, \boldsymbol{x}) + \boldsymbol{w}^{(0,k)}(\varepsilon^{-1}\boldsymbol{x}, z), \\ \Lambda^{(k)}(\varepsilon) &\sim \Lambda_0^{(k)}(z). \end{aligned} \quad (5.2.10)$$

Substituting (5.2.10) into (5.2.7) and neglecting terms of order $O(\varepsilon^{1-\delta})$ we obtain the following formulations for vectors $\boldsymbol{u}^{(0,k)}$:

$$\boldsymbol{L}(\frac{\partial}{\partial \boldsymbol{x}})\boldsymbol{u}^{(0,k)}(\boldsymbol{x}) = -\Lambda_0^{(k)}(z)(\boldsymbol{u}^{(0,k)}(\boldsymbol{x}) + z(z^{-1}\boldsymbol{v}^{(k)}(\boldsymbol{x})$$

$$+ a^{-1}(T^{(k)}(\boldsymbol{x}) + \mathfrak{C}^{(k)}))), \ \boldsymbol{x} \in \Omega;$$

$$\boldsymbol{\sigma}^{(n)}(\boldsymbol{u}^{(0,k)};\boldsymbol{x}) = -za^{-1}\boldsymbol{\sigma}^{(n)}(T^{(k)};\boldsymbol{x}), \ \ \boldsymbol{x} \in \partial\Omega. \tag{5.2.11}$$

Also, the solution $\boldsymbol{u}^{(0,k)}$ of (5.2.11) is assumed to satisfy the additional relation

$$\boldsymbol{u}^{(0,k)}(\mathbf{0}) = \mathbf{0}.$$

The solvability conditions for (5.2.11) yield the equality

$$\Lambda_0^{(k)}(z) = |z|a^{-1}\mathcal{R}_k^{-1},$$

where

$$\mathcal{R}_k = \text{mes}\Omega + \int_\Omega (u_k^{(0,k)}(\boldsymbol{x},z) + a^{-1}z(T_k^{(k)}(\boldsymbol{x}) + \mathfrak{C}_k^{(k)}))d\boldsymbol{x}.$$

The following representation holds (it follows from (5.2.11)),

$$\boldsymbol{u}^{(0,k)} = z\mathfrak{M}^{(k)}(z,\boldsymbol{u}^{(0,k)}),$$

where

$$\mathfrak{M}^{(k)}(t,\mathbf{W}) = \mathfrak{N}^{(k)}(a^{-1}\mathcal{R}_k^{-1}(\mathbf{W} + \boldsymbol{v}^{(k)} + ta^{-1} \times (T^{(k)}(\boldsymbol{x}) + \mathfrak{C}^{(k)})); -a^{-1}\boldsymbol{\sigma}^{(n)}(T^{(k)};\boldsymbol{x})),$$

while $\mathfrak{N}^{(k)} : (\mathfrak{F}^{(k)}, \boldsymbol{f}^{(k)}) \to \mathbf{W}$ is an operator which associates the vectors $\mathfrak{F}^{(k)}, \boldsymbol{f}^{(k)}$ with the vector $\mathbf{W}$ satisfying the boundary value problem

$$\boldsymbol{L}(\frac{\partial}{\partial \boldsymbol{x}})\mathbf{W}(\boldsymbol{x}) = \mathfrak{F}^{(k)}(\boldsymbol{x}), \ \ \boldsymbol{x} \in \Omega,$$

$$\boldsymbol{\sigma}^{(n)}(\mathbf{W};\boldsymbol{x}) = \boldsymbol{f}^{(k)}(\boldsymbol{x}), \ \ \boldsymbol{x} \in \partial\Omega,$$

and the additional condition $\mathbf{W}(\mathbf{0}) = \mathbf{0}$. The right-hand sides are assumed to satisfy the balance conditions

$$\int_\Omega \boldsymbol{v}^{(k)} \cdot \mathfrak{F}^{(k)} d\boldsymbol{x} = \int_{\partial\Omega} \boldsymbol{v}^{(k)} \cdot \boldsymbol{f}^{(k)} ds, \quad k = 1,2.$$

The solution $\boldsymbol{u}^{(0,k)}(\boldsymbol{x},z)$ is unique, and it depends analytically on z (see [56], [41]).

The following asymptotic relations are valid:

$$\boldsymbol{u}^{(0,k)}(\boldsymbol{x},z) = z\mathfrak{N}^{(k)}(a^{-1}(\text{mes}\Omega)^{-1}\boldsymbol{v}^{(k)}; -a^{-1}\boldsymbol{\sigma}^{(n)}(T^{(k)};\boldsymbol{x}))$$

$$+O(z^2), \tag{5.2.12}$$

$$\Lambda_0^{(k)}(z) = a^{-1}\{|z|(\text{mes}\Omega)^{-1} + z^2(\text{mes}\Omega)^{-2}$$

$$\times [\int_\Omega \mathfrak{N}_k^{(k)}(a^{-1}(\text{mes}\Omega)^{-1}\boldsymbol{v}^{(k)}; -a^{-1}\boldsymbol{\sigma}^{(n)}(T^{(k)};\boldsymbol{x}))d\boldsymbol{x}$$

$$+a^{-1}\int_\Omega (T_k^{(k)} + \mathfrak{C}_k^{(k)})d\boldsymbol{x}]\} + O(z^3),\ k=1,2. \tag{5.2.13}$$

Now, consider $k=3$. For the function $\boldsymbol{w}^{(1)}(\boldsymbol{\xi})$ we obtain

$$\boldsymbol{L}(\frac{\partial}{\partial\boldsymbol{\xi}})\boldsymbol{w}^{(1)}(\boldsymbol{\xi}) = \boldsymbol{0}, \quad \boldsymbol{\xi}\in\mathbb{R}^2\backslash\overline{\omega}, \quad \boldsymbol{w}^{(1)}(\boldsymbol{\xi}) = -\boldsymbol{v}^{(3)}(\boldsymbol{\xi}), \quad \boldsymbol{\xi}\in\partial\omega.$$

As $||\boldsymbol{\xi}||\to\infty$, the asymptotic formula holds,

$$\boldsymbol{w}^{(1)}(\boldsymbol{\xi}) = \Theta\boldsymbol{v}^{(3)}(\partial/\partial\boldsymbol{\xi})T(\boldsymbol{\xi}) + O(||\boldsymbol{\xi}||^{-2}),$$

with

$$\Theta = -\int_{\partial\omega}\boldsymbol{w}^{(1)}(\boldsymbol{\xi})\cdot\boldsymbol{\sigma}^{(n)}(\boldsymbol{w}^{(1)};\boldsymbol{\xi})ds, \quad \Theta<0.$$

The vector-valued function $\boldsymbol{U}^{(2)}$ satisfies the boundary value problem

$$\boldsymbol{L}(\frac{\partial}{\partial\boldsymbol{x}})\boldsymbol{U}^{(2)}(\boldsymbol{x}) + \Lambda_2^{(3)}\boldsymbol{v}^{(3)}(\boldsymbol{x}) = \boldsymbol{0}, \quad \boldsymbol{x}\in\Omega;$$

$$\boldsymbol{\sigma}^{(n)}(\boldsymbol{U}^{(2)};\boldsymbol{x}) = -\boldsymbol{\sigma}^{(n)}(\Theta\boldsymbol{v}^{(3)}(\partial/\partial\boldsymbol{x})T;\boldsymbol{x}), \quad \boldsymbol{x}\in\partial\Omega.$$

The solvability conditions yield

$$\Lambda_2^{(3)} = \Theta A^{-1},$$

where

$$A = \frac{1}{2}\int_\Omega (x_1^2 + x_2^2)d\boldsymbol{x},$$

and, therefore,

$$\Lambda^{(3)}(\varepsilon) = \varepsilon^2\Theta A^{-1} + O(\varepsilon^3). \tag{5.2.14}$$

Thus, in a small neighbourhood of the point $\Lambda = 0$ there exist three eigenvalues $\Lambda^{(k)}(\varepsilon)(k = 1,2,3)$ of problem (5.2.7), and asymptotic formulae (5.2.12), (5.2.14) hold.

5.2.3 Flexural vibration of a plate

Consider the problem of vibrations of a plate with a small hole:

$$\Delta^2 u(\varepsilon, \boldsymbol{x}) - \Lambda(\varepsilon) u(\varepsilon, \boldsymbol{x}) = 0, \quad \boldsymbol{x} \in \Omega_\varepsilon, \tag{5.2.15}$$

$$u(\varepsilon, \boldsymbol{x}) = (\partial u/\partial n)(\varepsilon, \boldsymbol{x}) = 0, \quad \boldsymbol{x} \in \partial\Omega_\varepsilon. \tag{5.2.16}$$

Here u is the deflection of the centre surface of the plate; $\Lambda(\varepsilon) = \rho h \vartheta(\varepsilon)^2 D^{-1}$, h is the thickness of the plate; D is the cylindrical rigidity; and $\vartheta(\varepsilon)$ is the vibrational frequency.

We take the solution U_s, Λ_s of the Sobolev problem as the leading order approximation for $u(\varepsilon, \boldsymbol{x})$, $\Lambda(\varepsilon)$:

$$\Delta^2 U_s(\boldsymbol{x}) - \Lambda_s U_s(\boldsymbol{x}) = 0, \quad \boldsymbol{x} \in \Omega \backslash O,$$

$$U_s(\boldsymbol{0}) = 0, \quad U_s(\boldsymbol{x}) = \frac{\partial U_s}{\partial n}(\boldsymbol{x}) = 0, \quad \boldsymbol{x} \in \partial\Omega.$$

The function U_s leaves discrepancies of orders $O(\varepsilon)$ and $O(1)$ in the boundary conditions (5.2.16). To compensate them, introduce the function $w^{(1)}$ of the boundary layer type such that

$$\Delta_\xi^2 w^{(1)}(\boldsymbol{\xi}) = 0, \quad \boldsymbol{\xi} \in \mathbb{R}^2 \backslash \overline{\omega}, \tag{5.2.17}$$

$$w^{(1)}(\boldsymbol{\xi}) = -\boldsymbol{\xi} \cdot \nabla U_s(\boldsymbol{0}), \quad \frac{\partial w^{(1)}}{\partial n}(\boldsymbol{\xi}) - \frac{\partial \boldsymbol{\xi}}{\partial n} \cdot \nabla U_s(\boldsymbol{0}), \quad \boldsymbol{\xi} \in \partial\omega. \tag{5.2.18}$$

One can specify two linearly independent solutions $W^{(j)}$, $j = 1, 2$, of the homogeneous problem (5.2.17), (5.2.18) which admit the following asymptotic representations at infinity:

$$W^{(j)}(\boldsymbol{\xi}) = \xi_j \ln \|\boldsymbol{\xi}\| + l_j(\boldsymbol{\xi}) + b_j \ln \|\boldsymbol{\xi}\| + O(\|\boldsymbol{\xi}\|^{-1}), \quad \|\boldsymbol{\xi}\| \to \infty,$$

where $l_j(\boldsymbol{\xi}) = l_j^{(1)}\xi_1 + l_j^{(2)}\xi_2 + d_j$, $(j = 1, 2)$; $l_j^{(1)}$, $l_j^{(2)}$, d_j, b_j are constants. Since the hole is small, at a distance of order $O(1)$ from ω_ε we can write that

$$|u(\varepsilon, \boldsymbol{x}) - U_s(\boldsymbol{x})| \ll 1.$$

Consequently,

$$w^{(1)}(\boldsymbol{\xi}) = -\boldsymbol{\xi} \cdot \nabla U_s(\boldsymbol{0}) + \sum_{j=1}^{2} a_j W^{(j)}(\boldsymbol{\xi}),$$

where the coefficients a_j are determined from the conditions of decay at infinity and admit the representation

$$a_j = z\frac{\partial U_s}{\partial x_j}(\mathbf{0}).$$

We seek the solution $u(\varepsilon, \boldsymbol{x})$, $\Lambda(\varepsilon)$ in the form

$$u(\varepsilon, \boldsymbol{x}) = U_s(\boldsymbol{x}) + zU^{(1)}(\boldsymbol{x}) + \varepsilon w^{(1)}(\varepsilon^{-1}\boldsymbol{x}) + O(z^2\|\boldsymbol{x}\|),$$

$$\Lambda(\varepsilon) = \Lambda_s + z\Lambda_1 + O(z^2),\ z = (\ln\varepsilon)^{-1}.$$

Here the function $U^{(1)}$ satisfies the eigenvalue problem

$$\Delta^2 U^{(1)}(\boldsymbol{x}) - \Lambda_c U^{(1)}(\boldsymbol{x}) - \Lambda_1 U_s(\boldsymbol{x}) - \Lambda_c V(\boldsymbol{x}) = 0,\ \boldsymbol{x} \in \Omega, \qquad (5.2.19)$$

$$U^{(1)}(\boldsymbol{x}) = -V(\boldsymbol{x}),\ (\partial U^{(1)}/\partial n)(\boldsymbol{x}) = -(\partial V/\partial n)(\boldsymbol{x}),\ \boldsymbol{x} \in \partial\Omega, \quad (5.2.20)$$

where $V(\boldsymbol{x}) = \sum_{j=1}^{2}(\partial U_s/\partial x_j)(\mathbf{0})(x_j \ln\|\boldsymbol{x}\| + l_j(\boldsymbol{x}))$. The solvability condition for (5.2.19), (5.2.20) yields $\Lambda_1 = 4\pi\|\nabla U_s(\mathbf{0})\|^2$, and hence $\Lambda(\varepsilon) = \Lambda_c + 4\pi z\|\nabla U_s(\mathbf{0})\|^2 + O(z^2)$. Thus, the fundamental frequency of oscillation $\Lambda(\varepsilon)$ of a plate with a fastened hole is greater than the frequency Λ_s of a plate with a stationary point.

In addition, it follows from Rayleigh's principle that the vibrational frequency Λ_0 of a plate without a hole is less than Λ_c. As a result, we obtain the relation $\Lambda_0 < \Lambda_c < \Lambda(\varepsilon)$.

5.3 Quasistatic interaction of a crack with small defects

Small imperfections (such as cavities or inclusions) exist in most engineering components and structures. Some are present from manufacture while others develop during service. The presence of a crack may induce an enhanced population of small defects near its tip, which may either amplify the stress near the crack tip or shield it. It is important, therefore, to take into account the effect of interactions between defects and cracks in calculating stress intensity factors.

The problem of the interaction between a crack and a small inclusion or cavity can be studied by means of an integral equation formulation. In this way, Rubinstein [112] performed an investigation of the stress-strain state in an elastic plane containing a crack and imperfections; results for stress-intensity factors at the crack tip were obtained by means of numerical methods for integral equations, and these were used to assess the effect of interactions between defects in an elastic body. Romalis and Tamuzh ([110], [111]) studied stress-intensity factors at the crack tip in a plane containing a group of small defects. Their papers [110], [111] use a numerical procedure together with an asymptotic analysis for integral equations. In papers [73], [74] a perturbation produced by small inhomogeneities is modelled with the aid of the integral characteristics introduced in Section 5.1.

In the present section we consider a crack in a plane containing small defects. It is known from experimental observations that the stress-intensity factor at the tip of the crack is affected by the presence of a defect or a set of defects. The value of K_I may change, and, also, the crack propagation may be accompanied by changes in direction. First, we analyze the problem on a semi-infinite crack in a plane with the finite number of defects. Here, the emphasis is made on the derivation of the crack path due to the interaction with a small defect. Next, we consider the quasistatic growth of a Mode-I crack in an elastic plane with a periodic structure. The use of the Pólya-Szegö tensor allows us to obtain explicit asymptotic formulae that can be quite efficient in mechanics of dilute composites.

5.3.1 A semi-infinite crack in a plane with a small defect

Let the isotropic elastic plane contain a semi-infinite crack

$$M_\varepsilon(a) = \{\boldsymbol{x} \in \mathbb{R}^2 : x_1 \le a,\ x_2 = \varepsilon^2 h(x_1)\}, \tag{5.3.1}$$

with $0 < \varepsilon \ll 1$, and h being smooth. Assume that the elastic medium contains a small defect, which may be represented by a small cavity or an elastic inclusion. The defect occupies a region Ω_ε specified by

$$\Omega_\varepsilon = \{\boldsymbol{x} : \varepsilon^{-1}(x_1, x_2 - b) \in \Omega\}, \tag{5.3.2}$$

where b is a real quantity such that $\varepsilon \ll |b|$, and $\bar{\Omega}$ is compact in $\mathbb{R}^2$. Assume that $b = 1$, diam $\Omega = 1$.

The function h, defining the crack path, satisfies the following condition at infinity:

$$h(x_1) = O(|x_1|^{-1}), \quad x_1 \to -\infty. \tag{5.3.3}$$

The plane is subjected to the plane-strain deformation, and the displacement field $\boldsymbol{u}$ is assumed to satisfy the homogeneous Lamé system

$$\boldsymbol{L}\left(\frac{\partial}{\partial \boldsymbol{x}}\right)\boldsymbol{u}(\varepsilon, a, \boldsymbol{x}) = \boldsymbol{0}, \quad \boldsymbol{x} \in \mathbb{R}^2 \backslash (\overline{\Omega}_\varepsilon \cup M_\varepsilon(a)), \tag{5.3.4}$$

and the homogeneous traction boundary condition

$$\sigma_i^{(n)} := \sigma_{ij} n_j = 0, \quad \boldsymbol{x} \in M_\varepsilon(a), \tag{5.3.5}$$

where n_j, $j = 1, 2$, are components of the unit normal vector with respect to M_ε. For the case of cavity we prescribe zero tractions on $\partial\Omega_\varepsilon$:

$$\boldsymbol{\sigma}^{(n)}(\boldsymbol{u}; \boldsymbol{x}) = \boldsymbol{0}, \quad \boldsymbol{x} \in \partial\Omega_\varepsilon. \tag{5.3.6}$$

If Ω_ε is occupied by an elastic inclusion, the corresponding field $\boldsymbol{u}^\circ$ in Ω_ε satisfies the homogeneous system

$$\boldsymbol{L}^\circ\left(\frac{\partial}{\partial \boldsymbol{x}}\right)\boldsymbol{u}^\circ(\varepsilon, a, \boldsymbol{x}) = \boldsymbol{0}, \quad \boldsymbol{x} \in \Omega_\varepsilon; \tag{5.3.7}$$

the conditions of ideal physical contact are specified by

$$\boldsymbol{u}(\boldsymbol{x}) = \boldsymbol{u}^\circ(\boldsymbol{x}), \ \boldsymbol{\sigma}^{(n)}(\boldsymbol{u}; \boldsymbol{x}) = \boldsymbol{\sigma}^{\circ,(n)}(\boldsymbol{u}^\circ; \boldsymbol{x}), \quad \boldsymbol{x} \in \partial\Omega_\varepsilon. \tag{5.3.8}$$

At infinity the displacement vector admits the following asymptotic form:

$$\boldsymbol{u}(\varepsilon, a; \boldsymbol{x}) = K_I^0(a) r^{1/2} \boldsymbol{\Phi}^I(\varphi) + O(r^{-1/2}), \ r \to \infty, \tag{5.3.9}$$

where (r, φ) are the polar coordinates with the centre $(a, 0)$, $|\varphi| \leq \pi$, and the angular part $\boldsymbol{\Phi}^I$ corresponds to the Mode-I load and is defined by the relations

$$\Phi_r^I(\varphi) = \frac{1}{4\mu\sqrt{2\pi}}\left((2\varkappa - 1)\cos\frac{\varphi}{2} - \cos\frac{3\varphi}{2}\right),$$

$$\Phi_\varphi^I(\varphi) = \frac{1}{4\mu\sqrt{2\pi}}\left(\sin\frac{3\varphi}{2} - (2\varkappa + 1)\sin\frac{\varphi}{2}\right), \tag{5.3.10}$$

with $\varkappa = (\lambda + 3\mu)(\lambda + \mu)^{-1}$. The coefficient K_I^0 is the Mode-I stress-intensity factor corresponding to the unperturbed stress state (elastic plane with a semi-infinite straight crack). Let $K_I^\varepsilon(a)$, $K_{II}^\varepsilon(a)$ denote the Mode-I and Mode-II stress intensity factors for the crack $M_\varepsilon(a)$. The unperturbed stress state corresponds to the Mode-I load. Consequently, $K_{II}^0 = 0$. The criteria of Erdogan and Sih (see [23]) of quasistatic crack growth implies

$$K_{II}^\varepsilon(a) = 0. \tag{5.3.11}$$

It turns out that the Mode-II stress-intensity factor admits the asymptotic form

$$K_{II}^\varepsilon(a) = \varepsilon^2 k_{II}(a) + O(\varepsilon^4), \quad \varepsilon \to 0. \tag{5.3.12}$$

The objective is to obtain the quantity k_{II}. Here, we use the integral characteristics of a defect described above.

The displacement field outside a neighbourhood of Ω_ε is specified by

$$\boldsymbol{u}(\varepsilon, a; \boldsymbol{x}) \sim \boldsymbol{v}(a; \boldsymbol{x}) + \varepsilon^2 \boldsymbol{w}(a; \boldsymbol{x}), \tag{5.3.13}$$

where

$$\boldsymbol{v}(a; \boldsymbol{x}) \sim K_I^0(a) r^{1/2} \boldsymbol{\Phi}^I(\varphi), \tag{5.3.14}$$

and it corresponds to the unperturbed stress state in an elastic plane with a semi-infinite straight crack.

The field (5.3.14) leaves a discrepancy in traction boundary conditions (5.3.6) (or (5.3.8)) on $\partial\Omega_\varepsilon$, which should be compensated for by the boundary layer constructed in the vicinity of the defect. The boundary layer decays at infinity, and it was described in Section 5.1. The perturbation term $\varepsilon^2 \boldsymbol{w}$ satisfies the following nonhomogeneous system:

$$\boldsymbol{L}\left(\frac{\partial}{\partial \boldsymbol{x}}\right) \boldsymbol{w}(a, \boldsymbol{x}) = -\sum_{j,k=1}^{3} l_j(\boldsymbol{v}; \boldsymbol{x}^0) m_{jk}(\Omega)$$

$$\times \boldsymbol{V}^{(k)}\left(\frac{\partial}{\partial \boldsymbol{x}}\right) \delta(\boldsymbol{x} - \boldsymbol{x}^0), \boldsymbol{x} \in \mathbb{R}^2 \backslash M_0(a), \tag{5.3.15}$$

where δ is the Dirac function, $\boldsymbol{x}^0$ is the defect centre, $l_j(\boldsymbol{v}; \boldsymbol{x}^0) = (\boldsymbol{V}^{(j)}(\frac{\partial}{\partial \boldsymbol{x}}))^T \boldsymbol{v}(\boldsymbol{x})\big|_{\boldsymbol{x}=\boldsymbol{x}^0}$; the definition of the Pólya-Szegö matrix $(m_{jk})_{j,k=1}^3$ was presented in Section 5.1. The displacement field $\boldsymbol{v}$ leaves a discrepancy $O(\varepsilon^4 r^{-3/2})$ in the boundary condition on the crack surface, and the term $\varepsilon^2 \boldsymbol{w}$ should satisfy the homogeneous traction boundary condition on the limit surface M_0,

$$\sigma_{12}(\boldsymbol{v};\boldsymbol{x}) = \sigma_{22}(\boldsymbol{v};\boldsymbol{x}) = 0, \quad \boldsymbol{x} \in M_0(a). \tag{5.3.16}$$

It should be noted that the polar coordinate systems, related to the limit contour $M_0(a)$ and to the curved crack $M_\varepsilon(a)$, are different. To change the system, one has to make the translation $(0, \varepsilon h(a))$ and the rotation by the small angle $\varepsilon h'(a)$. The asymptotic analysis of local stability of a crack with a small perturbation of the contour was presented by Rice [108] and by Movchan, Nazarov and Polyakova [75]. These results show that the field $\boldsymbol{w}$ is singular at $(a, 0)$:

$$\boldsymbol{w}(a;\boldsymbol{x}) = -\frac{1+\varkappa}{4\mu} h(a) K_I^0(a) r^{-1/2} \boldsymbol{\Psi}^{II}(\varphi) + k_I(a) r^{1/2} \boldsymbol{\Phi}^I(\varphi)$$

$$+ (k_{II}(a) - \frac{1}{2} K_I^0(a) h'(a)) r^{1/2} \boldsymbol{\Phi}^{II}(\varphi) + O(r|\log r|), \quad \text{as } r \to 0. \tag{5.3.17}$$

The angular part $\boldsymbol{\Phi}^{II}$ (it corresponds to the Mode-II opening) has the components

$$\Phi_r^{II}(\varphi) = \frac{1}{4\mu\sqrt{2\pi}} \left\{ (1-2\varkappa)\sin\frac{\varphi}{2} + 3\sin\frac{3\varphi}{2} \right\},$$

$$\Phi_\varphi^{II}(\varphi) = \frac{1}{4\mu\sqrt{2\pi}} \left\{ 3\cos\frac{3\varphi}{2} - (1+2\varkappa)\cos\frac{\varphi}{2} \right\}. \tag{5.3.18}$$

The vector $\boldsymbol{\Psi}^{II}$ represents the angular part of the Mode-II weight function (see Bueckner [17], Maz'ya and Plamenevskii [62]):

$$\Psi_r^{II}(\varphi) = \frac{1}{\sqrt{8\pi}(1+\varkappa)} \left\{ (1+2\varkappa)\sin\frac{3\varphi}{2} - \sin\frac{\varphi}{2} \right\},$$

$$\Psi_\varphi^{II}(\varphi) = \frac{1}{\sqrt{8\pi}(1+\varkappa)} \left\{ (2\varkappa-1)\cos\frac{3\varphi}{2} - \cos\frac{\varphi}{2} \right\}. \tag{5.3.19}$$

Using the weight function

$$\boldsymbol{\zeta}^{II}(a;\boldsymbol{x}) = r^{-1/2} \boldsymbol{\Psi}^{II}(\varphi) \tag{5.3.20}$$

for the calculation of the Mode-II stress-intensity factor, one can derive that

$$k_{II} = K_I^0(a)[\frac{1}{2}h'(a) - \sum_{j,k=1}^{3} m_{jk}(\Omega)$$

$$\times\, l_j(r^{-1/2}\Psi^{II}; \boldsymbol{x}^0) l_k(r^{1/2}\Phi^I; \boldsymbol{x}^0)]. \qquad (5.3.21)$$

With the use of (5.3.11), (5.3.12) and (5.3.21) the equation for the crack deflection can be written in the form

$$h'(a) = 2\sum_{j,k=1}^{3} m_{jk}(\Omega) l_j(\zeta^{II}; \boldsymbol{x}^0) l_k(\boldsymbol{z}^I; \boldsymbol{x}^0) \qquad (5.3.22)$$

with

$$\boldsymbol{z}^I = r^{1/2}\boldsymbol{\Phi}^I(\varphi).$$

The solution of (5.3.22), which vanishes as $a \to -\infty$, is given by

$$h(a) = -4\mu(1+\varkappa)^{-1}(a(1+a^2)^{-1/2}J(\varphi_a) + J(0)), \qquad (5.3.23)$$

where

$$\varphi_a = \cos^{-1}(-a(1+a^2)^{-1/2}),\, J(\varphi) = \boldsymbol{l}^{(1)}(\varphi)\cdot \boldsymbol{m}(\Omega)\boldsymbol{l}^{(1)}(\varphi),$$

$$l_1^{(1)}(\varphi) = (4\mu)^{-1}(2\pi)^{-1/2}\cos(\varphi/2)(\varkappa - 1 - 2\sin(\varphi/2)\sin(3\varphi/2)),$$

$$l_2^{(1)}(\varphi) = (4\mu)^{-1}(2\pi)^{-1/2}\cos(\varphi/2)(\varkappa - 1 + 2\sin(\varphi/2)\sin(3\varphi/2)),$$

$$l_3^{(1)}(\varphi) = (4\mu)^{-1}\pi^{-1/2}\sin\varphi\cos(3\varphi/2).$$

Here $\boldsymbol{m}(\Omega)$ is the Pólya-Szegö matrix of the defect.

Note that the principal term $\varepsilon^2 h$ of the function H, which describes the path of the crack, satisfies the condition

$$h(0) = h(+\infty) = -4\mu(1+\varkappa)^{-1}J(0)$$

$$= -4\mu(1+\varkappa)^{-1}\boldsymbol{l}^{(1)}(0)\cdot \boldsymbol{m}\boldsymbol{l}^{(1)}(0). \qquad (5.3.24)$$

If the inclusion is sufficiently rigid (soft), then the matrix $\boldsymbol{m}$ is positive (negative) definite. Consequently, a soft inclusion will attract the crack, whereas the rigid one should act in the opposite direction. Also, one can see that the principal terms $\varepsilon^2 h$ of the asymptotic representation of the crack path deflections $h(0)$ and $h(+\infty)$ are invariant under a rotation of the set Ω_ε.

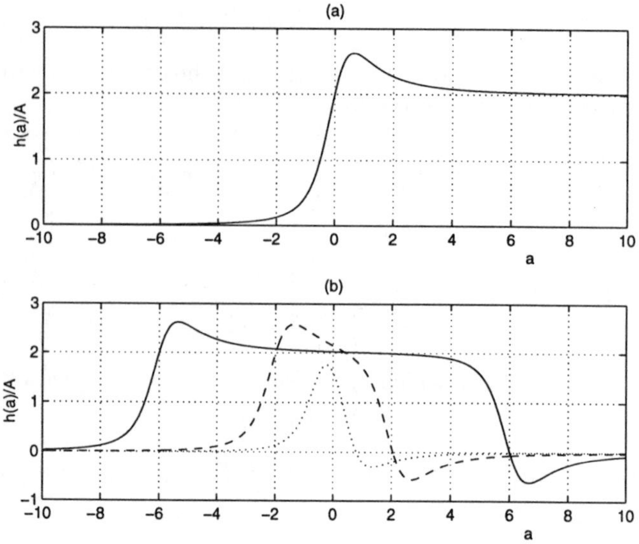

Fig. 5.1: The crack path deflection:
(a) a single circular cavity,
(b) a pair of circular cavities,
$d = 6$ (solid line), $d = 2$ (dashed line),
$d = 0.5$ (dotted line).

Using the representation (5.1.21) we can calculate the principal term of the crack path deflection for any small cavity in an elastic plane. In particular, for a circular cavity with the radius ε we obtain

$$h(a) = A(j_0(+\infty) - j_0(-a)),$$

where $j_0(t) = Z(t)(Z(t) - Z(t)^2 + 2), Z(t) = t(1+t^2)^{-1/2}, A = -(\lambda + 2\mu)^2(\lambda+\mu)^{-2}$. The plot of the function $A^{-1}h$ is given in Fig. 5.1(a).

Note that due to the linearity of the problem, the interaction between a crack and any number of defects, distances between which are much larger than ε, can be determined by superposing the interactions between the crack and each one of the defects.

Fig. 5.1 (b) shows the plots of the function $A^{-1}h(a)$, where $h(a) = A(j_0(-a+d) - j_0(-a-d))$, in the case of two circular cavities with centres at $(-d, 1), (d, -1)$ for different values of d.

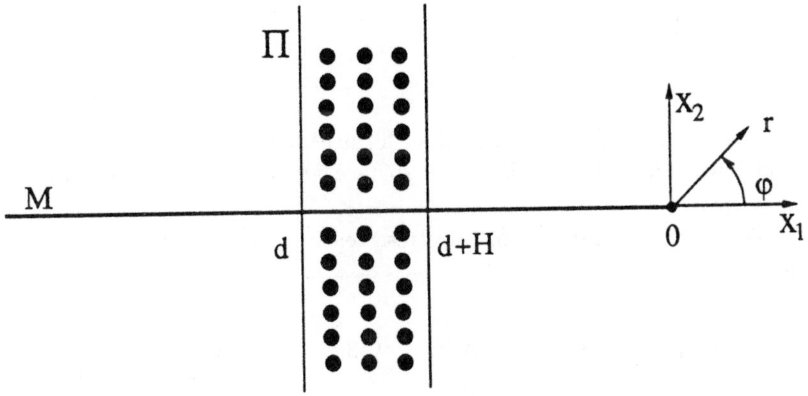

Fig. 5.2: A semi-infinite crack and a composite strip.

5.3.2 Interaction of a semi-infinite crack with a composite strip

The results of the previous section give a simple tool of analysis of stress-intensity factors for a crack in an elastic plane with a finite number of small inclusions or cavities.

Here, we show how to use the integral characteristics of defects for the stress-intensity factor analysis in a dilute periodic composite.

Let a nonhomogeneous elastic plane consist of two isotropic homogeneous half-planes and a composite strip

$$\Pi = \left\{ x \in \mathbb{R}^2 : x_1 \in (d, d+H) \right\}, \tag{5.3.25}$$

where H denotes the thickness of the strip.

A semi-infinite crack

$$M = \left\{ \boldsymbol{x} \in \mathbb{R}^2 : x_2 = 0,\ x_1 < 0 \right\} \tag{5.3.26}$$

is assumed to be perpendicular to $\partial\Pi$ (see Fig. 5.2), and the remote load is applied in such a way that the displacement vector admits the asymptotic representation

$$\boldsymbol{u}(\boldsymbol{x}) \sim K_I^0 r^{1/2} \boldsymbol{\Phi}^I(\varphi), \quad \text{as } r \to 0. \tag{5.3.27}$$

Here (r, φ) are polar coordinates with the centre at the origin. The upper and lower parts $M_\pm$ of the crack surface are specified by the equalities $\varphi = \pm\pi$.

The leading order part of the stress tensor components can be written as follows:

$$(\sigma_{rr}, \sigma_{\varphi\varphi}, \sqrt{2}\sigma_{r\varphi}) \sim K_I^0 r^{-1/2} (32\pi)^{-1/2} \left(5\cos\frac{\varphi}{2} - \cos\frac{3\varphi}{2}, \right.$$

$$\left. 3\cos\frac{\varphi}{2} + \cos\frac{3\varphi}{2},\ \sqrt{2}(\sin\frac{\varphi}{2} + \sin\frac{3\varphi}{2}) \right), \quad \text{as } r \to \infty. \tag{5.3.28}$$

We are studying the Mode-I crack, and the intensity of remote load is characterized by the coefficient K_I^0.

Consider the composite region Π. Let Q_h denote the square with the side h and with the centre at the origin. Assume that a simply connected closed set g belongs to Q_1, and introduce the following notation:

$$g_h = \left\{ \boldsymbol{x} : h^{-1}\boldsymbol{x} \in g \subset Q_1 \right\}. \tag{5.3.29}$$

Assuming that H/h is integer and $H/h \gg 1$, cover the strip Π by small squares $Q_h^{(m,n)}$ obtained by means of shifts of Q_h,

$$mhe^{(1)} + nhe^{(2)}, \quad m, n \in \mathbb{Z}, \tag{5.3.30}$$

along the Ox_1 and Ox_2 axes. Introduce the notation

$$G_h = \bigcup_{m,n} g_h^{(m,n)},$$

where $g_h^{(m,n)}$ denotes the corresponding shift of the set $Q_h^{(m,n)}$, and assume that

$$\Omega_h = (\bar{\Pi}\backslash G_h) \cup (\mathbb{R}^2\backslash\Pi)$$

is occupied by isotropic linearly elastic material with the Lamé constants λ and μ, whereas G_h corresponds to a different material characterized by the Lamé constants λ°, μ°. The ideal contact conditions are prescribed on the interface boundary.

For the sake of convenience introduce the vectors

$$\boldsymbol{\sigma} = (\sigma_{11}, \sigma_{22}, \sqrt{2}\sigma_{12})^T, \ \boldsymbol{\varepsilon} = (\varepsilon_{11}, \varepsilon_{22}, \sqrt{2}\varepsilon_{12})^T \qquad (5.3.31)$$

and matrices

$$\boldsymbol{D}(\xi_1, \xi_2) = \begin{pmatrix} \xi_1 & 0 \\ 0 & \xi_2 \\ 2^{-1/2}\xi_2 & 2^{-1/2}\xi_1 \end{pmatrix},$$

$$\boldsymbol{A} = \begin{pmatrix} 2\mu+\lambda & \lambda & 0 \\ \lambda & 2\mu+\lambda & 0 \\ 0 & 0 & 2\mu \end{pmatrix}; \qquad (5.3.32)$$

the quantities, related to elastic inclusions, will have the superscript "∘."

The homogeneous equilibrium equations and linear constitutive relations can be written in the form

$$\boldsymbol{D}^T\left(\frac{\partial}{\partial \boldsymbol{x}}\right)\boldsymbol{\sigma}(\boldsymbol{x}) = \boldsymbol{0}, \quad \boldsymbol{x} \in \Omega_h\backslash M, \qquad (5.3.33)$$

$$\boldsymbol{D}^T\left(\frac{\partial}{\partial \boldsymbol{x}}\right)\boldsymbol{\sigma}^\circ(\boldsymbol{x}) = \boldsymbol{0}, \quad \boldsymbol{x} \in G_h, \qquad (5.3.34)$$

$$\boldsymbol{\sigma}(\boldsymbol{x}) = \boldsymbol{A}\boldsymbol{\varepsilon}(\boldsymbol{x}), \ \boldsymbol{x} \in \Omega_h\backslash M; \ \boldsymbol{\sigma}^\circ(\boldsymbol{x}) = \boldsymbol{A}^\circ\boldsymbol{\varepsilon}^\circ(\boldsymbol{x}), \ \boldsymbol{x} \in G_h, \qquad (5.3.35)$$

and

$$\boldsymbol{\varepsilon}(\boldsymbol{x}) = \boldsymbol{D}\left(\frac{\partial}{\partial \boldsymbol{x}}\right)\boldsymbol{u}(\boldsymbol{x}), \quad \boldsymbol{x} \in \Omega_h;$$

$$\boldsymbol{\varepsilon}^\circ(\boldsymbol{x}) = \boldsymbol{D}\left(\frac{\partial}{\partial \boldsymbol{x}}\right)\boldsymbol{u}^\circ(\boldsymbol{x}), \quad \boldsymbol{x} \in G_h. \qquad (5.3.36)$$

Here and further in the text we shall use the Cartesian components of stress and strain. The differential operator $\boldsymbol{D}\left(\frac{\partial}{\partial \boldsymbol{x}}\right)$ is defined by (5.3.32), where ξ_i should be replaced by $\partial/\partial x_i$.

The contact conditions on the interface boundary are

$$\boldsymbol{u}(\boldsymbol{x}) = \boldsymbol{u}^\circ(\boldsymbol{x}), \ \boldsymbol{D}^T(\boldsymbol{n}(\boldsymbol{x}))\boldsymbol{\sigma}(\boldsymbol{x}) = \boldsymbol{D}^T(\boldsymbol{n}(\boldsymbol{x}))\boldsymbol{\sigma}^\circ(\boldsymbol{x}), \ \boldsymbol{x} \in \partial G_h, \quad (5.3.37)$$

where $\boldsymbol{n}$ is the unit outward normal vector with respect to ∂G_h.

The homogeneous traction boundary conditions are prescribed on the crack surfaces

$$\boldsymbol{D}^T(0,1)\boldsymbol{\sigma}(\boldsymbol{x}) = \boldsymbol{0}, \quad \boldsymbol{x} \in \Gamma^\pm. \quad (5.3.38)$$

The objective is to obtain the asymptotic approximation for the stress-intensity factor K_I due to the presence of dilute composite. The composite structure is assumed to be symmetric with respect to Ox_1, and the remote load is applied in such a way that the Mode-II stress-intensity factor is equal to zero.

First, consider the case when the *composite strip is located in front of the crack*. Let $d \gg h$. In this case (see, for example, Bensoussan, Lions and Papanicolaou [12]) the solution of the boundary value problem (5.3.33)–(5.3.38) can be approximated (up to terms of order $O(h)$) by the displacement field in a plane with a strip Π occupied by the homogeneous elastic material (not necessarily isotropic) with the matrix $\boldsymbol{A}^*$ such that

$$\boldsymbol{A}^* = \boldsymbol{A} + \beta^2 \boldsymbol{m} + O(\beta^3); \quad (5.3.39)$$

here β characterizes the diameter of the defect compared to the diameter of an elementary cell, $\beta \ll 1$, and $\boldsymbol{m}$ denotes the Pólya-Szegö matrix of the inclusion g.

The equality (5.3.39) holds for the case of dilute composite. Also, assume that the averaged composite medium is orthotropic, with $A^*_{13} = A^*_{23} = 0$. The homogenized displacement field is specified by $\boldsymbol{v}, \boldsymbol{v}^*$ which satisfy the following equations:

$$\boldsymbol{L}\boldsymbol{v} = \boldsymbol{0} \text{ in } \mathbb{R}^2 \backslash (\Pi \cup M); \quad (5.3.40)$$

$$\boldsymbol{L}^*\boldsymbol{v}^* = \boldsymbol{0} \text{ in } \Pi \backslash M; \quad (5.3.41)$$

and the interface boundary conditions

$$\boldsymbol{v} = \boldsymbol{v}^*, \ \boldsymbol{D}^T(1,0)\boldsymbol{\sigma} = \boldsymbol{D}^T(1,0)\boldsymbol{\sigma}^* \text{ on } \partial\Pi. \quad (5.3.42)$$

The following notations are used:

$$\boldsymbol{\varepsilon}^* = \boldsymbol{D}\left(\frac{\partial}{\partial \boldsymbol{x}}\right)\boldsymbol{v}^*; \ \boldsymbol{\sigma}^* = \boldsymbol{A}^*\boldsymbol{\varepsilon}^*; \ \boldsymbol{L}^* = \boldsymbol{D}^T\boldsymbol{A}^*\boldsymbol{D}.$$

The solution of problem (5.3.38), (5.3.40)–(5.3.42) admits the asymptotic form in a neighbourhood of the crack tip

$$v(x) = K_I r^{1/2} \Phi^I(\varphi) + O(r),$$

$$\sigma(v; x) = K_I r^{-1/2} \Sigma^I(\varphi) + O(1), \quad \text{as } r \to +0, \qquad (5.3.43)$$

where $\Sigma^I = (32\pi)^{-1/2} \left(5\cos\frac{\varphi}{2} - \cos\frac{3\varphi}{2},\ 3\cos\frac{\varphi}{2} + \cos\frac{3\varphi}{2}, \sqrt{2}\left(\sin\frac{\varphi}{2} + \sin\frac{3\varphi}{2}\right)\right)^T$. To calculate the stress-intensity factor K_I, we use the method of weight functions introduced by Bueckner [17] and developed further by Maz'ya and Plamenevskii [62]. Take the Mode-I weight function

$$\zeta^I(x) = r^{-1/2} \Psi^I(\varphi), \qquad (5.3.44)$$

where

$$\Psi_r^I(\varphi) = -\frac{1}{(1+\varkappa)\sqrt{8\pi}} \left((2\varkappa + 1)\cos\frac{3\varphi}{2} - 3\cos\frac{\varphi}{2} \right),$$

$$\Psi_\varphi^I(\varphi) = -\frac{1}{(1+\varkappa)\sqrt{8\pi}} \left((1 - 2\varkappa)\sin\frac{3\varphi}{2} + 3\sin\frac{\varphi}{2} \right), \qquad (5.3.45)$$

and apply the Betti formula to v, v^* and ζ^I in the region

$$\Xi_\delta = \left\{ x : \delta < \|x\| < \delta^{-1} \right\} \qquad (5.3.46)$$

for a small positive quantity δ.

Since the auxiliary field ζ^I satisfies the homogeneous equilibrium equations in $(\mathbb{R}^2 \backslash \Pi) \backslash M$ and homogeneous boundary conditions (5.3.38) on $\Gamma^\pm$, the integration by parts gives

$$0 = \int_{\Xi_\delta \cap \Pi} D(\partial/\partial x) v^*(x) \cdot (A - A^*) D(\partial/\partial x) \zeta(x) dx$$

$$+ \int_{S_\delta} \sigma^{(n)}(v; x) \cdot \zeta(x) - v(x) \cdot \sigma^{(n)}(\zeta; x) ds$$

$$+ \int_{S_{1/\delta} \backslash \Pi} \sigma^{(n)}(v; x) \cdot \zeta(x) - v(x) \cdot \sigma^{(n)}(\zeta; x) ds$$

$$+\int_{S_{1/\delta}\cap\Pi} \boldsymbol{\sigma}^{*(n)}(\boldsymbol{v}^*;\boldsymbol{x})\cdot\boldsymbol{\zeta}(\boldsymbol{x}) - \boldsymbol{v}^*(\boldsymbol{x})\cdot\boldsymbol{\sigma}^{(n)}(\boldsymbol{\zeta};\boldsymbol{x})ds. \quad (5.3.47)$$

Here S_R is a circle with the centre O and radius R; $\boldsymbol{\sigma}^{(n)} = \boldsymbol{D}(\boldsymbol{n})\boldsymbol{\sigma}$. Denote the integrals in the right-hand side (5.3.47) by $I_j,\ j = 1,2,3,4$.

Direct calculations show that

$$I_2 = K_I + O(\delta^{1/2}), \quad \text{as } \delta \to 0, \qquad (5.3.48)$$

(also, see Parton and Perlin [103], p. 318).

The length of arc $S_{1/\delta}\cap\Pi$ is the quantity of order $O(1)$. As $\delta\to 0$, the components of vectors $\boldsymbol{\zeta}(\boldsymbol{x})$ and $\boldsymbol{v}(\boldsymbol{x})$ have the order $O(\delta^{-1/2})$ and $O(\delta^{1/2})$, respectively. Consequently, one has

$$I_3 = -K_I^0 + O(\delta^{1/2}),\ I_4 = O(\delta). \qquad (5.3.49)$$

Taking the limit $\delta\to 0$ in (5.3.47) we obtain

$$K_I = K_I^0 + \int_\Pi \boldsymbol{D}\left(\frac{\partial}{\partial \boldsymbol{x}}\right)\boldsymbol{v}^*(\boldsymbol{x})\cdot(\boldsymbol{A}^* - \boldsymbol{A})\boldsymbol{D}\left(\frac{\partial}{\partial \boldsymbol{x}}\right)\boldsymbol{\zeta}(\boldsymbol{x})d\boldsymbol{x}. \quad (5.3.50)$$

Due to (5.3.39), the following asymptotic formula holds:

$$K_I = K_I^0\left(1 + \int_\Pi r^{-2}Zd\boldsymbol{x} + O(\beta^3)\right), \qquad (5.3.51)$$

where $r^{-2}Z = \boldsymbol{D}\left(\frac{\partial}{\partial \boldsymbol{x}}\right)(r^{1/2}\boldsymbol{\Phi})\cdot(\boldsymbol{A}^* - \boldsymbol{A})\boldsymbol{D}\left(\frac{\partial}{\partial \boldsymbol{x}}\right)\boldsymbol{\zeta}$.

Let J denote the integral in (5.3.51). With the use of polar coordinates it can be rewritten in the form

$$J = \int_{-\pi/2}^{\pi/2}\left\{\int_{d/\cos\varphi}^{(d+H)/\cos\varphi} r^{-1}dr\right\}Z(\varphi)d\varphi$$

$$= 2\ln\left(\frac{d+H}{d}\right)\int_0^{\pi/2} Z(\varphi)d\varphi = -2\beta^2 I(\boldsymbol{m}), \qquad (5.3.52)$$

where

$$I(\boldsymbol{m}) = \frac{1}{32\mu\pi(1+\varkappa)}\left\{\varkappa^2(m_{11} + 2m_{12} + m_{22})\right.$$
$$\left. - 4\varkappa(m_{12} + m_{11}) + 2m_{33} + 4m_{11}\right\}.$$

Finally, the approximation (5.3.51) can be written as follows:

$$K_I = K_I^0\{1 - 2\beta^2 I(m) \ln \frac{d+H}{d} + O((\beta^3 + \frac{h}{H})(1 + \frac{H}{d}))\}. \quad (5.3.53)$$

Exercise. *For the case of a strip* Π *perforated by small cracks or circular cavities show that*

$$I(m) = -\frac{a^2}{16}(m^2 + 2m + 2),$$

where βa *is the diameter of the cavity;* $m = 1$, $m = -1$ *correspond to longitudinal and transversal cracks, respectively;* $m = 0$ *corresponds to circular cavities.*

Similar analysis can be performed for a *semi-infinite crack which has crossed the strip* Π. In this case we assume that $d + H < 0$, $|d + H| \gg h$. Equations (5.3.40)–(5.3.42) do not change. The boundary condition (5.3.38) should be replaced by

$$\boldsymbol{D}^T(0,1)\boldsymbol{\sigma}(\boldsymbol{x}) = \boldsymbol{0}, \ \boldsymbol{x} \in \Gamma^{\pm} \backslash \Pi;$$

$$\boldsymbol{D}^T(0,1)\boldsymbol{\sigma}^*(\boldsymbol{x}) = \boldsymbol{0}, \ \boldsymbol{x} \in \Gamma^{\pm} \cap \Pi. \quad (5.3.54)$$

The integral J from (5.3.51) takes the form

$$J = 2 \ln \left(\frac{|d|}{|d| - H} \right) \int_{\pi/2}^{\pi} Z(\varphi) d\varphi = 2\beta^2 I(m) \ln \left(\frac{|d|}{|d| - H} \right),$$

and the asymptotic approximation for the stress-intensity factor is given by

$$K_I \sim K_I^0 \left(1 + 2\beta^2 I(m) \ln \frac{|d|}{|d| - H} \right). \quad (5.3.55)$$

5.3.3 Semi-infinite crack with the tip located within the composite strip

Assume that

$$d < 0, \ d + H > 0 \ \text{and} \ |d| \gg h, \ H + d \gg h.$$

This problem is more complicated than the one described above. The region, occupied by a composite material, has a non-smooth boundary (the crack tip), and, therefore, the averaging procedure does not work for the evaluation of the stress-intensity factor. However, the averaged formulation can be used for description of the displacement field outside a neighbourhood of the crack tip. Up to terms of order $O(\beta^3)$ the vectors $\boldsymbol{v}, \boldsymbol{v}^*$ can be approximated as follows:

$$\boldsymbol{v}(\boldsymbol{x}) \sim K_I^0 r^{1/2} \boldsymbol{\Phi}^I(\varphi) + \beta^2 \boldsymbol{w}(\boldsymbol{x}),$$

$$\boldsymbol{v}^*(\boldsymbol{x}) \sim K_I^0 r^{1/2} \Phi^I(\varphi) + \beta^2 \boldsymbol{w}^*(\boldsymbol{x}), \tag{5.3.56}$$

where the fields $\boldsymbol{w}, \boldsymbol{w}^*$ satisfy the boundary value problem

$$\boldsymbol{L}\left(\frac{\partial}{\partial \boldsymbol{x}}\right) \boldsymbol{w} = \boldsymbol{0} \text{ in } \mathbb{R}^2 \backslash (\Pi \cup M); \tag{5.3.57}$$

$$\boldsymbol{L}^*\left(\frac{\partial}{\partial \boldsymbol{x}}\right) \boldsymbol{w}^* = K_I^0 \boldsymbol{D}^T\left(\frac{\partial}{\partial \boldsymbol{x}}\right) \boldsymbol{F} \text{ in } \Pi \backslash M, \tag{5.3.58}$$

$$\boldsymbol{D}^T(1,0)\boldsymbol{\sigma}(\boldsymbol{w}^* - \boldsymbol{w}) = K_I^0 \boldsymbol{D}^T(1,0)\boldsymbol{F}; \ \boldsymbol{w} = \boldsymbol{w}^* \text{ on } \partial\Pi, \tag{5.3.59}$$

$$\boldsymbol{D}^T(0,1)\boldsymbol{\sigma}(\boldsymbol{w}) = \boldsymbol{0} \text{ on } \Gamma^\pm \backslash \Pi; \tag{5.3.60}$$

$$\boldsymbol{D}^T(0,1)\boldsymbol{\sigma}(\boldsymbol{w}^*) = K_I^0 \boldsymbol{D}^T(0,1)\boldsymbol{F} \quad \text{on } \Gamma^\pm \cap \Pi, \tag{5.3.61}$$

where

$$\boldsymbol{F} = -m\boldsymbol{D}\left(\frac{\partial}{\partial \boldsymbol{x}}\right) r^{1/2} \Phi^I(\varphi).$$

Thus, the terms $\boldsymbol{w}, \boldsymbol{w}^*$ compensate the leading order discrepancy left by principal part of (5.3.56) in equations (5.3.40)–(5.3.42), (5.3.54).

To find the leading order non-smooth part of the field $\boldsymbol{w}^*$ in a neighbourhood of the crack tip, we need a particular solution

$$\boldsymbol{W}(\boldsymbol{x}) = K_I^0 r^{1/2} \boldsymbol{\Upsilon}(\varphi), \tag{5.3.62}$$

of the system (5.3.57), (5.3.58) in $\mathbb{R}^2 \backslash M$ with the boundary conditions (5.3.60), (5.3.61) on $\Gamma^\pm$.

The vector function $\boldsymbol{\Upsilon}$ satisfies a certain boundary value problem for a system of ordinary differential equations on the interval $(-\pi, \pi)$ and has the form (the detailed analysis is presented by Kondrat'ev [38])

$$\boldsymbol{\Upsilon} = \boldsymbol{\Upsilon}^0 + C_\Upsilon \boldsymbol{\Phi}; \ C_\Upsilon = \text{const}; \tag{5.3.63}$$

$$\Upsilon_r^0 = \mu^{-1}(32\pi)^{-1/2}(k_1 \cos\frac{1}{2}\varphi + k_2\cos\frac{3}{2}\varphi + k_3\cos\frac{5}{2}\varphi + k_4\cos\frac{9}{2}\varphi);$$

$$\Upsilon_\varphi^0 = \mu^{-1}(32\pi)^{-1/2}(n_1 \sin\frac{1}{2}\varphi + n_2\sin\frac{3}{2}\varphi + n_3\sin\frac{5}{2}\varphi + n_4\sin\frac{9}{2}\varphi);$$

$$k_1 = [2(3\lambda+7\mu)]^{-1}[m_{11}(5-2\varkappa) - 4\varkappa m_{12} + m_{22}(3-2\varkappa) + 4m_{33}];$$

$$k_2 = [64\mu(3\lambda^2+13\lambda\mu+14\mu^2)]^{-1}\{m_{11}[30\lambda^2+258\lambda\mu+412\mu^2$$
$$-\varkappa(33\lambda^2+190\lambda\mu+253\mu^2)] + 2m_{12}[3\lambda^2+10\lambda\mu+7\mu^2$$
$$-32\mu\varkappa(\lambda+2\mu)] + m_{22}[-36\lambda^2-150\lambda\mu-170\mu^2+\varkappa(33\lambda^2$$
$$+126\lambda\mu+125\mu^2)] + 2m_{33}(3\lambda^2+42\lambda\mu+71\mu^2)\};$$

$$k_3 = [128\mu(\lambda+2\mu)]^{-1}\{m_{11}[3(5\lambda+17\mu) - 2\varkappa(5\lambda+13\mu)]$$
$$-2(5\lambda+13\mu)(m_{12}+m_{33}) + m_{22}[2\varkappa(5\lambda+13\mu) - 5(\lambda+5\mu)]\};$$

$$k_4 = (3\lambda+7\mu)[128\mu(\lambda+2\mu)]^{-1}[2(m_{12}+m_{33}) - m_{11} - m_{22}];$$

$$n_1 = 0;\; n_2 = -k_2;\; n_3 = [128\mu(\lambda+2\mu)]^{-1}\{m_{11}$$
$$\times[2\varkappa(3\lambda+11\mu) - 9\lambda - 53\mu] + 2(m_{33}+m_{12})(3\lambda+11\mu)$$
$$+ m_{22}[3\lambda+31\mu-2\varkappa(3\lambda+11\mu)]\};$$
$$n_4 = -(\lambda+5\mu)(3\lambda+7\mu)^{-1}k_4.$$

The constant C_Υ in (5.3.63) can be chosen in such a way that the vector function $\boldsymbol{\Upsilon}$ satisfies the orthogonality condition

$$\int_{S_\delta}\left\{\zeta^I \boldsymbol{D}^T(\boldsymbol{n}(\boldsymbol{x}))\boldsymbol{A}\boldsymbol{D}\left(\frac{\partial}{\partial \boldsymbol{x}}\right) r^{1/2}\boldsymbol{\Upsilon}\right.$$
$$\left. - r^{1/2}\boldsymbol{\Upsilon}\boldsymbol{D}^T(\boldsymbol{n}(\boldsymbol{x}))\boldsymbol{A}\boldsymbol{D}\left(\frac{\partial}{\partial \boldsymbol{x}}\right)\zeta^I\right\} ds$$
$$+\int_{S_\delta}\zeta^I \boldsymbol{D}^T(\boldsymbol{n}(\boldsymbol{x}))\boldsymbol{m}\boldsymbol{D}\left(\frac{\partial}{\partial \boldsymbol{x}}\right) r^{1/2}\boldsymbol{\Phi} ds = 0. \qquad (5.3.64)$$

Direct calculations give

$$C_\Upsilon = [128\mu(3\lambda^3+16\lambda^2\mu+27\lambda\mu^2+14\mu^3)]^{-1}[-m_{11}(51\lambda^3$$
$$+125\lambda^2\mu+81\lambda\mu^2+71\mu^3) + 2m_{12}(51\lambda^3+205\lambda^2\mu+117\lambda\mu^2-41\mu^3)$$

$$- m_{22}(51\lambda^3 + 285\lambda^2\mu + 593\lambda\mu^2 + 423\mu^3)$$

$$+ 2m_{33}(3\lambda^3 + 29\lambda^2\mu + 49\lambda\mu^2 + 23\mu^3)].$$

In particular, for the case of a strip perforated by small cavities (see exercise in the previous section),

$$C_\Upsilon = \pi a^2[8(3\lambda + 7\mu)]^{-1}[m^2(5\lambda + 9\mu) + 2m(5\lambda + 11\mu) + 19\mu + 11\lambda].$$

Introduce the vector-valued functions $\boldsymbol{y}$ and $\boldsymbol{y}^*$ specified by

$$\boldsymbol{y} = \boldsymbol{w} \text{ in } \mathbb{R}^2\backslash(\Pi \cup M); \;\; \boldsymbol{y}^* = \boldsymbol{w}^* - \boldsymbol{W} \text{ in } \Pi\backslash M.$$

The field $\boldsymbol{y}^*$ satisfies the homogeneous equations (5.3.58), (5.3.61), and, consequently,

$$\boldsymbol{y}^* = K_I^0 k r^{1/2}\boldsymbol{\Phi}^I(\varphi) + O(r), \text{ as } r \to 0.$$

Using the Bueckner method and taking into account the contact conditions on $\partial\Pi$ we obtain

$$k = -2\lim_{R\to\infty}\int_0^R \{r^{-1/2}\boldsymbol{\Psi}^I(\varphi)\cdot \boldsymbol{D}^T(1,0)(\boldsymbol{A}\boldsymbol{D}(\partial/\partial\boldsymbol{x})r^{1/2}\boldsymbol{\Upsilon}(\varphi)$$

$$+ \boldsymbol{m}\boldsymbol{D}(\partial/\partial\boldsymbol{x})r^{1/2}\boldsymbol{\Phi}^I(\varphi)) - r^{1/2}$$

$$\times\boldsymbol{\Upsilon}(\varphi)\cdot \boldsymbol{D}^T(1,0)\boldsymbol{A}\boldsymbol{D}(\partial/\partial\boldsymbol{x})r^{-1/2}\boldsymbol{\Psi}^I(\varphi)\}\Big|_{x_1=d}^{x_1=d+H} dx_2. \qquad (5.3.65)$$

Integration by parts and equality (5.3.64) give

$$k = -\lim_{\substack{R\to\delta\\ \delta\to 0}}\int_{\Pi_R\backslash\{D_\delta\cup M\}} \boldsymbol{D}\left(\frac{\partial}{\partial\boldsymbol{x}}\right)\boldsymbol{\zeta}^I\cdot \boldsymbol{m}\boldsymbol{D}\left(\frac{\partial}{\partial\boldsymbol{x}}\right) r^{1/2}\boldsymbol{\Phi}^I d\boldsymbol{x}$$

$$= [32\pi\mu(\lambda^3 + 4\lambda^2\mu + 5\lambda\mu^2 + 2\mu^3)]^{-1}\{\ln|d(d+H)^{-1}|$$

$$\times\; [m_{11}(\lambda^3 - \lambda^2\mu - \lambda\mu^2 + \mu^3) + m_{22}(\lambda^3 + 7\lambda^2\mu + 15\lambda\mu^2 + 9\mu^3)$$

$$+2m_{12}(\lambda\mu^2 + 3\mu^3 - \lambda^3 - 3\lambda^2\mu) + 2m_{33}(\lambda+\mu)^3]$$

$$+\; \pi[m_{22}(3\lambda^3 + 17\lambda^2\mu + 29\lambda\mu^2 + 15\mu^3) + m_{11}(3\lambda^3 + \lambda^2\mu - 3\lambda\mu^2 - \mu^3)$$

$$+ 2m_{12}(\mu^3 - 3\lambda^3 - 9\lambda^2\mu - 5\lambda\mu^2) - 6m_{33}(\lambda+\mu)^3]\}. \qquad (5.3.66)$$

For the case of a strip Π, perforated by small cavities, the expression (5.3.66) is simple:

$$k = -\frac{a^2}{8}\left\{\ln\left|\frac{d}{d+H}\right|(m^2+2m+2)+\pi(m^2+4m+1)\right\}$$

(the values of m are chosen in the same way as in the exercise of the previous section).

Now, we can write a simple asymptotic approximation for the vector function $\boldsymbol{w}^*$

$$\boldsymbol{w}^*(\boldsymbol{x}) \sim K_I^0 r^{1/2}\left\{\boldsymbol{\Upsilon}(\varphi)+k\boldsymbol{\Phi}^I(\varphi)\right\}, \quad \text{as } r \to 0. \qquad (5.3.67)$$

5.3.4 Boundary layer near the crack tip

We noted that the approximation (5.3.56), (5.3.67) can be used outside a small neighbourhood of the crack tip. However, as one approaches the point O, the crack should be treated as a cut in a material with the Hooke matrix $\boldsymbol{A}$ (corresponding to unperturbed state). Thus, the displacement field should have the form

$$\boldsymbol{u} \sim K_I r^{1/2}\boldsymbol{\Phi}^I(\varphi) + O(r), \qquad (5.3.68)$$

which is different from (5.3.67).

To obtain the stress-intensity factor K_I, introduce nondimensional variables

$$\boldsymbol{\xi} = (\xi_1, \xi_2) = \left(\frac{x_1}{h}+\tau,\ \frac{x_2}{h}\right),$$

where $\tau \in [0,1)$ characterizes the location of the crack tip, and $d = -h(q+\tau); q \in \mathbb{Z}$. The geometry of the composite structure with a crack is shown in Fig. 5.3.

Taking the limit $h \to 0$, one has the boundary value problem corresponding to the displacement field near a semi-infinite crack

$$M_\tau = \{\boldsymbol{\xi} : \xi_2 = 0, \xi_1 < \tau\}$$

in a nonhomogeneous elastic plane with a periodic system of defects (elastic inclusions or cavities):

$$\boldsymbol{L}\left(\frac{\partial}{\partial \boldsymbol{\xi}}\right)\boldsymbol{z}(\boldsymbol{\xi}) = \boldsymbol{0}, \ \boldsymbol{\xi} \in \mathbb{R}^2 \backslash (G_1 \cup M_\tau), \qquad (5.3.69)$$

$$L^{\circ}\left(\frac{\partial}{\partial\boldsymbol{\xi}}\right)\boldsymbol{z}^{\circ}(\boldsymbol{\xi})=\mathbf{0},\ \boldsymbol{\xi}\in G_1, \tag{5.3.70}$$

$$\boldsymbol{z}(\boldsymbol{\xi})-\boldsymbol{z}^{\circ}(\boldsymbol{\xi}),\ \boldsymbol{D}^T(\boldsymbol{n}(\boldsymbol{\xi}))\boldsymbol{\sigma}(\boldsymbol{z};\boldsymbol{\xi})=\boldsymbol{D}^T(\boldsymbol{n}(\boldsymbol{\xi}))\boldsymbol{\sigma}^{\circ}(\boldsymbol{z}^{\circ};\boldsymbol{\xi}),$$

$$\boldsymbol{\xi}\in\partial G_1, \tag{5.3.71}$$

$$\boldsymbol{D}^T(0,1)\boldsymbol{\sigma}(\boldsymbol{z};\boldsymbol{\xi})=\mathbf{0},\ \boldsymbol{\xi}\in\Gamma_\tau^{\pm}. \tag{5.3.72}$$

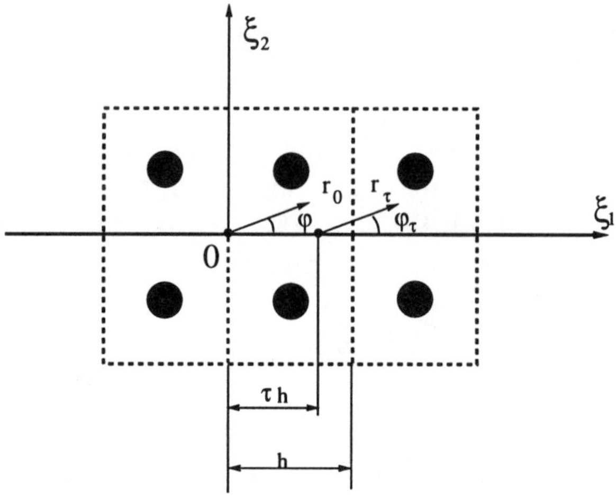

Fig. 5.3: A composite region corresponding to the boundary layer.

We also have to set certain conditions at infinity to provide matching of the boundary layer with the "outer solution."

As usual (see, for example, Bakhvalov and Panasenko [8], or Nazarov [92]), the fields $\boldsymbol{z}, \boldsymbol{z}^{\circ}$ are characterized at infinity by the asymptotic relations

$$\boldsymbol{z}(\boldsymbol{\xi}) = \text{const}\left\{\boldsymbol{X}(\boldsymbol{\xi})+\boldsymbol{X}^{(1)}(\boldsymbol{\xi};\boldsymbol{\xi})+\boldsymbol{X}^{(2)}(\boldsymbol{\xi};\boldsymbol{\xi})\right\}+\ldots,$$

$$\boldsymbol{z}^{\circ}(\boldsymbol{\xi}) = \text{const}\left\{\boldsymbol{X}(\boldsymbol{\xi})+\boldsymbol{X}^{\circ,(1)}(\boldsymbol{\xi};\boldsymbol{\xi})+\boldsymbol{X}^{\circ,(2)}(\boldsymbol{\xi};\boldsymbol{\xi})\right\}+\ldots,$$

$$\text{as } ||\boldsymbol{\xi}||\to\infty. \tag{5.3.73}$$

Here,

$$\boldsymbol{X}(\boldsymbol{\xi}) = ||\boldsymbol{\xi}||^{\Lambda}\boldsymbol{F}(\varphi),$$

$$\boldsymbol{X}^{(j)}(\boldsymbol{\xi};\boldsymbol{\eta}) = ||\boldsymbol{\xi}||^{\Lambda-j}\boldsymbol{F}^{(j)}(\varphi,\boldsymbol{\eta}), \quad j=1,2, \tag{5.3.74}$$

φ is the polar angle, $\varphi \in (-\pi, \pi)$; the exponent Λ and vector functions $\boldsymbol{F}, \boldsymbol{F}^{(j)}$ are to be specified. The dependence on $\boldsymbol{\eta} = (\eta_1, \eta_2)$ is assumed to be periodic with the period which is equal to 1.

Take the expressions from braces in (5.3.73) and substitute into equations (5.3.69)–(5.3.71). Let us note that when we write $\boldsymbol{X}^{(j)}(\boldsymbol{\xi};\boldsymbol{\eta})$, it is assumed that $\boldsymbol{\xi}$ is the "slow" variable, and $\boldsymbol{\eta}$ is the "fast" variable. In formulae (5.3.73) we use $\boldsymbol{\eta} = \boldsymbol{\xi}$. The differentiation with respect to the first argument reduces the exponent of power of $||\boldsymbol{\xi}||$ by 1, whereas the differentiation with respect to the second argument does not change this exponent. Collecting the terms of order $O(||\boldsymbol{\xi}||^{\Lambda-1})$, $O(||\boldsymbol{\xi}||^{\Lambda-2})$, one can obtain the boundary value problems with respect to $\boldsymbol{X}^{(1)}, \boldsymbol{X}^{(2)}$ on the unit cell Q_1. The solvability conditions for these problems allow one to obtain the equations with respect to $\boldsymbol{X}$ (this is the standard procedure in problems of homogenization).

The fields $\boldsymbol{X}^{(1)}, \boldsymbol{X}^{\circ,(1)}$ satisfy the following problem:

$$\boldsymbol{L}(\partial/\partial\boldsymbol{\eta})\boldsymbol{X}^{(1)}(\boldsymbol{\xi},\boldsymbol{\eta}) = \boldsymbol{0}; \; \boldsymbol{\eta} \in Q_1\backslash\overline{g};$$

$$\boldsymbol{L}^{\circ}(\partial/\partial\boldsymbol{\eta})\boldsymbol{X}^{\circ,(1)}(\boldsymbol{\xi},\boldsymbol{\eta}) = \boldsymbol{0}; \; \boldsymbol{\eta} \in g; \tag{5.3.75}$$

$$\boldsymbol{D}^T[\boldsymbol{n}(\boldsymbol{\eta})][\boldsymbol{\sigma}(\boldsymbol{X}^{(1)};\boldsymbol{\xi},\boldsymbol{\eta}) - \boldsymbol{\sigma}^{\circ}(\boldsymbol{X}^{\circ,(1)};\boldsymbol{\xi},\boldsymbol{\eta})$$
$$+ (\boldsymbol{A}-\boldsymbol{A}^{\circ})\boldsymbol{D}(\partial/\partial\boldsymbol{\xi})\boldsymbol{X}(\boldsymbol{\xi})] = \boldsymbol{0}, \; \boldsymbol{\eta} \in \partial g; \tag{5.3.76}$$

$$\boldsymbol{X}^{(1)}(\boldsymbol{\xi},\boldsymbol{\eta}) = \boldsymbol{X}^{\circ,(1)}(\boldsymbol{\xi},\boldsymbol{\eta}), \; \boldsymbol{\eta} \in \partial g. \tag{5.3.77}$$

We seek a solution in the class of functions, periodic in $\boldsymbol{\eta}$; $\boldsymbol{\xi}$ is treated as a parameter.

The following orthogonality conditions hold:

$$\int_{\partial g} \boldsymbol{D}^T(\boldsymbol{n}(\boldsymbol{\eta}))(\boldsymbol{A}-\boldsymbol{A}^{\circ})\boldsymbol{D}\left(\frac{\partial}{\partial\boldsymbol{\xi}}\right)\boldsymbol{X}(\boldsymbol{\xi})ds_\eta = \boldsymbol{0},$$

$$\int_{\partial g} \left(\boldsymbol{\eta} \times \boldsymbol{D}^T(\boldsymbol{n}(\boldsymbol{\eta}))(\boldsymbol{A}-\boldsymbol{A}^{\circ})\boldsymbol{D}\left(\frac{\partial}{\partial\boldsymbol{\xi}}\right)\boldsymbol{X}(\boldsymbol{\xi})\right)\cdot \boldsymbol{e}^{(3)}ds_\eta = 0.$$

The first relation corresponds to the principal force vector, the second equality corresponds to the third component of the principal moment. Thus, the problem (5.3.75)–(5.3.77) is solvable. The solution admits the form

$$\boldsymbol{X}^{(1)}(\boldsymbol{\xi},\boldsymbol{\eta}) = \Xi(\boldsymbol{\eta})\boldsymbol{D}(\partial/\partial\boldsymbol{\xi})\boldsymbol{X}(\boldsymbol{\xi});$$

$$\boldsymbol{X}^{\circ,(1)}(\boldsymbol{\xi},\boldsymbol{\eta}) = \Xi^{\circ}(\boldsymbol{\eta})\boldsymbol{D}(\partial/\partial\boldsymbol{\xi})\boldsymbol{X}(\boldsymbol{\xi}). \tag{5.3.78}$$

Here, Ξ, Ξ° are the 2×3 matrices which columns

$$\Xi^{(k)}, \Xi^{\circ,(k)}, \quad k = 1, 2,$$

satisfy the problem (5.3.75)–(5.3.77) with $\boldsymbol{D}(\partial/\partial\boldsymbol{\xi})\boldsymbol{X}(\boldsymbol{\xi})$ being replaced by $\boldsymbol{e}^{(k)}$, $k = 1, 2$.

Next, the boundary value problem for $\boldsymbol{X}^{(2)}, \boldsymbol{X}^{\circ,(2)}$ can be written as follows:

$$\boldsymbol{L}(\partial/\partial\boldsymbol{\eta})\boldsymbol{X}^{(2)}(\boldsymbol{\xi},\boldsymbol{\eta}) = -[\boldsymbol{D}^{T}(\partial/\partial\boldsymbol{\xi})\boldsymbol{A}\boldsymbol{D}(\partial/\partial\boldsymbol{\eta}) \tag{5.3.79}$$

$$+ \boldsymbol{D}^{T}(\partial/\partial\boldsymbol{\eta})\boldsymbol{A}\boldsymbol{D}(\partial/\partial\boldsymbol{\xi})]\boldsymbol{X}^{(1)}(\boldsymbol{\xi},\boldsymbol{\eta}) - \boldsymbol{L}(\partial/\partial\boldsymbol{\xi})\boldsymbol{X}(\boldsymbol{\xi}), \boldsymbol{\eta} \in Q_1\backslash\overline{g};$$

$$\boldsymbol{L}^{\circ}(\partial/\partial\boldsymbol{\eta})\boldsymbol{X}^{\circ,(2)}(\boldsymbol{\xi},\boldsymbol{\eta}) = -[\boldsymbol{D}^{T}(\partial/\partial\boldsymbol{\xi})\boldsymbol{A}^{\circ}\boldsymbol{D}(\partial/\partial\boldsymbol{\eta})$$

$$+ \boldsymbol{D}^{T}(\partial/\partial\boldsymbol{\eta})\boldsymbol{A}^{\circ}\boldsymbol{D}(\partial/\partial\boldsymbol{\xi})]\boldsymbol{X}^{\circ,(1)}(\boldsymbol{\xi},\boldsymbol{\eta}) - \boldsymbol{L}^{\circ}(\partial/\partial\boldsymbol{\xi})\boldsymbol{X}(\boldsymbol{\xi}), \boldsymbol{\eta} \in g;$$

$$\boldsymbol{X}^{(2)}(\boldsymbol{\xi},\boldsymbol{\eta}) = \boldsymbol{X}^{\circ,(2)}(\boldsymbol{\xi},\boldsymbol{\eta});\ \boldsymbol{D}^{T}(\boldsymbol{n}(\boldsymbol{\eta}))[\boldsymbol{A}\boldsymbol{D}(\partial/\partial\boldsymbol{\eta})\boldsymbol{X}^{(2)}(\boldsymbol{\xi},\boldsymbol{\eta}) \tag{5.3.80}$$

$$+ (\boldsymbol{A} - \boldsymbol{A}^{\circ})\boldsymbol{D}(\partial/\partial\boldsymbol{\xi})\boldsymbol{X}^{(1)}(\boldsymbol{\xi},\boldsymbol{\eta})$$

$$- \boldsymbol{A}^{\circ}\boldsymbol{D}(\partial/\partial\boldsymbol{\eta})\boldsymbol{X}^{\circ,(2)}(\boldsymbol{\xi},\boldsymbol{\eta})] = \boldsymbol{0},\ \boldsymbol{\eta} \in \partial g.$$

The objective is to derive conditions for solvability of (5.3.79), (5.3.80) in the class of periodic functions.

First, integrate by parts equations (5.3.79), (5.3.80)

$$I := \int_{Q_1\backslash\overline{g}} \boldsymbol{L}(\partial/\partial\boldsymbol{\eta})\boldsymbol{X}^{(2)}(\boldsymbol{\xi},\boldsymbol{\eta})d\boldsymbol{\eta} + \int_{g} \boldsymbol{L}^{\circ}(\partial/\partial\boldsymbol{\eta})\boldsymbol{X}^{\circ,(2)}(\boldsymbol{\xi},\boldsymbol{\eta})d\boldsymbol{\eta}$$

$$= -[(1 - \mathrm{mes} g)\boldsymbol{L}(\partial/\partial\boldsymbol{\xi}) + \mathrm{mes} g\boldsymbol{L}^{\circ}(\partial/\partial\boldsymbol{\xi})]\boldsymbol{X}(\boldsymbol{\xi}) + I_1 + I_2;$$

$$I_1 = \boldsymbol{D}^{T}(\partial/\partial\boldsymbol{\xi}) \int_{\partial g} (\boldsymbol{A} - \boldsymbol{A}^{\circ})\boldsymbol{D}(\boldsymbol{n}(\boldsymbol{\eta}))\boldsymbol{X}^{(1)}(\boldsymbol{\xi};\boldsymbol{\eta})ds_{\eta}$$

$$= \boldsymbol{D}^T(\partial/\partial\boldsymbol{\xi}) \int_{\partial g} (\boldsymbol{A} - \boldsymbol{A}^\circ)\boldsymbol{D}(\boldsymbol{n}(\boldsymbol{\eta}))\Xi(\boldsymbol{\eta}) ds_\eta \boldsymbol{D}(\partial/\partial\boldsymbol{\xi})\boldsymbol{X}(\boldsymbol{\xi});$$

$$I_2 = \int_{\partial g} \boldsymbol{D}^T(\boldsymbol{n}(\boldsymbol{\eta}))(\boldsymbol{A} - \boldsymbol{A}^\circ)\boldsymbol{D}(\partial/\partial\boldsymbol{\xi})\boldsymbol{X}^{(1)}(\boldsymbol{\xi};\boldsymbol{\eta}) ds_\eta. \tag{5.3.81}$$

Due to the Betti formula and the interface boundary conditions (5.3.80)

$$I = -\int_{\partial g} \boldsymbol{D}^T(\boldsymbol{n}(\boldsymbol{\eta})) \left[\boldsymbol{A}\boldsymbol{D}\left(\frac{\partial}{\partial \boldsymbol{\eta}}\right) \boldsymbol{X}^{(2)}(\boldsymbol{\xi};\boldsymbol{\eta}) \right.$$

$$\left. - \boldsymbol{A}^\circ \boldsymbol{D}\left(\frac{\partial}{\partial \boldsymbol{\eta}}\right) \boldsymbol{X}^{\circ,(2)}(\boldsymbol{\xi};\boldsymbol{\eta})\right] ds_\eta = I_2. \tag{5.3.82}$$

The integral I_1 can be rewritten in the form

$$I_1 = \boldsymbol{D}^T\left(\frac{\partial}{\partial \boldsymbol{\xi}}\right) \{\boldsymbol{A} - \boldsymbol{A}^* + \text{mes} g(\boldsymbol{A}^\circ - A)\} \boldsymbol{D}\left(\frac{\partial}{\partial \boldsymbol{\xi}}\right) \boldsymbol{X}(\boldsymbol{\xi}),$$

and then relations (5.3.81), (5.3.82) show that the problem (5.3.79), (5.3.80) has a periodic solution if and only if

$$\boldsymbol{D}^T\left(\frac{\partial}{\partial \boldsymbol{\xi}}\right) \boldsymbol{A}^*\boldsymbol{D}\left(\frac{\partial}{\partial \boldsymbol{\xi}}\right) \boldsymbol{X}(\boldsymbol{\xi}) = \boldsymbol{0}. \tag{5.3.83}$$

Taking into account the boundary conditions (5.3.72), one has

$$\boldsymbol{D}^T(0,1)\boldsymbol{A}^*\boldsymbol{D}\left(\frac{\partial}{\partial \boldsymbol{\xi}}\right) \boldsymbol{X}(\xi_1, \pm 0) = \boldsymbol{0}, \ \xi_1 < 0. \tag{5.3.84}$$

The standard procedure can be used for determining Λ and $\boldsymbol{F}$ from (5.3.83), (5.3.84) (see, for example, Kondrat'ev [38]). Direct substitution of (5.3.74) into (5.3.83), (5.3.84) and elementary calculations show that

$$\Lambda = 1/2, \ \boldsymbol{F}(\varphi) = \boldsymbol{\Phi}^I(\varphi) + \beta^2\boldsymbol{\Upsilon}(\varphi) + O(\beta^3). \tag{5.3.85}$$

Using (5.3.73), (5.3.67) and (5.3.56), we obtain the following condition at infinity for the vector field $\boldsymbol{z}$:

$$\boldsymbol{z}(\boldsymbol{\xi}) = K_I^0 \{1 + \beta^2 k + O(\beta^3)\} ||\boldsymbol{\xi}||^{1/2} \Big\{ \boldsymbol{\Phi}^I(\varphi) + \beta^2 \boldsymbol{\Upsilon}(\varphi)$$

$$+O(\beta^3)\Bigg\} + O(\|\boldsymbol{\xi}\|^{-1/2}), \text{ as } \|\boldsymbol{\xi}\| \to \infty. \tag{5.3.86}$$

The matching conditions for the "outer" solutions (5.3.56) and for the boundary layer $h^{1/2}\boldsymbol{z}(\boldsymbol{\xi})$ are satisfied.

The displacement and stress, corresponding to (5.3.69)–(5.3.72), (5.3.86), admit the asymptotic representation

$$\boldsymbol{z}(\boldsymbol{\xi}) = K_z(\tau) r_\tau^{1/2} \boldsymbol{\Phi}^I(\varphi_\tau) + O(r_\tau);$$

$$\boldsymbol{\sigma}(\boldsymbol{z};\boldsymbol{\xi}) = K_z(\tau) r_\tau^{-1/2} \boldsymbol{\Sigma}^I(\varphi_\tau) + O(1), \tag{5.3.87}$$

near the end $(\tau, 0)$ of the crack M_τ. Here, (r_τ, φ_τ) are polar coordinates with the centre at $(\tau, 0)$. As $\beta \to 0$,

$$K_I \sim K_z(\tau) = K_I^0 \Bigg\{ 1 + \beta^2 k$$

$$-\beta^2 \sum_{i,j=-\infty}^{\infty} \left[\int_{Q^{(ij)}} \boldsymbol{\varepsilon}(\boldsymbol{\zeta}^I;\boldsymbol{\xi}) \cdot \boldsymbol{m}\boldsymbol{D}\left(\frac{\partial}{\partial \boldsymbol{\xi}}\right) r_0^{1/2} \boldsymbol{\Phi}^I(\varphi_0) d\boldsymbol{\xi} \right.$$

$$\left. - \boldsymbol{\varepsilon}(\boldsymbol{\zeta}^I; \boldsymbol{q}^{(ij)}) \cdot \boldsymbol{m}\boldsymbol{D}\left(\frac{\partial}{\partial \boldsymbol{\xi}}\right) r_\tau^{1/2} \boldsymbol{\Phi}^I(\varphi_\tau)\Big|_{\boldsymbol{\xi}=\boldsymbol{q}^{(ij)}} \right]\Bigg)\Bigg\} + O(\beta^3), \tag{5.3.88}$$

where $Q^{(ij)} = \{\boldsymbol{\xi} : |\xi_1 - (i+1/2)| < 1/2,\ |\xi_2 - (j+1/2)| < 1/2\}$, $\boldsymbol{q}^{(ij)} = (i+1/2, j+1/2)$.

Solved exercise. *Prove the formula (5.3.88). For the reader's convenience we present the detailed derivation of the equality (5.3.88) for K_z.*

Represent the displacement fields $\boldsymbol{z}, \boldsymbol{z}^\circ$ in the form

$$\boldsymbol{z}(\boldsymbol{\xi}) = (K_I^0 + \beta^2 k)\{r_\tau^{1/2} \boldsymbol{\Phi}^I(\varphi_\tau)$$

$$+ \beta \sum_{i,j} \chi(\boldsymbol{\eta}) \boldsymbol{W}(\beta^{-1}\boldsymbol{\eta}) \varepsilon^{(ij)}(\tau)\} + \boldsymbol{Y}(\boldsymbol{\xi}); \tag{5.3.89}$$

$$\boldsymbol{z}^\circ(\boldsymbol{\xi}) = (K_I^\circ + \beta^2 k)\{r_\tau^{1/2} \boldsymbol{\Phi}^I(\varphi_\tau)$$

$$+ \beta \sum_{i,j} \boldsymbol{W}^\circ(\beta^{-1}\boldsymbol{\eta}) \varepsilon^{(ij)}(\tau)\} + \boldsymbol{Y}^\circ(\boldsymbol{\xi}). \tag{5.3.90}$$

Here, the summation is taken with respect to squares $Q^{(ij)}$ that cover a plane; $\boldsymbol{q}^{(ij)}$ is the centre of $Q^{(ij)}$, and

$$\boldsymbol{\eta} = \boldsymbol{\xi} - \boldsymbol{q}^{(ij)};$$

χ is a smooth cut-off function which is equal to 1 in a neighbourhood $\boldsymbol{\eta} = \boldsymbol{0}$, and it vanishes as one approaches the boundary of the cell. The notations $\boldsymbol{W}, \boldsymbol{W}^\circ$ are used for the 2×3 matrices with columns $\boldsymbol{W}^{(k)}, \boldsymbol{W}^{\circ,(k)}$ which solve the auxiliary boundary value problem in the plane with the inclusion g,

$$\boldsymbol{L}\left(\frac{\partial}{\partial \boldsymbol{x}}\right) \boldsymbol{W}(\boldsymbol{x}) = \boldsymbol{0}, \quad \boldsymbol{x} \in \mathbb{R}^2 \backslash \overline{g}; \tag{5.3.91}$$

$$\boldsymbol{L}^\circ\left(\frac{\partial}{\partial \boldsymbol{x}}\right) \boldsymbol{W}^\circ(\boldsymbol{x}) = \boldsymbol{0}, \quad \boldsymbol{x} \in g; \tag{5.3.92}$$

$$\boldsymbol{D}^T(\boldsymbol{n}(\boldsymbol{x}))\left\{\boldsymbol{\sigma}(\boldsymbol{W};\boldsymbol{x}) - \boldsymbol{\sigma}^\circ(\boldsymbol{W}^\circ;\boldsymbol{x}) + \boldsymbol{A} - \boldsymbol{A}^\circ\right\} = \boldsymbol{0}, \quad \boldsymbol{x} \in \partial g. \tag{5.3.93}$$

The vector $\boldsymbol{\varepsilon}^{(ij)}(\tau)$ of deformations is calculated as follows:

$$\boldsymbol{\varepsilon}^{(ij)}(\tau) = \boldsymbol{D}\left(\frac{\partial}{\partial \boldsymbol{\xi}}\right) r_\tau^{1/2} \boldsymbol{\Phi}^I(\varphi_\tau)\Big|_{\boldsymbol{\xi}=\boldsymbol{q}^{(ij)}}. \tag{5.3.94}$$

Near the centre $\boldsymbol{q}^{(ij)}$ of a square we can use the expansion

$$r^{1/2}\boldsymbol{\Phi}^I(\varphi_\tau) = \mathbf{C} + \boldsymbol{D}^T(\boldsymbol{\xi} - \boldsymbol{q}^{(ij)})\boldsymbol{\varepsilon}^{(ij)}(\tau) + \boldsymbol{\Theta}^{(ij)}(\boldsymbol{\xi} - \boldsymbol{q}^{(ij)}) + O(\|\boldsymbol{\xi} - \boldsymbol{q}^{(ij)}\|^3),$$

where $\mathbf{C}$ is a constant vector, and $\boldsymbol{\Theta}^{(ij)}$ denotes the second-order vector polynomial which solves the homogeneous Lamé system $\boldsymbol{L}\left(\frac{\partial}{\partial \boldsymbol{\xi}}\right)\boldsymbol{\Theta}^{(ij)} = \boldsymbol{0}$. The fields $\boldsymbol{Y}, \boldsymbol{Y}^\circ$ satisfy the boundary value problem

$$\boldsymbol{L}\left(\frac{\partial}{\partial \boldsymbol{\xi}}\right) \boldsymbol{Y}(\boldsymbol{\xi}) = -\beta^2 K_I^0$$

$$\times \sum_{i,j} \left[\boldsymbol{L}\left(\frac{\partial}{\partial \boldsymbol{\eta}}\right), \chi(\boldsymbol{\eta})\right] \left(m\boldsymbol{D}\left(\frac{\partial}{\partial \boldsymbol{\eta}}\right) \boldsymbol{T}(\boldsymbol{\eta})\right)^T \boldsymbol{\varepsilon}^{(ij)}(\tau)$$

$$+ O(\beta^3), \quad \boldsymbol{\xi} \in \mathbb{R}^2 \backslash (G_1 \cup M_\tau), \tag{5.3.95}$$

$$\boldsymbol{L}^{\circ}\left(\frac{\partial}{\partial\boldsymbol{\xi}}\right)\boldsymbol{Y}^{\circ}(\boldsymbol{\xi}) = -K_I^0\boldsymbol{L}^{\circ}\left(\frac{\partial}{\partial\boldsymbol{\xi}}\right)\sum_{i,j}\boldsymbol{\Theta}^{(ij)}(\boldsymbol{\xi}-\boldsymbol{q}^{(ij)})$$

$$+O(\beta^3),\ \boldsymbol{\xi}\in G_1, \qquad (5.3.96)$$

$$\boldsymbol{Y}(\boldsymbol{\xi}) = \boldsymbol{Y}^{\circ}(\boldsymbol{\xi}),\quad \boldsymbol{D}^T(\boldsymbol{n}(\boldsymbol{\xi}))(\boldsymbol{\sigma}(\boldsymbol{Y};\boldsymbol{\xi})-\boldsymbol{\sigma}^{\circ}(\boldsymbol{Y}^{\circ};\boldsymbol{\xi}))$$

$$= \boldsymbol{D}^T(\boldsymbol{n}(\boldsymbol{\xi}))K_I^0(\boldsymbol{A}^{\circ}-\boldsymbol{A})\boldsymbol{D}\left(\frac{\partial}{\partial\boldsymbol{\xi}}\right)\sum_{i,j}\boldsymbol{\Theta}^{(ij)}(\boldsymbol{\xi}-\boldsymbol{q}^{(ij)})$$

$$+O(\beta^3),\ \boldsymbol{\xi}\in\partial G_1, \qquad (5.3.97)$$

$$\boldsymbol{D}^T(0,1)\boldsymbol{\sigma}(\boldsymbol{Y};\boldsymbol{\xi}) = \boldsymbol{0},\ \boldsymbol{\xi}\in\Gamma_\tau^{\pm}, \qquad (5.3.98)$$

where $[\boldsymbol{L},\chi] = \boldsymbol{L}\chi - \chi\boldsymbol{L}$ *denotes the commutator of the cut-off function. Here,* $\boldsymbol{T}$ *is the Somigliana tensor.*

Introduce a "large" set

$$Q_{N,\delta} = \{\boldsymbol{\xi} : |\xi_j| < N,\ j=1,2\}\setminus(M_\tau\cup\{\boldsymbol{\xi}: r_\tau<\delta\}),$$

where $N \gg 1$, $N\in\mathbb{N}$, *and* $0<\delta\ll 1$. *It is assumed that the circle of a small radius* δ *with the centre at* $(\tau,0)$ *does not intersect any of the elastic inclusions.*

Multiply (5.3.95), (5.3.96) by the weight function $\boldsymbol{\zeta}^I$ *and integrate by parts in* $Q_{N,\delta}$*:*

$$-K_I^0\sum_{i,j}\left\{\beta^2\int_{Q_1}\boldsymbol{\zeta}^I\cdot[\boldsymbol{L},\chi]\left(\boldsymbol{m}\boldsymbol{D}\left(\frac{\partial}{\partial\boldsymbol{\xi}}\right)\boldsymbol{T}\right)^T\boldsymbol{\varepsilon}^{(ij)}(\tau)d\boldsymbol{\xi}\right.$$

$$\left.+\int_g\boldsymbol{\zeta}^I\cdot\boldsymbol{L}^{\circ}\boldsymbol{\Theta}^{(ij)}d\boldsymbol{\xi}+\int_{\partial g}\boldsymbol{\zeta}^I\cdot\boldsymbol{D}^T(\boldsymbol{n}(\boldsymbol{\xi}))(\boldsymbol{A}-\boldsymbol{A}^{\circ})\boldsymbol{D}^T\left(\frac{\partial}{\partial\boldsymbol{\xi}}\right)\boldsymbol{\Theta}^{(ij)}ds_\xi\right\}$$

$$=\int_{\partial Q_{N,\delta}}\left\{\boldsymbol{\zeta}^I\cdot\boldsymbol{\sigma}^{(n)}(\boldsymbol{Y})-\boldsymbol{Y}\cdot\boldsymbol{\sigma}^{(n)}(\boldsymbol{\zeta}^I)\right\}ds_\xi+O(\beta^3). \qquad (5.3.99)$$

The summation is restricted by those squares which centres belong to $Q_{N,\delta}$. *Let* J *denote the expression in braces in the left-hand side of (5.3.99). Then, one can write*

$$J = J_1 + J_2,$$

where

$$J_1 = \beta^2 \int_{Q_1} \boldsymbol{L}\boldsymbol{\zeta}^I \cdot \chi \left(\boldsymbol{m}\boldsymbol{D}\left(\frac{\partial}{\partial \boldsymbol{\xi}}\right)\boldsymbol{T}\right)^T \boldsymbol{\varepsilon}^{(ij)}(\tau) \tag{5.3.100}$$

$$+\boldsymbol{\zeta}^I \cdot \left(\boldsymbol{m}\boldsymbol{D}\left(\frac{\partial}{\partial \boldsymbol{\xi}}\right)(\delta(\boldsymbol{\xi})\boldsymbol{I})\right)^T \boldsymbol{\varepsilon}^{(ij)}(\tau) d\boldsymbol{\xi}$$

$$= \beta^2 \int_{Q_1} \boldsymbol{\zeta}^I \cdot \left(\boldsymbol{m}\boldsymbol{D}\left(\frac{\partial}{\partial \boldsymbol{\xi}}\right)(\delta(\boldsymbol{\xi})\boldsymbol{I})\right)^T \boldsymbol{\varepsilon}^{(ij)}(\tau) d\boldsymbol{\xi}$$

$$= -\boldsymbol{\varepsilon}(\boldsymbol{\zeta}^I; \boldsymbol{q}^{(ij)}) \cdot \boldsymbol{m}\boldsymbol{\varepsilon}^{(ij)}(\tau),$$

$$J_2 = -\int_j \boldsymbol{\Theta}^{(ij)} \cdot \boldsymbol{L}^\circ \boldsymbol{\zeta}^I d\boldsymbol{\xi} + \int_{\partial g} \boldsymbol{\Theta}^{(ij)} \cdot (\boldsymbol{\sigma}^{(n)}(\boldsymbol{\zeta}^I) - \boldsymbol{\sigma}^{\circ,(n)}(\boldsymbol{\zeta}^I)) ds_\xi$$

$$= O(\beta^3).$$

*Consider the integral in the right-hand side of (*5.3.99*). It is equal to*

$$I_N + I_\delta,$$

*where I_N is the integral over the boundary of the large square, and I_δ is the integral over a circle of a small radius δ. Using the asymptotic formula (*5.3.87*) and (*5.3.89*), (*5.3.90*) one can write*

$$I_\delta = K_z(\tau) - (K_I^0 + \beta^2 k + O(\beta^3)) + O(\delta^{1/2}). \tag{5.3.101}$$

*For calculation of the integral I_N, formulae (*5.3.89*), (*5.3.90*) and (*5.3.86*) give*

$$I_N = (K_I^0 + \beta^2 k + O(\beta^3))\{\int_{\Gamma_N} \boldsymbol{\zeta}^I \cdot \boldsymbol{D}^T(\boldsymbol{n}(\boldsymbol{\xi}))\boldsymbol{A}\boldsymbol{D}\left(\frac{\partial}{\partial \boldsymbol{\eta}}\right) \boldsymbol{X}^{(1)}\Big|_{\boldsymbol{\eta}=\boldsymbol{\xi}} ds_\xi$$

$$+\beta^2 \int_{\Gamma_N} \left(\boldsymbol{\sigma}^{(n)}(r_0^{1/2}\boldsymbol{\Upsilon}(\varphi_0))\right) \cdot \boldsymbol{\zeta}^I(\boldsymbol{\xi}) - r_0^{1/2}\boldsymbol{\Upsilon}(\varphi_0) \cdot \boldsymbol{\sigma}^{(n)}(\boldsymbol{\zeta}^I) ds_\xi$$

$$+O(N^{-1})\}. \tag{5.3.102}$$

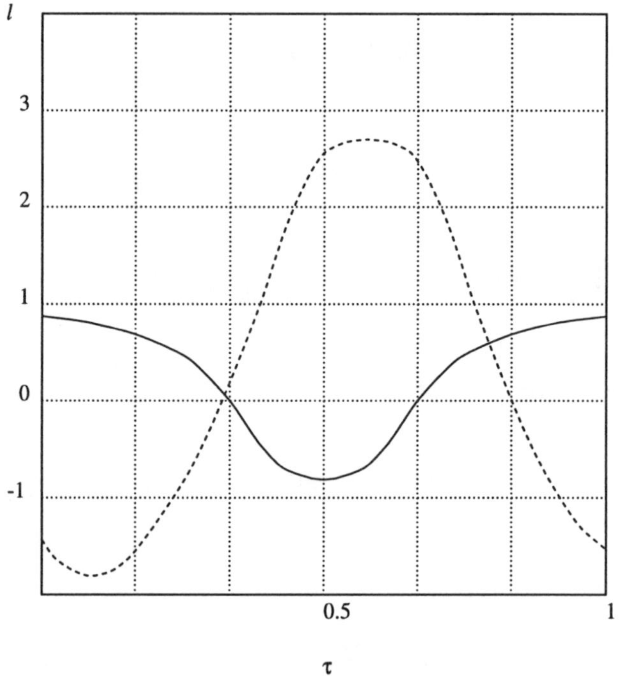

Fig. 5.4: Correction term for the stress-intensity factor:
$m = -1$ (transversal cracks, solid line),
$m = 1$ (longitudinal cracks, dashed line).

Up to terms of order $O(N^{-1})$, *the expression from the braces can be represented in the form*

$$\beta^2 \int_{\Gamma_N} r_0^{-1/2} \boldsymbol{\Psi}^I(\varphi_0) \cdot \boldsymbol{D}^T(\boldsymbol{n}(\boldsymbol{\xi})) \left\{ \boldsymbol{A}\boldsymbol{D}\left(\frac{\partial}{\partial \boldsymbol{\xi}}\right) r_0^{1/2} \boldsymbol{\Upsilon}(\varphi_0) \right.$$

$$\left. + \boldsymbol{m}\boldsymbol{D}\left(\frac{\partial}{\partial \boldsymbol{\xi}}\right) r_0^{1/2} \boldsymbol{\Phi}^I(\varphi_0) \right\} ds_\xi$$

$$- \beta^2 \int_{\Gamma_N} r_0^{1/2} \boldsymbol{\Upsilon}(\varphi_0) \cdot \boldsymbol{D}^T(\boldsymbol{n}(\boldsymbol{\xi})) \boldsymbol{\sigma}(r_0^{-1/2} \boldsymbol{\Psi}^I(\varphi_0)) ds_\xi$$

$$= \beta^2 \int_{Q_N} \{ \boldsymbol{\varepsilon}(r_0^{-1/2} \boldsymbol{\Psi}^I(\varphi_0)) \cdot (\boldsymbol{A}\boldsymbol{\varepsilon}(r_0^{1/2} \boldsymbol{\Upsilon}(\varphi_0)))$$

$$+ \boldsymbol{\varepsilon}(r_0^{-1/2} \boldsymbol{\Psi}^I(\varphi_0)) \cdot (\boldsymbol{m}\boldsymbol{\varepsilon}(r_0^{1/2} \boldsymbol{\Phi}^I(\varphi_0))) - \boldsymbol{\varepsilon}(r_0^{1/2} \boldsymbol{\Upsilon}) \cdot (\boldsymbol{A}\boldsymbol{\varepsilon}(r_0^{-1/2} \boldsymbol{\Psi}^I(\varphi_0))) \} d\boldsymbol{\xi}$$

$$= \beta^2 \int_{Q_N} \varepsilon(r_0^{-1/2} \Psi^I(\varphi_0)) \cdot (m\varepsilon(r_0^{1/2} \Phi^I(\varphi_0))) d\xi. \qquad (5.3.103)$$

Substituting (5.3.100)–(5.3.103) into (5.3.99) and taking the limit $N \to \infty$, $\delta \to 0$, one obtains the asymptotic approximation (5.3.88).

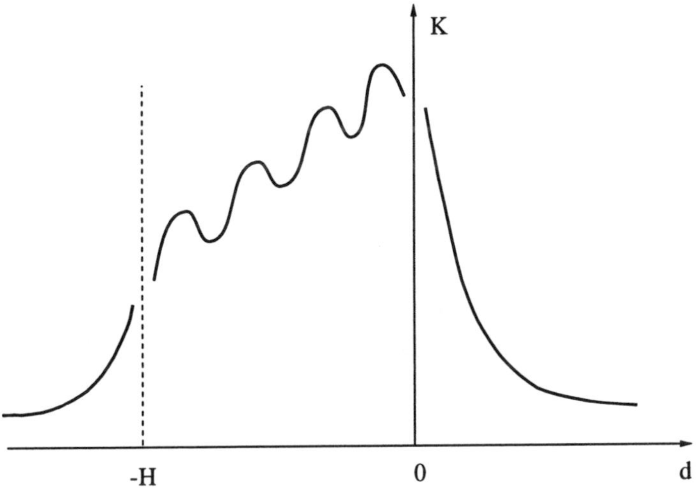

Fig. 5.5: Qualitative picture for the stress-intensity factor.

Example. The explicit asymptotic formula (5.3.88) can be rewritten in the form

$$K_z(\tau) = K_I^0(1 + \beta^2 k + \beta^2 l(\tau) + O(\beta^3)).$$

In Fig. 5.4 we present the plots of the correction term $l(\tau)$ for the case of a medium perforated by small cracks (longitudinal and transversal cracks). It can be observed that for transversal remote loading, applied at infinity, the longitudinal cracks give the greatest influence on the stress-intensity factor.

Fig 5.5 shows the qualitative behavior of the Mode–I stress intensity factor as a function of the coordinate of the crack tip. The gaps correspond to the regions, which are close to the sides of the strip perforated by small holes. First, the perforated region attracts the crack, then a certain oscillation occurs while the crack is propagating through the composite. At the end, when the array of small holes is located behind the crack front, one has the effect of unloading.

The REDUCE program was generated on the basis of present analysis. The source code, presented in the Appendix, can be used to calculate the correction term $l(\tau)$ for other types of defects (small inclusions or cavities) with the given matrix of Pólya and Szegö.

5.4 Exercises

1. Follow the algorithm of Section 5.3.1 and consider the interaction of a small micro crack and a semi–infinite crack propagating along the Ox_1 axis. Determine the perturbation of the crack path, based on the formula (5.3.23) including the Pólya-Szegö matrix (5.1.22).

2. Calculate the perturbation of the stress intensity factor for a semi–infinite crack propagating in an elastic plane perforated by a periodic system of small circular cavities. Use the formula (5.3.88) from Section 5.3.4 and the REDUCE program from the Appendix.

APPENDIX A. REDUCE program on the asymptotic algorithm in thin domains

Here we present the listing of the test program, written in REDUCE, for the asymptotic derivation of the equations of thin beams on the basis of the asymptotic algorithm presented in Section 3.3. For the sake of simplicity we consider just an elementary example of a beam with constant thickness. However, the same program works well for beams of variable thickness, and for the case of thin plates. First, we specify the space dimension and the order of the system of equations

```
N:=2$
COMMENT SPACE DIMENSION;
M:=2$
COMMENT SYSTEM ORDER;
```

Then we define a certain set of matrices

```
MATRIX (DOP(M,M),LL(M,M),MI(M,M),MO(M,M),
UU(M,1),VV(M,1),XX(M,1),N1(N,1),N2(N,1),
BB1(M,M),BB2(M,M),S1(M,1));
```

The leading components of the displacement field are assumed to depend on the longitudinal variable (variables) only.

```
FOR J:=1:M DO <<DEPEND U(J),X(1),X(2),X(3);
NODEPEND U(J),X(N+1),X(N+2)>>;
*** U declared operator
*** X declared operator
```

We also introduce a set of eigenvectors and generalized eigenvectors for the spectral problem in a cylinder

```
FOR J1:=1:M DO FOR J2:=1:M DO <<DEPEND
SV(J1,J2),X(1),X(2);
NODEPEND SV(J1,J2),X(N)>>;
*** SV declared operator
FOR J1:=1:M DO FOR J2:=1:M DO FOR J3:=1:M DO
<<DEPEND V1(J1,J2,J3),X(1),X(2),X(3);NODEPEND
V1(J1,J2,J3),X(N+1),X(N+2)>>;
*** V1 declared operator
```

The functions H1 and H2 define the profile of the upper and lower surfaces of the thin layer.

```
DEPEND HN1,X(1),X(2);
NODEPEND HN1,X(N);
H1:=P*HN1$
DEPEND HN2,X(1),X(2);
NODEPEND HN2,X(N);
H2:=P*HN2$
```

```
LET HN=(H1+H2)/P;
```

In the present example the layer is assumed to have a constant thickness

```
LET HN1=h/2,HN2=h/2;
```

Here we define components of the normal vector to the surface

```
SS:=FOR J:=1:N-1 SUM DF(H1,X(J))**2;
SS := 0
FOR J:=1:N-1 DO
N1(J,1):=-DF(H1,X(J))/SQRT(SS+1);
N1(N,1):=1/SQRT(SS+1);
N1(2,1) := 1
SS:=FOR J:=1:N-1 SUM DF(H2,X(J))**2;
SS := 0
FOR J:=1:N-1 DO
N2(J,1):=-DF(H2,X(J))/SQRT(SS+1);
N2(N,1):=-1/SQRT(SS+1);
N2(2,1) := -1
```

Description of the Jordan chains is included as well; in the present example we have two chains, and their lengths are 2 (for the longitudinal displacements) and 4 (for transversal displacements)

```
FOR ALL K1,K2,K3 SUCH THAT K2<0 LET
V1(K1,K2,K3)=0;
FOR ALL K1,K2,K3 SUCH THAT K2=0 LET
V1(K1,K2,K3)=SV(K1,K3);
FOR ALL K1,K2 SUCH THAT (K1-K2)**2>0 LET
SV(K1,K2)=0;
COMMENT ------------------------------------;
NK:=4$
ARRAY NKK(M);
NKK(1):=2$
NKK(2):=4$
```

Then we need to define the elastic moduli of the elastic medium; in the present example the material is assumed to be homogeneous and isotropic

```
ARRAY (A(M,M),A1(M,M,N),A2(M,M,N,N));
COMMENT LM:=E*NV/((1+NV)*(1-2*NV)),
MU:=E/(2*(1+NV));
A2(1,1,1,1):=A2(2,2,2,2):=LM+2*MU$
A2(1,1,2,2):=A2(2,2,1,1):=MU$
A2(1,2,1,2):=A2(2,1,2,1):=LM$
A2(1,2,2,1):=A2(2,1,1,2):=MU$
```

The following are the setting parameters for the output

```
FACTOR P;
ON RAT,DIV;
```

ON RATIONAL;

Now, we introduce the matrix differential operators of the traction boundary conditions and of the Lamé system

```
FOR ALL L,K,J LET
B(L,K,J)=A1(L,K,J)-FOR J1:=1:N SUM (A2(L,K,J,J1)
*S(J1));
*** B declared operator
FOR ALL K1,K2 LET
L1(K1,K2)=FOR J:=1:N SUM DF(B(K1,K2,J),X(J)),
L2(K1,K2)=FOR J:=1:N SUM (B(K1,K2,J)*S(J)),
L3(K1,K2)=FOR J:=1:N SUM (A1(K1,K2,J)*S(J));
*** L1 declared operator
*** L2 declared operator
*** L3 declared operator
FOR K1:=1:M DO
FOR K2:=1:M DO
LL(K1,K2):=L1(K1,K2)+L2(K1,K2)+L3(K1,K2)+A(K1,
K2)*DE;
*** S declared operator
FOR ALL L,K LET
B1(L,K)=FOR J:=1:N SUM (B(L,K,J)*N1(J,1)),
B2(L,K)=FOR J:=1:N SUM (B(L,K,J)*N2(J,1));
*** B1 declared operator
*** B2 declared operator
FOR L:=1:M DO FOR K:=1:M DO
<<BB1(L,K):=SUB(X(N)=H1,B1(L,K));
BB2(L,K):=SUB(X(N)=-H2,B2(L,K))>>;
```

One can print the following matrix forms of the differential operators

LL;

$$-\begin{pmatrix} (S(2)^2*MU+S(1)^2*(LM+2*MU)) & (S(2)*S(1))*(LM+MU) \\ (S(2)*S(1))*(LM+MU) & (S(2)^2*(LM+2*MU)+S(1)^2*MU) \end{pmatrix}$$

BB1;

$$\begin{pmatrix} -S(2)*MU & -S(1)*MU \\ -S(1)*LM & -S(2)*(LM+2*MU) \end{pmatrix}$$

BB2;

$$\begin{pmatrix} S(2)*MU & S(1)*MU \\ S(1)*LM & S(2)*(LM+2*MU) \end{pmatrix}$$

These are some auxiliary differential operators.

```
ON NAT;
FOR ALL F LET DFN(F)=DF(F,X(N))/P,
DFNN(F)=DF(F,X(N),2)/P**2;
*** DFN declared operator
*** DFNN declared operator
```

Next we introduce subroutines that allow one to deal with the symbolic form of matrix differential operators.

```
PROCEDURE CFR(F,X,J);
BEGIN SCALAR FM,FN,FF,S,FD;
FN:=NUM(F); FD:=DEN(F);
IF DEG(FD,X)>0 THEN WRITE "ERROR";
IF DEG(FD,X)>0 THEN RETURN 0;
FM:=DEG(FN,X);
IF FM=0 AND J=0 THEN RETURN FN/FD;
IF FM<J THEN RETURN 0;
ARRAY CCF(FM); COEFF(FN,X,CCF);
S:=CCF(J)/FD; RETURN S
END;
CFR
PROCEDURE DDOP1(N,M);
BEGIN SCALAR WW;
COMMENT DOP(M,M) - FORMAL REPRESENTATION
OF DIFFERENTIAL OPERATOR,
U(M)- ORIGINAL OPERATOR-FUNCTION,
N - SPACE DIMENSION,
VV(M,1) - OUTPUT VECTOR-FUNCTION;
FOR JI:=1:M DO << VV(JI,1):=0;WW:=0;
FOR J:=1:M DO << DOP(JI,J):=SUB(S(N)
=S(N)/P,DOP(JI,J));
FOR ALL K1,K2,N1,N2 LET
S(K1)**N1*S(K2)**N2=DF(U(J),X(K1),N1,X(K2),N2);
FOR ALL K1,K2 LET
S(K1)*S(K2)=DF(U(J),X(K1),1,X(K2),1);LET DE=U(J);
WW:=WW+DOP(JI,J);
FOR ALL K1 LET
S(K1)**2=DF(U(J),X(K1),2);
NN:=N-1;
FOR K1:=1:N DO WW:=SUB(S(K1)=DF(U(J),
X(K1)),WW);
VV(JI,1):=WW;
FOR ALL K1,K2,N1,N2 CLEAR
S(K1)**N1*S(K2)**N2,DE,S(K1)**2,
```

```
S(K1)*S(K2)>>;>>
END;
```
DDOP1
```
PROCEDURE MDOP(N,MT);
<<BEGIN ;
N-SPACE DIMENSION,VV(MT,1) - INPUT OPERATOR-
FUNCTION,
MT- SYSTEM DIMENSION, U(MT) - ORIGINAL
VECTOR-FUNCTION,
DOP(MT,MT)- OUTPUT DIFFERENTIAL OPERATOR
MATRIX;
FOR JI:=1:MT DO
FOR J:=1:MT DO
<< COMMENT;
FOR ALL K1,K2,N1,N2 LET
DF(U(J),X(K1),N1)=D(K1)**N1,
DF(U(J),X(K1),N1,X(K2),N2)=D(K1)*N1*D(K2)**N2,
DF(U(J),X(K1))=D(K1),DF(U(J),X(K1),X(K2))=D(K1)
*D(K2);
VW(JI):=VV(JI,1);
FOR K:=1:MT DO
IF (K-J)**2>0 THEN
VW(JI):=SUB(U(K)=0,VW(JI));
FOR K1:=1:N DO
VW(JI):=SUB(D(K1)=S(K1),VW(JI));
VW(JI):=SUB(U(J)=DE,VW(JI));
DOP(JI,J):=VW(JI);
FOR ALL K1,K2,N1,N2 CLEAR
DF(U(J),X(K1),N1),DF(U(J),X(K1),N1,X(K2),N2),
DF(U(J),X(K1)),DF(U(J),X(K1),X(K2))>> END>>;
```
MDOP
```
PROCEDURE MDOP1(N,MT);
<<BEGIN ;
COMMENT N-SPACE DIMENSION,VV(MT,1) - INPUT
OPERATOR-FUNCTION,
MT- SYSTEM DIMENSION, U(MT) - ORIGINAL
VECTOR-FUNCTION,
DOP(MT,MT)- OUTPUT DIFFERENTIAL OPERATOR
MATRIX;
FOR JI:=1:MT DO
FOR J:=1:MT DO
<< COMMENT;
FOR ALL K1,K2,N1,N2 LET
```

```
DF(SV(J,J),X(K1),N1)=D(K1)**N1,
DF(SV(J,J),X(K1),N1,X(K2),N2)=D(K1)*N1*D(K2)**N2,
DF(SV(J,J),X(K1))=D(K1),DF(SV(J,J),X(K1),X(K2))=
D(K1)*D(K2);
VW(JI):=VV(JI,1);
FOR K:=1:MT DO
IF (K-J)**2>0 THEN
VW(JI):=SUB(SV(K,K)=0,VW(JI));
COMMENT;
FOR K1:=1:N DO
VW(JI):=SUB(D(K1)=S(K1),VW(JI));
COMMENT WRITE VW(JI):=SUB(DZ=SZ,VW(JI));
VW(JI):=SUB(SV(J,J)=DE,VW(JI));
DOP(JI,J):=VW(JI);
FOR ALL K1,K2,N1,N2 CLEAR
DF(SV(J,J),X(K1),N1),DF(SV(J,J),X(K1),N1,X(K2),N2),
DF(SV(J,J),X(K1)),DF(SV(J,J),X(K1),X(K2)) >>
END>>;
```

MDOP1

```
PROCEDURE DDOP(N,M);
BEGIN SCALAR WW;
FOR JI:=1:M DO << VV(JI,1):=0;WW:=0;
FOR J:=1:M DO
<< FOR ALL K1,K2,N1,N2 LET
S(K1)**N1*S(K2)**N2=DF(U(J),X(K1),N1,X(K2),N2),
DE=U(J);
FOR ALL K1,K2 LET S(K1)*S(K2)=DF(U(J),X(K1),
1,X(K2),1);
WW:=WW+DOP(JI,J);FOR ALL K1 LET S(K1)**2=
DF(U(J),X(K1),2);
COMMENT SZ**2=DF(U(J),Z,2);
FOR K1:=1:N DO WW:=SUB(S(K1)=DF(U(J),X(K1)),
WW);
VV(JI,1):=WW;
FOR ALL K1,K2,N1,N2 CLEAR
S(K1)**N1*S(K2)**N2,DE,S(K1)**2,S(K1)*S(K2) >>; >>
END;
```

DDOP

```
DOP:=LL;
```

The following output demonstrates the change of the transversal variable and the splitting up procedure.

$$DOP :=$$

$$-\begin{pmatrix} (S(2)^2*MU+S(1)^2*(LM+2*MU)) & (S(2)*S(1))*(LM+MU) \\ (S(2)*S(1))*(LM+MU) & (S(2)^2*(LM+2*MU)+S(1)^2*MU) \end{pmatrix}$$

DDOP1(N,M);
MDOP(N,M);
**** D declared operator*
**** VW declared operator*
LL:=DOP;

$$LL :=$$

$$-\begin{pmatrix} S(1)^2*(LM+2*MU)+P^{-2}*S(2)^2*MU & (P^{-1}*S(2)*S(1))*(LM+MU) \\ (P^{-1}*S(2)*S(1))*(LM+MU) & S(1)^2*MU+(P^{-2}*S(2)^2)*(LM+2*MU) \end{pmatrix}$$

DOP:=BB1;

$$DOP := \begin{pmatrix} -S(2)*MU & -S(1)*MU \\ -S(1)*LM & -S(2)*(LM+2*MU) \end{pmatrix}$$

DDOP1(N,M);
MDOP(N,M);
BB1:=DOP;

$$BB1 := \begin{pmatrix} -P^{-1}*S(2)*MU & -S(1)*MU \\ -S(1)*LM & -(P^{-1}*S(2))*(LM+2*MU) \end{pmatrix}$$

DOP:=BB2;

$$DOP := \begin{pmatrix} S(2)*MU & S(1)*MU \\ S(1)*LM & S(2)*(LM+2*MU) \end{pmatrix}$$

DDOP1(N,M);
MDOP(N,M);
BB2:=-DOP;

$$BB2 := \begin{pmatrix} -P^{-1}*S(2)*MU & -S(1)*MU \\ -S(1)*LM & -(P^{-1}*S(2))*(LM+2*MU) \end{pmatrix}$$

ON NAT;
PROCEDURE TAYLOR(F,P,NK);
SUB(P=0,F)+FOR K:=1:NK SUM (SUB(P=0,DF(F,P, K))*PK/FOR J:=1:K PRODUCT J);**
TAYLOR

```
FOR K1:=1:M DO FOR K2:=1:M DO
<<LL(K1,K2):=TAYLOR(LL(K1,K2)*P**2,P,NK);
BB1(K1,K2):=TAYLOR(BB1(K1,K2)*P,P,NK);
BB2(K1,K2):=TAYLOR(BB2(K1,K2)*P,P,NK)>>;
PROCEDURE MTL(P,J);
BEGIN ARRAY CC(J);
IF J=0 THEN GOTO M1;
COMMENT MI(M,M) - INPUT MATRIX,
MO(M,M) - OUTPUT MATRIX;
ON RATIONAL;
ON RAT;
ON DIV; MI:=MI;
FOR JL:=1:M DO FOR JK:=1:M DO
MO(JL,JK):=CFR(MI(JL,JK),P,J) ;
COMMENT CLEAR P**(J+1);
GOTO M2;
M1: FOR JL:=1:M DO FOR JK:=1:M DO
MO(JL,JK):=SUB(P=0,MI(JL,JK));
M2: END;
MTL
OPERATOR INTG;
FOR ALL F,Z SUCH THAT DF(F,Z)=0 LET
INTG(F,Z)=F*Z;
FOR ALL N LET INTG(Z**N,Z)=Z**(N+1)/(N+1);
LET INTG(Z,Z)=Z**2/2;
LINEAR INTG;
PROCEDURE GGG1(R);
BEGIN
MATRIX GS1(M,1);
MI:=BB1;
FOR JM:=1:R DO
<< MTL(P,JM); DOP:=MO;
RM:=R-JM;
FOR J:=1:M DO
U(J):=V1(K,RM,J);
DDOP(N,M);
FOR J:=1:M DO <<
S1(J,1):=SUB(X(N)=H1/P,VV(J,1));
GS1(J,1):=GS1(J,1)-S1(J,1)>>;>>
END;
GGG1
PROCEDURE GGG2(R);
BEGIN
```

```
MATRIX GS2(M,1);
MI:=BB2;
FOR JM:=1:R DO
<< MTL(P,JM); DOP:=MO;
RM:=R-JM;
FOR J:=1:M DO
U(J):=V1(K,RM,J);
DDOP(N,M);
FOR J:=1:M DO <<
S1(J,1):=SUB(X(N)=-H2/P,VV(J,1));
GS2(J,1):=GS2(J,1)-S1(J,1)>>;>>
END;
GGG2
PROCEDURE FFF(R);
BEGIN
MATRIX FS(M,1);
MI:=LL;
FOR JM:=1:R DO <<MTL(P,JM);DOP:=MO;
FOR J:=1:M DO
U(J):=V1(K,R-JM,J);
DDOP(N,M);
FOR J:=1:M DO
FS(J,1):=FS(J,1)-VV(J,1) >>END;
FFF
PROCEDURE VP(L,F,G1,G2);
BEGIN SCALAR S;
F:=SUB(X(N)=Z,F);
G1:=SUB(X(N)=Z,G1);
G2:=SUB(X(N)=Z,G2);
LET RFS=-HN2;
IN1:=INTG((Z1-Z)*F,Z);
INT1:=IF IN1=0 THEN 0 ELSE SUB(Z=Z1,IN1)
-SUB(Z=RFS,IN1);
INT1:=IF IN1=0 THEN 0 ELSE SUB(Z1=Z,INT1);
IN2:=INTG((HN1**2/2-HN1*Z+Z**2/2)*F,Z);
INT2:=IF IN2=0 THEN 0 ELSE -HN**(-1)
*(SUB(Z=HN1,IN2)-SUB(Z=RFS,IN2));
IN3:=INTG(F,Z);
INT3:=IF IN3=0 THEN 0 ELSE
SUB(Z=HN1,IN3)-SUB(Z=RFS,IN3);
XX(L,1):=G2-G1+INT3;
S:=-SUB(P=0,Z=0,A2(L,L,N,N)**(-1))*(INT1+Z
*G2+INT2-XX(L,1)*(Z**2/2+Z*HN2+HN2*(HN2-HN1)
```

```
/3-HN1**2/6)-(HN1-HN2)*G2/2);
S:=SUB(Z=X(N),S);
RETURN S
END;
VP
FOR J1:=1:M DO FACTOR U(J1),SV(J1,J1);
FOR J1:=1:M DO FOR J3:=1:NK DO FOR J2:=1:N-1 DO
FACTOR DF(U(J1),X(J2),J3),DF(SV(J1,J1),X(J2),J3);
FOR J1:=1:M DO FACTOR DF(U(J1),X(1),1,X(2),1);
FOR KK:=1:M DO
FOR R:=1:NKK(KK) DO << LET K=KK;
WRITE "================================";
WRITE K:=K; WRITE R:=R; GGG1(R);
GGG2(R);FFF(R);
FOR LJ:=1:M DO << F:=FS(LJ,1);
G1:=GS1(LJ,1);
G2:=GS2(LJ,1); FACTOR X(1), X(2);
WRITE V1(KK,R,LJ):=VP(LJ,F,G1,G2);
REMFAC X(1),X(2)>>;
VV:=XX; MDOP1(N,M); FACTOR S(1), S(2);
WRITE DOP; REMFAC S(1), S(2) >>;
```

This is the printing which shows the generalized eigenvectors and the differential operators of equations with respect to leading order terms of the longitudinal and transversal displacements

================================

$$K := 1$$

$$R := 1$$

$$V1(1,1,1) := 0$$

$$V1(1,1,2) := -\frac{DF(SV(1,1),X(1)) * X(2) * LM}{LM + 2 * MU}$$

$$\begin{pmatrix} 0 & 0 \\ 0 & 0 \end{pmatrix}$$

The presence of the zero matrix means that the boundary value problem for the corresponding generalized eigenfunction is solvable, and one can proceed further.

================================

$$K := 1$$

$$R := 2$$

$$V1(1,2,1) :=$$

$$\frac{2 * DF(SV(1,1), X(1), 2) * X(2)^2 * (H * (LM + MU) - \frac{3}{4} * LM - MU)}{LM + 2 * MU}$$

$$+\frac{2 * DF(SV(1,1), X(1), 2) * X(2) * H * (H * (LM + MU) - LM - MU)}{LM + 2 * MU}$$

$$-\frac{(\frac{1}{6} * DF(SV(1,1), X(1), 2) * H^2) * (H * (LM + MU) - \frac{3}{4} * LM - MU)}{LM + 2 * MU}$$

$$V1(1,2,2) := 0$$

$$\begin{pmatrix} \frac{4*S(1)^2*H*MU*(LM+MU)}{LM+2*MU} & 0 \\ 0 & 0 \end{pmatrix}$$

The (11) component of the above matrix corresponds to the second-order differential operator acting on the longitudinal component of the displacement vector with the thin rectangle.
Now we look at the transversal mode of the deformation.

==

$$K := 2$$

$$R := 1$$

$$V1(2,1,1) := -DF(SV(2,2), X(1)) * X(2)$$

$$V1(2,1,2) := 0$$

==

$$K := 2$$

$$R := 2$$

$$V1(2,2,1) := 0$$

$$V1(2,2,2) := \frac{\frac{1}{2} * DF(SV(2,2), X(1), 2) * X(2)^2 * LM}{LM + 2 * MU}$$

$$-\frac{\frac{1}{24} * DF(SV(2,2), X(1), 2) * H^2 * LM}{LM + 2 * MU}$$

$$\begin{pmatrix} 0 & 0 \\ 0 & 0 \end{pmatrix}$$

==

$$K := 2$$

$$R := 3$$

$$V1(2,3,1) := \frac{\frac{1}{2} * DF(SV(2,2), X(1), 3) * X(2)^3 * (LM + \frac{4}{3} * MU)}{LM + 2 * MU}$$

$$- \frac{(\frac{11}{24} * DF(SV(2,2), X(1), 3) * X(2) * H^2) * (LM + \frac{12}{11} * MU)}{LM + 2 * MU}$$

$$V1(2,3,2) := 0$$

$$\begin{pmatrix} 0 & 0 \\ 0 & 0 \end{pmatrix}$$

==

$$K := 2$$

$$R := 4$$

$$V1(2,4,1) := 0$$

$$V1(2,4,2) :=$$

$$- \frac{(\frac{1}{8} * DF(SV(2,2), X(1), 4) * X(2)^4)}{(LM + 2 * MU)^2} * (LM^2 + \frac{8}{3} * LM * MU$$

$$+ \frac{4}{3} * MU^2) - ((\frac{1}{6} * DF(SV(2,2), X(1), 4) * X(2)^2 * H^2)$$

$$*(H * LM * MU + H * MU^2 - \frac{11}{8} * LM^2 - 3 * LM * MU$$

$$- \frac{3}{2} * MU^2))/(LM + 2 * MU)^2 - \frac{(\frac{1}{6} * DF(SV(2,2), X(1), 4) * X(2))}{(LM + 2 * MU)^2}$$

$$*H^3 * MU * (H * LM + H * MU - LM - MU) + (\frac{1}{72} * DF(SV(2,2), X(1), 4)$$

$$*H^4 * (H * LM * MU + H * MU^2 - \frac{101}{80} * LM^2 - \frac{27}{10} * LM * MU$$

$$-\frac{27}{20} * MU^2))/(LM + 2 * MU)^2$$

$$\begin{pmatrix} 0 & 0 \\ 0 & -\frac{\frac{1}{3}*S(1)^4*H^3*MU*(LM+MU)}{LM+2*MU} \end{pmatrix}$$

END;

The (22) component of the last matrix represents the fourth-order differential operator that is acting on the transversal component of the displacement vector; this is the differential operator corresponding to the well–known beam equation.

APPENDIX B. Cracks in composite media. REDUCE program

In this part of the Appendix we present the source code of the REDUCE program for the evaluation of the stress–intensity factor at the tip of a crack in a periodic composite media. The asymptotic analysis of this problem was presented in Chapter 4.

First, we introduce the operator of integration

```
OPERATOR INTGR;
FOR ALL N SUCH THAT DF(N,T)=0
LET INTGR(N,T)=N*T,INTGR(COS(T),T)=SIN(T),
INTGR(COS(N*T),T)=SIN(N*T)/N;
LINEAR INTGR;
OPERATOR PPP;
F:=2*M/(L+M);
COMMENT AL- crack length;
```

We consider an example of elastic medium perforated by small elliptical cavities of different orientation (the numerical results in Chapter 4 have presented for the case of small cracks). Here we specify components of the polarization matrix

```
COMMENT CCP:=PI*AL**2*(L+2*M)/(2*M);
FOR ALL S,M0 LET PPP(S,M0)=
<< BEGIN SCALAR SS; SS:=SUB(A11=-CCP*((L
+M)*(M0*(M0-F)+1)+M*(F-2*M0))/2,
A22=-CCP*((L+M)*(M0*(M0+F)+1)+
M*(F+2*M0))/2,A12=-CCP*((L+M)*(M0*(M0+F)+1)
-M*(F+2*M0))/2,A33=-CCP*M*F,S); RETURN SS;
END>>;
```

The following is the differential operator of the Lamé system

```
PROCEDURE LAME(M);
<< BEGIN
FD:=ROT*FF;WRITE "FD=",FD;
D1:=COS(T)*DF(FD(1,1),R)-SIN(T)*DF(FD(1,1),T)/R;
D2:=SIN(T)*DF(FD(2,1),R)+COS(T)*DF(FD(2,1),T)/R;
D3:=(COS(T)*DF(FD(2,1),R)-
SIN(T)*DF(FD(2,1),T)/R+
SIN(T)*DF(FD(1,1),R)+
COS(T)*DF(FD(1,1),T)/R)/SQRT(2);
DFD:=MAT((D1),(D2),(D3)); WRITE "DFD=",DFD;
IF M=4 THEN GOTO M3;
ADF:=AP*DFD; WRITE "ADF=", ADF;
IF M=3 THEN GOTO M3;
```

```
IF M=2 THEN GOTO M2;
IF M=1 THEN GOTO M1;
DT1:=COS(T)*DF(ADF(1,1),R)-SIN(T)*DF(ADF(1,
1),T)/R+(SIN(T)*DF(ADF(3,1),R)+COS(T)*
DF(ADF(3,1),T)/R)/SQRT(2);
DT2:=SIN(T)*DF(ADF(2,1),R)+COS(T)*DF(ADF(2,1),T)
/R+(COS(T)*DF(ADF(3,1),R)-SIN(T)*DF(ADF(3,1),T)
/R)/SQRT(2);
LR:=MAT((DT1),(DT2)); WRITE "LR=",LR; GOTO M3;
M1: DT10:=MAT((1,0,0),(0,0,1./SQRT(2)));
BR1:=DT10*ADF; WRITE "BR1=",BR1;
GOTO M3;
M2: DT01:=MAT((0,0,1/SQRT(2)),(0,1,0));
BR2:=DT01*ADF; WRITE "BR2=",BR2;
M3: END>>;
```

The elastic material is assumed to have the orthotropic symmetry. For the isotropic and orthotropic media the formulae for the asymptotics of the stress–intensity factor have been presented in Chapter 4. Also, in this part of the program we have to define some elementary rules of trigonometric transformations.

```
A:=MAT((2*M+L,L,0),
(L,2*M+L,0),
(0,0,2*M));
AA:=MAT((A11,A12,0),
(A12,A22,0),
(0,0,A33));
ROT:=MAT((COS(T),-SIN(T)),(SIN(T),COS(T)));
LET COS(PI/2)=0,SIN(PI/2)=1,
COS(3*PI/2)=0,SIN(3*PI/2)=-1,
COS(5*PI/2)=0,SIN(5*PI/2)=1,
COS(7*PI/2)=0,SIN(7*PI/2)=-1,
COS(9*PI/2)=0,SIN(9*PI/2)=1,
COS(11*PI/2)=0,SIN(11*PI/2)=-1,
COS(13*PI/2)=0,SIN(13*PI/2)=1;
FOR ALL X,Y LET
COS(X)*COS(Y)=(COS(X+Y)+COS(X-Y))/2,
SIN(X)*SIN(Y)=(COS(X-Y)-COS(X+Y))/2,
SIN(X)*COS(Y)=(SIN(X+Y)+SIN(X-Y))/2,
COS(X)**2=(1+COS(2*X))/2,
SIN(X)**2=(1-COS(2*X))/2;
```

The following part of the program provides the calculation of the stress–intensity factor for the case where the crack tip is located outside the composite strip.

```
FP:=MAT(((2*P-1)*COS(T/2)-COS(3*T/2)),
(-(2*P+1)*SIN(T/2)+SIN(3*T/2)))*CF;
PSP:=-MAT(((2*P+1)*COS(3*T/2)-3*COS(T/2)),((1-2*
P)*SIN(3*T/2)+3*SIN(T/2)))*CP;
ZFP:=R**(1/2)*FP;
ZPSP:=R**(-1/2)*PSP;
FF:=ZPSP;AP:=AA;LAME(3);DDZ:=ADF;FF:=ZFP;
LAME(4);
RR:=DFD(1,1)*DDZ(1,1)+DFD(2,1)*DDZ(2,1)+DFD(3,1)
*DDZ(3,1);
RR:=RR*R**2;
RRI:=INTGR(RR,T);
RRS:=SUB(T=PI/2,RRI)-SUB(T=0,RRI);
LET CP=1/((1+P)*SQRT(8*PI)),
CF=1/(M*SQRT(32*PI));
RRS;R1:=PPP(RRS,M0);R2:=PPP(RRS,1);
R3:=PPP(RRS,-1);R4:=PPP(RRS,0);
LET P=(L+3*M)/(L+M);RRS;R1;R2;R3;R4;
CLEAR P,CP,CF;
```

Next, we present the REDUCE code for the asymptotic analysis of the stress–strain state near the crack tip located within a composite strip. This part involves the boundary layer corresponding to a boundary value problem posed in a periodic composite plane with a semi–infinite crack.

```
IPSP:=MAT((K1*COS(T/2)+K2*COS(3*T/2)
+K3*COS(5*T/2)+K4*COS(9*T/2)),
(N1*SIN(T/2)+N2*SIN(3*T/2)+N3*SIN(5*T/2)+N4*SIN
(9*T/2)))*CF;
ZIPSP:=R**(1/2)*IPSP;ZIPSD:=ROT*ZIPSP;ZPSD:=
ROT*ZPSP;
FF:=ZIPSP;AP:=A;
LAME(3); ADSP:=ADF;
FF:=ZFP;AP:=AA;
LAME(3);AAFI:=ADF;
FF:=ZPSP;AP:=A;
LAME(1);DAPSI:=BR1;
RQ:=ADSP+AAFI;DRQ:=DT10*RQ;
RS:=(ZPSD(1,1)*DRQ(1,1)+ZPSD(2,1)*DRQ(2,1)-
ZIPSD(1,1)*DAPSI(1,1)-ZIPSD(2,1)*DAPSI(2,1))*R;
CONST:=SUB(COS(T)=0,COS(2*T)=0,COS(3*T)=0,
COS(4*T)=0,COS(5*T)=0,COS(6*T)=0,COS(7*T)=0,
RS);
RS1:=RS-CONST; LG:=LOG(DD2/DD1);
RS2:=SUB(COS(2*T)=-LG,COS(4*T)=LG,COS(6*T)=
```

```
-LG,COS(T)=PI,COS(3*T)=-PI,COS(5*T)=PI,
COS(7*T)=-PI,RS1);
KRS:=RS2+CONST*LG;
FACTOR K1,K2,K3,K4,N1,N2,N3,N4,LG,PI;KRS;
LET P=(L+3*M)/(L+M),CF=1/(M*SQRT(32*PI)),
CP=1/((1+P)*SQRT(8*PI));
KRS;
COMMENT ———————————————————————————-;
FF:=ZIPSP; AP:=A; LAME(0);
LIP1:=LR(1,1);LIP2:=LR(2,1);
LAME(2);BIP1:=BR2(1,1);BIP2:=BR2(2,1);
FF:=ZFP;AP:=AA;LAME(0);
LFI1:=LR(1,1);LFI2:=LR(2,1);
LAME(2);BFI1:=BR2(1,1);BFI2:=BR2(2,1);
ARRAY LP1(10),LP2(10),LF1(10),LF2(10),
LS1(10),LS2(10);
ROT1:=TP(ROT);
LIPP:=ROT1*MAT((LIP1),(LIP2));LFIP:=ROT1*MAT
((LFI1),(LFI2));
FOR ALL K LET COS(K*T/2)=TC**K,SIN(K*T/2)
=TS**K,
COS(T/2)=TC,SIN(T/2)=TS;FACTOR TC,TS;
LIP1:=LIPP(1,1);LIP2:=LIPP(2,1);LFI1:=LFIP(1,1);
LFI2:=LFIP(2,1);
NLIP1:=NUM(LIP1);NLIP2:=NUM(LIP2);
NLFI1:=NUM(LFI1);NLFI2:=NUM(LFI2);
DLIP1:=LIP1/NLIP1;DLIP2:=LIP2/NLIP2;
DLFI1:=LFI1/NLFI1;DLFI2:=LFI2/NLFI2;
COEFF(NLIP1,TC,LP1);COEFF(NLFI1,TC,LF1);
COEFF(NLIP2,TS,LP2);COEFF(NLFI2,TS,LF2);
FOR K:=0:9 DO
BEGIN WRITE LP1(K)*DLIP1; WRITE LF1(K)*DLFI1;
WRITE LP2(K)*DLIP2; WRITE LF2(K)*DLFI2;
WRITE LS1(K):=LP1(K)*DLIP1+LF1(K)*DLFI1;
WRITE LS2(K):=LP2(K)*DLIP2+LF2(K)*DLFI2; END;
FOR ALL K CLEAR COS(K*T/2),SIN(K*T/2),
COS(T/2),SIN(T/2);
BS1:=SUB(T=PI,BIP1+BFI1);
BS2:=SUB(T=PI,BIP2+BFI2);
COMMENT CALCULATIONS OF COEFFICIENTS
K1,N1,K2,N2;
BI:=MAT((((2*M+L)*(-1/2)+L)*PSP(1,1)
+L*DF(PSP(2,1),T)),
```

```
(M*DF(PSP(1,1),T)+M*(-3/2)*PSP(2,1)));
BP:=MAT((((2*M+L)*(1/2)+L)*IPSP(1,1)+L*DF(IPSP
(2,1),T)),(M*DF(IPSP(1,1),T)+M*(-1/2)*IPSP(2,1)));
RZR:=BP(1,1)*PSP(1,1)+BP(2,1)*PSP(2,1)-BI(1,1)
*IPSP(1,1)-BI(2,1)*IPSP(2,1);
RZR1:=INTGR(RZR,T);RZR:=SUB(T=PI,RZR1)-SUB
(T=-PI,RZR1);
SOLVE(LST(LS1(5),LS2(5)),K3,N3);K3;N3; K3:=
SOLN(1,1);
N3:=SOLN(1,2);
SOLVE(LST(LS1(9),LS2(9)),K4,N4);K4;N4; K4:=
SOLN(1,1);
N4:=SOLN(1,2);
SOLVE(LST(LS1(1),LS1(3),BS1,RZR),K1,N1,K2,N2);
K1;N1;K2;N2;
K1:=SOLN(1,1);N1:=SOLN(1,2);K2:=SOLN(1,3);N2:=
SOLN(1,4);
K1N:=SUB(R=1,K1);K2N:=SUB(R=1,K2);
K3N:=SUB(R=1,K3);K4N:=SUB(R=1,K4);
N1N:=SUB(R=1,N1);N2N:=SUB(R=1,N2);
N3N:=SUB(R=1,N3);N4N:=SUB(R=1,N4);
KRS:=SUB(K1=K1N,K2=K2N,K3=K3N,K4=K4N,
N1=N1N,N2=N2N,N3=N3N,N4=N4N,KRS);
COMMENT--------------------------------------;
COMMENT STRESS INTENSITY FACTOR
INCREMENT INSIDE THE COMPOSITE STRIP;
FF:=ZFP;AP:=AA;LAME(3);ADF1:=ADF;
FF:=ZPSP;AP:=AA;LAME(4);RSI:=TP(DFD)*ADF1;
RSIM:=PPP(RSI(1,1),M0);
ARRAY CCF(2);
LET COS(4*T)=CST4;NRS:=NUM(RSI(1,1));DRS:=
RSI(1,1)/NRS;
COEFF(NRS,CST4,CCF);CCF(1):=CCF(1)*DRS;CLEAR
COS(4*T);
CCF(1):=SUB(R=1,CCF(1));
LET C=1;RSJ:=KRS+2*CCF(1)-PI*CCF(1)/2;RSJM:=
PPP(RSJ,M0);
LET P=(L+3*M)/(L+M);LET L=V/((1+V)*(1-2*V)),
M=1/(2*(1+V));
RSJM;RSIM;
```

On the final step we generate the FORTRAN code for further numerical calculations.

```
ON FORT;OUT FORTIL;
```

```
WRITE " FUNCTION RSI(R,T,M0,V)";
RSI:=RSIM;
WRITE " RETURN";
WRITE " END";
WRITE " FUNCTION RSJ(M0,V,DD1,DD2)";
RSJ:=SUB(LOG(DD2/DD1)=ALOG(DD2/DD1),RSJM);
WRITE " RETURN";
WRITE " END";
WRITE " FUNCTION RRS(M0,V)";
RRS:=R1;
WRITE " RETURN";
WRITE " END";
SHUT FORTIL; OFF FORT;END;
```

We do not present the complete listing of the results produced by this program, because it is quite long. The reader can try to run this program and to generalize the numerical results presented in Chapter 4 for the case of elliptical cavities in orthotropic media.

References

[1] Arutyunyan, N. Kh. and Abramyan, B. L., *Torsion of Elastic Bodies.* Fizmatgiz, Moscow, 1963 (in Russian).

[2] Arutyunyan, N. Kh., Movchan, A. B. and Nazarov, S. A., Correct formulations of Lekhnitskii problems. *Prikl. Mat. Mekh. (J. Appl. Math. Mech.),* 50(2), 237, 1986.

[3] Arutyunyan, N. Kh., Movchan, A. B. and Nazarov, S. A., Behaviour of solutions of elasticity problems in noncompact domains with parabolic and cylindrical inclusions or cavities. *Adv. Mech.,* 10(4), 3, 1987.

[4] Arutyunyan, N. Kh., Movchan, A. B. and Nazarov, S. A., Asymptotic interpretation of the Lekhnitskii problem. *Zh. Prikl. Mekh. Tekhn. Fiziki (J. Appl. Mech. Tech. Phys.),* 5, 123, 1989.

[5] Arutyunyan, N. Kh. and Nazarov, S. A., On singularities of the stress function at the angular points of the cross-section of a rod with a thin surface layer. *Prikl. Mat. Mekh. (J. Appl. Math. Mech.),* 47(1), 122, 1982.

[6] Atkin, R. J. and Fox, N., *An Introduction to the Theory of Elasticity.* Longman, London, 1980.

[7] Babich, V. M. and Ivanov, M. I., Long waves asymptotics in a scattering problems in elastic media. *Zap. Nauchn. Sem. Leningrad. Otdel. Mat. Inst. Steklov. (LOMI) (J. Soviet Math.),* 156, 3, 1986.

[8] Bakhvalov, N. S. and Panasenko, G. P., *Homogenization in Periodic Media.* Nauka, Moscow, 1984 (in Russian).

[9] Barenblatt, G. I., On the equilibrium cracks occured due to brittle

fracture, *Appl. Math. Mech. (PMM)*. 23(3), 434, 1959.

[10] Bazant, Z. P., Three–dimensional harmonic function near termination of intersection of gradient singularity method. *Int. J. Eng. Sci.*, 12(3), 221, 1974.

[11] Bazant, Z. P. and Keer, L. M., Singularities of elastic stresses and of harmonic functions at conical notches or inclusions. *Int. J. Solids Struct.*, 10(9), 957, 1974.

[12] Bensoussan, A., Lions, J.- L. and Papanicolaou, G., *Asymptotic Methods in Periodic Structures.* North-Holland, Amsterdam, 1978.

[13] Benthem, J. P., State of stress at the vertex of a quarter–infinite crack in a half–space. *Int. J. Solids Struct.*, 13(5), 479, 1977.

[14] Benthem, J. P., A quarter–infinite crack in a half–space; alternative and additional solutions. *Int. J. Solids Struct.*, 16(2), 119, 1980.

[15] Benthem, J. P. and Koiter, W. T., Asymptotic approximation to crack problems, in *Methods of Analysis and Solutions of Crack Problems.* Noordhoff, Holland, 1973.

[16] Berezhnitskii, L. T., Delyavskii, M. V. and Panasyuk, V. V., *Flexure of Thin Lates with Cracks.* Naukova Dumka, Kiev, 1979 (in Russian).

[17] Bueckner, H. F., A novel principle for the computation of stress intensity factor. *Z. Angew. Math. Mech.*, 50, 529, 1970.

[18] Caillerie, D., The effect of a thin inclusion of high rigidity in an elastic body. *Math. Meth. Appl. Sci.*, 2(3), 251, 1980.

[19] Cherepanov, G. P., *Mechanics of Brittle Failure.* Nauka, Moscow, 1974; Engl. transl., McGraw-Hill, 1980.

[20] Cherepanov, G. P., Kocharov, R.C. and Sotkilava, O.V., Elastic plane with a parabolic inclusion, in *Phyz. Chim. Processy Gorn. Proizv.* Moscow, 751, 1976.

[21] Craggs, I. W., On the propagation of a crack in elastic-brittle material. *J. Mech. Phys. Solids*, 8, 66, 1961.

[22] Dzhavadov, M. G., Asymptotics of solutions of a boundary value problem for second order elliptic equations in thin domains. *Differentzial'nye Uravneniya (Differential Equations)*, 4(10), 1901, 1968.

[23] Erdogan, F. and Sih, G. C., *J. Basic Eng.*, 85(4), 519, 1963.

[24] Eshelby, J., *Continuum Dislocation Theory.* Izd. Inostr. Lit., Moscow, 1963.

[25] Fedoryuk, M. V., Asymptotics of the solution of the Dirichlet problem for the Laplace and Helmholtz equations in the exterior of a thin cylinder. *Izv. Acad. Nauk SSSR. Ser. Mat. (Math. USSR-Izv.),* 45(1), 167, 1981.

[26] Galin, L. A., *Contact Problems of the Theory of Elasticity.* Gostekhizdat, Moscow, 1953 (in Russian).

[27] Geer, J. F. and Keller, J. B., Uniform asymptotic solutions for potential flow around a thin airfoil and the electrostatic potential about a thin conductor. *SIAM J. Appl. Math.*, 16(1), 75, 1968.

[28] Goldenveizer, A. L., The approximate theory of shells with the asymptotic integration of equations of the theory of elasticity. *Prikl. Mat. Mekh. (J. Appl. Math. Mech.),* 27(4), 593, 1963.

[29] Griffith, A. A., The phenomena of rupture and flow in solids. *Philos. Trans. R. Soc. London,* A221, 163, 1921.

[30] Griffith, A. A., The theory of rupture, in *Proc. First Int. Congr. Appl. Mech.*, Delft, 1924.

[31] Hartranft, R. J. and Sih, G., Alternating method applied to edge and surface crack problems, in *Methods of Analysis and Solutions of Crack Problems.* Noordhoff, Holland, 1973.

[32] Il'in, A. M., Boundary value problems for second-order elliptic equations in a domain with a narrow cavity. I. Two-dimensional case. *Mat. Sb. (Math. USSR-Sb.),* 99(4), 514, 1976.

[33] Il'in, A. M., Boundary value problems for second-order elliptic equations in a domain with a narrow cavity. II. A domain with a small void. *Mat. Sb. (Math. USSR-Sb.),* 103(2), 265, 1977.

[34] Irwin, G. R., Fracture, in *Handbuch der Physik.* Bd. 6, Berlin, Springer Verlag, 1958.

[35] Kanaun, S. K., A thin defect in a homogeneous elastic medium, in *Theoretical Principles of Analysis of Building Structures.* Build Eng. Inst. Press, Leningrad, 1983 (in Russian).

[36] Keer, L. M. and Parihar, K. S., Elastic stress singularity at conical inclusions. *Int. J. Solids Struct.*, 14(4), 261, 1978.

[37] Klarbring, A., Derivation of a model of adhesively bonded joints by the asymptotic expansion method. *Int. J. Eng. Sci.,* 29(4), 493, 1991.

[38] Kondrat'ev, V. A., Boundary-value problems for elliptic equations in domains with conical or angular points. *Trudy Moskov. Mat. Obshch. (Trans. Moscow Math. Soc.),* 16, 209, 1967.

[39] Kondrat'ev, V. A. and Oleinik, O. A., Boundary–value problems for partial differential equations in non–smooth domains. *Uspekhi Mat. Nauk (Russian Math. Surveys),* 38(2), 3, 1983.

[40] Kozlov, V. A., Maz'ya, V. G. and Movchan, A. B., Asymptotic representation of an elastic field in a multi–structure. Mathematics preprint LiTH-MAT-R-95-02, Linköping University; *Asymptotic Analysis,* 1995 (to appear).

[41] Krasnosel'skii, M. A., Vainikko, G. M., Zabreiko, P. P. et al., *Approximate Solution of Operator Equations.* Nauka, Moscow, 1969 (in Russian).

[42] Kucherenko, V. V. and Popov, V. A., On the high frequency oscillation of elastic plates. *Dokl. Acad. Nauk SSSR (Soviet Math. Dokl.),* 244(4), 819, 1979.

[43] Kuliev, V. D., Singular problem of the theory of elasticity for a semi–infinite rectangular void. *Prikl. Mat. Mekh. (J. Appl. Math. Mech.),* 44(5), 952, 1980.

[44] Kunetz, Ya. I., Axi-symmetric torsion of an elastic space with a thin elastic inclusion of a variable thickness, in *Mechanics of Inhomogeneous Structures.* Proc. 2nd USSR National Conf., Lvov, 1983.

[45] Landkof, N. S., *Principles of Modern Potential Theory.* Nauka, Moscow, 1966 (in Russian).

[46] Lekhnitskii, S. G., Symmetric deformation and torsion of an anisotropic solid with anisotropies of particular form. *Prikl. Mat. Mekh. (J. Appl. Math. Mech.),* 4(3), 43, 1940.

[47] Lekhnitskii, S. G., *Theory of Elasticity of an Anisotropic Body.* Nauka, Moscow, 1977 (in Russian).

[48] Leonov, M. Ya. and Panasyuk, V. V., Propagation of small cracks in a solid, *Prikl. Mekh. (Soviet Appl. Mech.),* 5(4), 391, 1969.

[49] Leora, S. N., Nazarov, S. A. and Proskura, A. V., Derivation of the limit equations of elliptic boundary value problems in thin domains. *Zh. Vych. Mat. i Mat. Fiziki,* 26(7), 1032, 1986.

[50] Lurie, A. I., *Theory of Elasticity.* Nauka, Moscow, 1970 (in Russian).

[51] Maz'ya, V. G. and Nazarov, S. A., Asymptotics of energy ranges for small perturbations of the boundary in the vicinity of nodal and conical points. *Trudy Moskov. Mat. Obshch. (Trans. Moscow Math. Soc.),* 50, 79, 1987.

[52] Maz'ya, V. G. and Nazarov, S. A., Singularities of solutions of the Neumann problem at a conical point. *Sibirsk. Mat. Zh. (Siberian Math. J.),* 30(3), 52, 1989.

[53] Maz'ya, V. G., Nazarov, S. A. and Plamenevskii, B. A., *Asymptotics of Solutions of Elliptic Boundary Value Problems in Singularly Perturbed Regions.* Izd. Tbil Univ., Tbilisi, 1981 (in Russian).

[54] Maz'ya, V. G., Nazarov, S. A. and Plamenevskii, B. A., On asymptotics of solutions to elliptic boundary value problems with singular perturbations of the boundary, in *Problemy Mat. Anal.,* vyp. 8. Izdat. Leningr. Univ., Leningrad, 1981 (in Russian).

[55] Maz'ya, V. G., Nazarov, S. A. and Plamenevskii, B. A., On the singularities of the solutions of the Dirichlet problem in the exterior of a thin cone. *Mat. Sb. (Math. USSR-Sb.),* 122(4), 435, 1983.

[56] Maz'ya, V. G., Nazarov, S. A. and Plamenevskii, B. A., Asymptotic expansions of eigenvalues of elliptic boundary value problems for the Laplace operator in domains with small voids. *Izv. Acad. Nauk SSSR. Ser. Mat. (Math. USSR-Izv.),* 48(2), 347, 1984.

[57] Maz'ya, V. G., Nazarov, S. A. and Plamenevskii, B. A., Elliptic boundary value problems for domains in the exterior of a cusp, in *Problemy Mat. Anal.,* vyp. 9. Izdat. Leningr. Univ., Leningrad, 1984 (in Russian).

[58] Maz'ya, V. G., Nazarov, S. A. and Plamenevskii, B. A., Dirichlet's problem in regions with thin bridges. *Sibirsk. Mat. Zh. (Siberian Math. J.),* 25(2), 161, 1984.

[59] Maz'ya, V. G., Nazarov, S. A. and Plamenevskii, B. A., *Asymptotische Theorie Elliptischer Randwertanfgaben in Singulär Gestörten Gebieten.* Vol. 1, 2. Akademie Verlag, Berlin, 1991.

[60] Maz'ya, V. G. and Plamenevskii, B. A., On the asymptotic behaviour of solutions of differential equations in Hilbert space. *Izv. Acad. Nauk SSSR. Ser. Mat. (Math. USSR-Izv.),* 36(5), 1080, 1972.

[61] Maz'ya, V. G. and Plamenevskii, B. A., Elliptic boundary value problems on manifolds with singularities, in *Problemy Mat. Anal.,* vyp. 6. Izdat. Leningr. Univ., Leningrad, 1977 (in Russian).

[62] Maz'ya, V. G. and Plamenevskii, B. A., On the coefficients in the asymptotic form of the solution of elliptic boundary-value problems in domains with conical points. *Math. Nachr.,* 76, 29, 1977.

[63] Morozov, N. F., *Mathematical Problems of the Theory of Cracks.* Nauka, Moscow, 1984 (in Russian).

[64] Morozov, N. F., Nazarov, S. A., On stress–strain state in the vicinity of a crack interacting with rigid inclusion, in *Issled. po uprugosti i plastichnosti,* vyp. 14. Izd. Leningr. Univ., 1982 (in Russian).

[65] Movchan, A. B., Brittle failure of an elastic plane with a thin rectangular noth. *Vestnik Leningr. Univ., Ser. I* , 1, 63, 1988.

[66] Movchan, A. B., Integral characteristics of elastic inclusions and cavities in the two–dimensional theory of elasticity. *Eur. J. Appl. Math.,* 3, 21, 1992.

[67] Movchan, A. B., Morozov, N. F. and Nazarov, S.A., Cracks with smoothly closing edges under plane deformation. *Prikl. Mat. Mekh. (J. Appl. Math. Mech.),* 51(1), 130, 1987.

[68] Movchan, A. B., Morozov, N. F. and Nazarov, S. A., A three–dimensional crack with smoothly closed edges. *Vestnik Leningr. Univ., Ser. I,* 3, 60, 1989.

[69] Movchan, A. B. and Nazarov, S. A., The asymptotic behaviour of a stress-strain state near the three-dimensional sharp inclusion. *Mekh. Kompozit. Mater. (Mech. Comp. Mater.),* 5, 792, 1985.

[70] Movchan, A. B. and Nazarov, S.A., Stress–strain state at the vertex of a sharp inclusion. *Izv. Acad. Nauk SSSR, Mekh. Tv. Tela (Mech. Solids),* 21(3), 155, 1986.

[71] Movchan, A. B. and Nazarov, S. A., Stress–strain state of a plane with a thin elastic inclusion. *Izv. Acad. Nauk SSSR, Mekh. Tv. Tela (Mech. Solids),* 22(1), 75, 1987.

[72] Movchan, A. B. and Nazarov, S. A., Stress–strain state near the tip of a perfectly rigid three–dimensional spike inserted into an elastic body. *Prikl. Mekh. (Soviet Appl. Mech.),* 25(12), 10, 1989.

[73] Movchan, A. B. and Nazarov, S. A., Cracks in composite materials. 1. A semiinfinite crack in an elastic plane with an orthotropic composite strip. *Mekh. Kompozit. Mater. (Mech. Comp. Mater.),* 5, 842, 1990.

[74] Movchan, A. B. and Nazarov, S. A., Cracks in composite materials. 2. A finite crack in an orthotropic composite plane. *Mekh. Kompozit. Mater. (Mech. Comp. Mater.),* 6, 1038, 1990.

[75] Movchan, A. B., Nazarov, S. A. and Polyakova, O. R., The quasistatic growth of a semi–infinite crack in a plane containing small defects. *C. R. Acad. Sci. Paris, Série II,* 313, 1223, 1991.

[76] Movchan, A. B. and Nazarov, S. A., Asymptotic behaviour of stress–strain state in the vicinity of sharp defects in an elastic body. *IMA J. Appl. Math.,* 49, 245, 1992.

[77] Movchan, A. B. and Morozov, N. F., Novozhilov's problem for stationary propagating cracks. *Fiziko–Khimich. Mekh. Mater. (Soviet Mater. Sci.),* 25(6), 33, 1989.

[78] Movchan, A. B. and Serkov, S. K., A longitudinal loading of rectangular notches. *Prikl. Mekh. (Soviet Appl. Mech.),* 27(10), 45, 1991.

[79] Movchan, N. V., Asymptotics of eigenvalues of three–dimensional elasticity problem for a body with a small inclusion. *Izv. Vyssh. Uchebn. Zaved. Mat. (Soviet Math.),* 1, 50, 1988.

[80] Movchan, N. V., Vibration of elastic solids with small holes. *Vestn. Leningr. Univ., Ser. I,* 1, 33, 1989.

[81] Movchan, N. V. and Nazarov, S. A., On singularity of stress at a vertex of a conical inclusion, in *Mechanics of Rheological Systems,* vyp. 87. Novosibirsk, 1988.

[82] Movchan, N. V. and Nazarov, S. A., Stress singularity exponents for non–symmetric thin conical inclusions, in *Mechanics of Inhomogeneous Structures.* Proc. 2nd USSR National Conf., Lvov, 1987.

[83] Movchan, N. V. and Nazarov, S. A., On state of stress and strain near cone apices. *Prikl. Mat. Mekh. (J. Appl. Math. Mech.),* 54(2),

231, 1990.

[84] Muskhelishvili, N. I., Sur l'integration approchée de *Comptes Rendus,* Paris, 185, 1184, 1927.

[85] Muskhelishvili, N. I., *Some Basic Problems of Mathematical Theory of Elasticity.* Noordhoff, Groningen, Holland, 1953.

[86] Murray, J. D., *Asymptotic Analysis.* Springer Verlag, New York, 1984.

[87] Myasnikov, V. P. and Fedoryuk, M. V., The Rayleigh approximation in the theory of elasticity. *Dokl. Akad. Nauk SSSR (Soviet. Math. Dokl.),* 254(3), 589, 1980.

[88] Nazarov, S. A., Stress–strain state at the concentration point of collinear microcracks. *Vestn. Leningr. Univ., Ser. I,* 13, 63, 1983.

[89] Nazarov, S. A., An introduction to the asymptotic methods of the theory of elasticity. Izd. Leningr. Univ., Leningrad, 1983.

[90] Nazarov, S. A., The structure of solutions of elliptic boundary value problems in slender domains. *Vestn. Leningr. Univ. Ser. I,* 15, 99, 1983.

[91] Nazarov, S. A., Asymptotics at infinity of a solution to the Neumann problem with the contact conditions in the angle. *Izv. Vyssh. Ucheb. Zaved. Mat. (Soviet Math.),* 1, 18, 1984.

[92] Nazarov, S. A., *Uspekhi Mat. Nauk (Russian Math. Surveys),* 40(5), 219, 1985.

[93] Nazarov, S. A., Local stability and instability of tensile mode cracks. *Izv. Akad. Nauk SSSR, Mekh. Tverd. Tela (Mech. Solids),* 3, 124, 1988.

[94] Nazarov, S. A. and Paukshto, M. V., *Discrete Models and Homogenization in Elasticity.* Izd. Leningr. Univ., Leningrad, 1984.

[95] Nazarov, S. A. and Romashev, Yu. A., Variation of the stress-intensity factor during the failure of a junction between two collinear cracks, *Izv. Acad. Nauk Armyan. SSR, Mekhanika,* 4, 30, 1982.

[96] Noble, B., *Methods Based on the Wiener-Hopf Technique, for the Solution of Partial Differential Equations.* Pergamon Press, New York, 1958.

[97] Novozhilov, V. V., The necessary and sufficient conditions of brittle failure. *Prikl. Mat. Mekh. (J. Appl. Math. Mech.),* 33(2), 212, 1969.

[98] Novozhilov, V. V., On principles of the theory of equilibrium cracks in elastic bodies. *Prikl. Mat. Mekh. (J. Appl. Math. Mech.),* 33(5), 797, 1969.

[99] Nowacki, W. *Elasticity Theory.* Mir, Moscow, 1975.

[100] Nuller, B. M., On a mixed problem on the torsion of an elastic cone. *Izv. Acad. Nauk SSSR, Mekh. Tverd. Tela (Mech. Solids),* 4, 146, 1967.

[101] Nuller, B. M., On the solution of an elasticity theory problem concerning a truncated hollow cone. *Izv. Acad. Nauk SSSR, Mekh. Tverd. Tela (Mech. Solids),* 5, 102, 1967.

[102] Panasyuk, V. V., Ed., *Mechanics of Fracturing and Strength of Materials,* Vol. 1, 2. Naukova Dumka, Kiev, 1988 (in Russian).

[103] Parton, V. Z. and Perlin, P. I., *Methods of Mathematical Elasticity Theory.* Nauka, Moscow, 1981 (in Russian).

[104] Pazy, A., Asymptotic expansions of solutions of ordinary differential equations in Hilbert space. *Arch. Rational Mech. Anal.,* 24, 193, 1967.

[105] Poddubnyak, A. P. and Kunetz, Ya. I., Axi-symmetric torsion of an elastic half-space with an elastic ring. *Prikl. Mech. (Soviet Appl. Mech.),* 19(7), 66, 1983.

[106] Podstrigach, Ya. S., The jump conditions for stresses and displacements across a thin elastic inclusion in a continuum. *Dokl. Acad. Nauk SSSR (Soviet Dokl.),* A12, 30, 1982.

[107] Pólya, G. and Szegö, G., *Isoperimetric Inequalities in Mathematical Physics.* Princeton Univ. Press, Princeton, 1951.

[108] Rice, J. R., Some remarks on elastic crack tip stress fields. *Int. J. Solids Struct.,* 8, 751, 1972.

[109] Rice, J. R., Mathematical analysis in the mechanics of fracture, in *Fracture,* Vol. 2, Liebowitz, H., Ed., Academic Press, New York, 1968.

[110] Romalis, N. B. and Tamuzh, V. P., A macrocrack propagation in a body with microcracks. *Mekh. Kompozit. Mater. (Mech. Comp. Mater.),* 1, 42, 1984.

[111] Romalis, N. B. and Tamuzh, V. P., Propagation of cracks in a porous material. *Mekh. Kompozit. Mater. (Mech. Comp. Mater.)*, 1, 146, 1985.

[112] Rubinstein, A. A., Macrocrack-microdefect interaction. *J. Appl. Mech.,* 53, 505, 1986.

[113] Rvachev, V. L., On a compression of an elastic half–space by a wedge shaped punch. *Prikl. Mat. Mekh. (J. Appl. Math. Mech.)*, 23(1), 169, 1959.

[114] Sedov, L. I., *A Course in Continuum Mechanics.* Vol. 4. Elastic and Plastic Solids and the Formation of Cracks. Wolter-Noordhoff, Groningen, 1972.

[115] Sherman, D. I., On a method of solution of the static plane problem of the theory of elasticity for multiply connected regions. *Trudy Seismologiczesk. Inst. Acad. Nauk SSSR,* 54, 1935.

[116] Sih, G. and Liebowitz, H., Mathematical theory of brittle fracture, in *Fracture,* Vol. 2, Liebowitz, H., Ed., Academic Press, New York, 1968.

[117] Sokolnikoff, I. S., *Mathematical Theory of Elasticity.* McGraw–Hill, New York, 1956.

[118] Stallybrass, M. P., A crack perpendicular to an elastic half-plane. *Int. J. Eng. Sci.,* 8, 351, 1970.

[119] Thompson, T. R. and Little, R. W., End effects in a truncated semi-infinite cone. *Q. J. Mech. Appl. Math.,* 23(2), 185, 1970.

[120] Timoshenko, S. P. and Goodier, J. N., *Theory of Elasticity.* McGraw–Hill, New York, 1951.

[121] Uflyand, Ya. S., *Integral Transformations in Problems of Elasticity Theory.* Nauka, Leningrad, 1967 (in Russian).

[122] Ulitko, A. F., *The Method of Vector Eigenfunctions in Spatial Problems of Elasticity Theory.* Naukova Dumka, Kiev, 1979 (in Russian).

[123] Van Dyke, M. D., *Perturbation Methods in Fluid Mechanics.* Academic Press, New York, 1964.

[124] Williams, M. L., Stress singularities, adhesion, and fracture, in *Proc. 5th U.S. National Congr. Appl. Mech.,* ASME, New York, 1966.

[125] Willis, J. R., The stress field around an elliptical crack in an anisotropic elastic medium. *Int. J. Eng. Sci,* 6, 253, 1968.

[126] Willis, J. R., A comparison of the fracture criteria of Barenblatt and Griffith. *J. Mech. Phys. Solids,* 15, 151, 1967.

[127] Zino, I. E. and Tropp, E. A., *Asymptotic Methods in Problems of Heat Conduction and Thermo-Elasticity.* Izd. Leningr. Univ., Leningrad, 1978 (in Russian).

[128] Zorin, I. S., On brittle failure of an elastic plane with a thin cut. *Vestn. Leningr. Univ. Ser. II,* 7, 11, 1982.

[129] Zorin, I. S., Brittle failure of an elastic plane with a thin cut under the biaxial load. *Izv. Acad. Nauk SSSR, Mekh. Tverd. Tela (Mech. Solids),* 1, 188, 1985.

[130] Zorin, I. S., Babich, V. M., Ivanov, M. I., Movchan, A. B. and Nazarov, S. A., *Integral Characteristics in Elasticity Problems.* Preprint LOMI P-6-89, Leningrad. Otdel. Mat. Inst. Steklov. (LOMI), Leningrad, 1989 (in Russian).

[131] Zorin, I. S., Movchan, A. B. and Nazarov, S. A., The use of the elastic polarization tensor in problems of the mechanics of cracks. *Izv. Acad. Nauk SSSR, Mekh. Tverd. Tela (Mech. Solids)*, 6, 128, 1988.

Index